Lecture [illegible] in Mathematics

[illegible]

Lecture Notes in Mathematics

For information about Vols. 1–758, please contact your bookseller or Springer-Verlag.

Vol. 759: R. L. Epstein, Degrees of Unsolvability: Structure and Theory. XIV, 216 pages. 1979.

Vol. 760: H.-O. Georgii, Canonical Gibbs Measures. VIII, 190 pages. 1979.

Vol. 761: K. Johannson, Homotopy Equivalences of 3-Manifolds with Boundaries. 2, 303 pages. 1979.

Vol. 762: D. H. Sattinger, Group Theoretic Methods in Bifurcation Theory. V, 241 pages. 1979.

Vol. 763: Algebraic Topology, Aarhus 1978. Proceedings, 1978. Edited by J. L. Dupont and H. Madsen. VI, 695 pages. 1979.

Vol. 764: B. Srinivasan, Representations of Finite Chevalley Groups. XI, 177 pages. 1979.

Vol. 765: Padé Approximation and its Applications. Proceedings, 1979. Edited by L. Wuytack. VI, 392 pages. 1979.

Vol. 766: T. tom Dieck, Transformation Groups and Representation Theory. VIII, 309 pages. 1979.

Vol. 767: M. Namba, Families of Meromorphic Functions on Compact Riemann Surfaces. XII, 284 pages. 1979.

Vol. 768: R. S. Doran and J. Wichmann, Approximate Identities and Factorization in Banach Modules. X, 305 pages. 1979.

Vol. 769: J. Flum, M. Ziegler, Topological Model Theory. X, 151 pages. 1980.

Vol. 770: Séminaire Bourbaki vol. 1978/79 Exposés 525–542. IV, 341 pages. 1980.

Vol. 771: Approximation Methods for Navier-Stokes Problems. Proceedings, 1979. Edited by R. Rautmann. XVI, 581 pages. 1980.

Vol. 772: J. P. Levine, Algebraic Structure of Knot Modules. XI, 104 pages. 1980.

Vol. 773: Numerical Analysis. Proceedings, 1979. Edited by G. A. Watson. X, 184 pages. 1980.

Vol. 774: R. Azencott, Y. Guivarc'h, R. F. Gundy, Ecole d'Eté de Probabilités de Saint-Flour VIII-1978. Edited by P. L. Hennequin. XIII, 334 pages. 1980.

Vol. 775: Geometric Methods in Mathematical Physics. Proceedings, 1979. Edited by G. Kaiser and J. E. Marsden. VII, 257 pages. 1980.

Vol. 776: B. Gross, Arithmetic on Elliptic Curves with Complex Multiplication. V, 95 pages. 1980.

Vol. 777: Séminaire sur les Singularités des Surfaces. Proceedings, 1976-1977. Edited by M. Demazure, H. Pinkham and B. Teissier. IX, 339 pages. 1980.

Vol. 778: SK_1 von Schiefkörpern. Proceedings, 1976. Edited by P. Draxl and M. Kneser. II, 124 pages. 1980.

Vol. 779: Euclidean Harmonic Analysis. Proceedings, 1979. Edited by J. J. Benedetto. III, 177 pages. 1980.

Vol. 780: L. Schwartz, Semi-Martingales sur des Variétés, et Martingales Conformes sur des Variétés Analytiques Complexes. XV, 132 pages. 1980.

Vol. 781: Harmonic Analysis Iraklion 1978. Proceedings 1978. Edited by N. Petridis, S. K. Pichorides and N. Varopoulos. V, 213 pages. 1980.

Vol. 782: Bifurcation and Nonlinear Eigenvalue Problems. Proceedings, 1978. Edited by C. Bardos, J. M. Lasry and M. Schatzman. VIII, 296 pages. 1980.

Vol. 783: A. Dinghas, Wertverteilung meromorpher Funktionen in ein- und mehrfach zusammenhängenden Gebieten. Edited by R. Nevanlinna and C. Andreian Cazacu. XIII, 145 pages. 1980.

Vol. 784: Séminaire de Probabilités XIV. Proceedings, 1978/79. Edited by J. Azéma and M. Yor. VIII, 546 pages. 1980.

Vol. 785: W. M. Schmidt, Diophantine Approximation. X, 299 pages. 1980.

Vol. 786: I. J. Maddox, Infinite Matrices of Operators. V, 122 pages. 1980.

Vol. 787: Potential Theory, Copenhagen 1979. Proceedings, 1979. Edited by C. Berg, G. Forst and B. Fuglede. VIII, 319 pages. 1980.

Vol. 788: Topology Symposium, Siegen 1979. Proceedings, 1979. Edited by U. Koschorke and W. D. Neumann. VIII, 495 pages. 1980.

Vol. 789: J. E. Humphreys, Arithmetic Groups. VII, 158 pages. 1980.

Vol. 790: W. Dicks, Groups, Trees and Projective Modules. IX, 127 pages. 1980.

Vol. 791: K. W. Bauer and S. Ruscheweyh, Differential Operators for Partial Differential Equations and Function Theoretic Applications. V, 258 pages. 1980.

Vol. 792: Geometry and Differential Geometry. Proceedings, 1979. Edited by R. Artzy and I. Vaisman. VI, 443 pages. 1980.

Vol. 793: J. Renault, A Groupoid Approach to C*-Algebras. III, 160 pages. 1980.

Vol. 794: Measure Theory, Oberwolfach 1979. Proceedings 1979. Edited by D. Kölzow. XV, 573 pages. 1980.

Vol. 795: Séminaire d'Algèbre Paul Dubreil et Marie-Paule Malliavin. Proceedings 1979. Edited by M. P. Malliavin. V, 433 pages. 1980.

Vol. 796: C. Constantinescu, Duality in Measure Theory. IV, 197 pages. 1980.

Vol. 797: S. Mäki, The Determination of Units in Real Cyclic Sextic Fields. III, 198 pages. 1980.

Vol. 798: Analytic Functions, Kozubnik 1979. Proceedings. Edited by J. Ławrynowicz. X, 476 pages. 1980.

Vol. 799: Functional Differential Equations and Bifurcation. Proceedings 1979. Edited by A. F. Izé. XXII, 409 pages. 1980.

Vol. 800: M.-F. Vignéras, Arithmétique des Algèbres de Quaternions. VII, 169 pages. 1980.

Vol. 801: K. Floret, Weakly Compact Sets. VII, 123 pages. 1980.

Vol. 802: J. Bair, R. Fourneau, Etude Géometrique des Espaces Vectoriels II. VII, 283 pages. 1980.

Vol. 803: F.-Y. Maeda, Dirichlet Integrals on Harmonic Spaces. X, 180 pages. 1980.

Vol. 804: M. Matsuda, First Order Algebraic Differential Equations. VII, 111 pages. 1980.

Vol. 805: O. Kowalski, Generalized Symmetric Spaces. XII, 187 pages. 1980.

Vol. 806: Burnside Groups. Proceedings, 1977. Edited by J. L. Mennicke. V, 274 pages. 1980.

Vol. 807: Fonctions de Plusieurs Variables Complexes IV. Proceedings, 1979. Edited by F. Norguet. IX, 198 pages. 1980.

Vol. 808: G. Maury et J. Raynaud, Ordres Maximaux au Sens de K. Asano. VIII, 192 pages. 1980.

Vol. 809: I. Gumowski and Ch. Mira, Recurences and Discrete Dynamic Systems. VI, 272 pages. 1980.

Vol. 810: Geometrical Approaches to Differential Equations. Proceedings 1979. Edited by R. Martini. VII, 339 pages. 1980.

Vol. 811: D. Normann, Recursion on the Countable Functionals. VIII, 191 pages. 1980.

Vol. 812: Y. Namikawa, Toroidal Compactification of Siegel Spaces. VIII, 162 pages. 1980.

Vol. 813: A. Campillo, Algebroid Curves in Positive Characteristic. V, 168 pages. 1980.

Vol. 814: Séminaire de Théorie du Potentiel, Paris, No. 5. Proceedings. Edited by F. Hirsch et G. Mokobodzki. IV, 239 pages. 1980.

Vol. 815: P. J. Slodowy, Simple Singularities and Simple Algebraic Groups. XI, 175 pages. 1980.

Vol. 816: L. Stoica, Local Operators and Markov Processes. VIII, 104 pages. 1980.

continuation on page 329

Lecture Notes in Mathematics

Edited by A. Dold and B. Eckmann

969

Combinatorial Theory

Proceedings of a Conference Held at
Schloss Rauischholzhausen, May 6-9, 1982

Edited by D. Jungnickel and K. Vedder

Springer-Verlag
Berlin Heidelberg New York 1982

Editors

Dieter Jungnickel
Klaus Vedder
Mathematisches Institut, Justus-Liebig-Universität Gießen
Arndtstr. 2, 6300 Gießen, Federal Republic of Germany

AMS Subject Classifications (1980): 05

ISBN 978-3-540-11971-5 Springer-Verlag Berlin Heidelberg New York
ISBN 978-0-387-11971-7 Springer-Verlag New York Heidelberg Berlin

2146/3140-543210

This volume contains the proceedings of a conference on Combinatorial Theory that took place at Schloss Rauischholzhausen in May 1982 to mark the 375th anniversary of the Universität Giessen. There were eight invited lectures and over twenty contributed talks. 21 of these papers are contained in this volume. In accordance with the aim of the conference, they cover the whole range of Combinatorics. We hope that the conference and this book will contribute to a better understanding of the various aspects of this fast developing and diverging field, as well as stimulate the exchange of ideas.

We would like to thank all the referees for their cooperation and, in particular, their prompt response. We are also indebted to Frau D. Begemann and to Frau R. Schmidt for helping with the organizational details of the conference, and to the Hochschulgesellschaft for financial support. Finally, we are very grateful to the secretaries of the Mathematisches Institut; without their help, the manuscript would not have been completed in time.

Giessen, October 1982

Dieter Jungnickel
Klaus Vedder

CONTENTS

* An asterisk indicates an invited speaker

CRITICAL PERFECT SYSTEMS OF DIFFERENCE SETS WITH THE MINIMUM START

Jaromir Abrham
Department of Industrial Engineering
University of Toronto
Toronto, Ontario, Canada

and

Anton Kotzig
C. R. M. A.
Universite de Montreal
Montreal, Quebec
Canada

The concept of a perfect system of difference sets has been introduced in [4] as a mathematical model of the following problem in radio-astronomy: A few movable antennas are used in several successive configurations to measure various spatial frequencies relative to some area of the sky. The distances between antennas determine the frequencies obtained. We do not want to miss any frequency, and want to avoid redundancy (repetition of the same spacing between antennas). For more details, the reader is referred to [4] and [5].

Let $c,m,p_1,\dots,p_m$ be positive integers, let $S_i = \{x_{0i}<x_{1i}<\dots<x_{p_i,i}\}$, $i = 1,\dots,m$ be sequences of integers, and let $D_i = \{x_{ji}-x_{ki},\ 1\le k < j\le p_i\}$, $i = 1,\dots,m$ be their difference sets. Then we say that the system $\{D_1,\dots,D_m\}$ is a perfect system of difference sets (PSDS) starting with c if $\bigcup_{i=1}^{m} D_i = \{c,c+1,\dots,c-1+\sum_{i=1}^{m}\binom{p_i+1}{2}\}$. Each set D_i is called a component of the PSDS $\{D_1,\dots,D_m\}$. The size of D_a is p_a, the half-size of D_a is $r_a = [p_a/2]$ where $[x]$ denotes the integer part of a real number x. Then $p_a = 2r_2+\delta_a$ where $\delta_a = 0$ or 1 according to whether p_a is even or odd. This notation will be used throughout the paper. The reader will observe that the size of a component is not the number of its elements; if the size of D_a is p_a then D_a has $1+2+\dots+p_a = \frac{1}{2}p_a(p_a+1)$ elements.

We will briefly review some earlier results concerning PSDS:

A PSDS is called regular if all its components have the same size. A regular PSDS with m components of size p, starting at c, will be called an (m,p,c)-system. In [4], the existence of (m,p,1)-systems has been related to graceful numberings of certain graphs, and some relations between m,p,c, necessary for the existence of an (m,p,c)-system, have been obtained. Further existence studies have been carried out in [7]; one of the results obtained here is that, if an (m,p,c)-system exists, then $p \le 4$. Without this result, a lot of time and money could have been spent in efforts aimed at finding (m,p,c)-systems with large values of p. A generalization of this result to the nonregular case has been obtained in [9]: Every PSDS contains at least one "small" component (a component of size ≤ 4). This has been further generalized in [2]: Every PSDS starting at c ($c \geq 1$) contains at least c small components. This follows immediately from the inequality (5) below. Proceeding from the inequality (2) it has been proved in [1] that, in a PSDS with m components with the half-sizes $r_1 \leq r_2 \leq \dots \leq r_m$, it is $r_m \leq K(\sqrt{m}+1)$ where K is a constant, and that the average of half-sizes of the components of any PSDS is bounded by a constant. The first result implies that the number of perfect systems of difference sets starting with a given c, which has a given number m of components, is finite. Moreover, it follows from the results in [2] that $c \leq m$. This means that the number of all PSDS with a given number of components and all possible starts c is finite.

Let us now denote (similarly as in [1, 2])

$$n = \frac{1}{2}\sum_{a=1}^{m}(2r_a+\delta_a)(2r_a+\delta_a+1)$$

$$s = \frac{1}{2}\sum_{a=1}^{m} r_a(3r_a+2\delta_a+1)$$

$$\ell' = \frac{1}{2}\sum_{a=1}^{m} r_a(r_a+1), \qquad \ell = n-\ell'$$

and let $S = \{c,c+1,\dots,c+s-1\}$, $L = \{c+\ell,\dots,c+n-1\}$, $M = \{c+s,\dots,c+\ell-1\}$. Furthermore, let us put $x_{j+k-1,a}-x_{j-1,a} = d_{ja}^k$, $j = 1,\dots,p_a$, $k = 1,\dots,p_a+1-j$, $a = 1,\dots,m$. Then the elements of D_a can be represented in the form of a difference triangle

$$\begin{array}{ccccc} & & d_{1a}^{p_a} & & \\ & \cdots & \cdots & \cdots & \\ & d_{1a}^2 & d_{2a}^2 & d_{p_a-1,a}^2 & \\ d_{1a}^1 & d_{2a}^1 & \cdots\cdots & & d_{p_a,a}^1 \end{array}$$

The top (bottom) r_a rows of this triangle will be referred to as its upper (lower) half. Then s and ℓ' denote the number of elements in the lower (upper) halves of all triangles corresponding to $\{D_1,\dots,D_m\}$, and n denotes the number of all elements in all such triangles. According to Proposition 1.1 in [4] we have

$$\sum_{j=1}^{k} d_{ja}^{p_a+1-k} = \sum_{j=1}^{p_a+1-k} d_{ja}^k, \quad k = 1,2,\dots,r_a, \quad a = 1,\dots,m$$

Adding over k and a we get

$$(1) \qquad \sum_{a=1}^{m}\sum_{k=1}^{r_a}\sum_{j=1}^{k} d_{ja}^{p_a+1-k} = \sum_{a=1}^{m}\sum_{k=1}^{r_a}\sum_{j=1}^{p_a+1-k} d_{ja}^k$$

in words: The sum of all elements in the upper halves of the difference triangles corresponding to $D_1,\dots,D_m$ is equal to the sum of all elements in the lower halves. If we replace the elements in the lower halves by $c,\dots,c+s-1$ in the middle rows by $c+s,\dots,c+\ell-1$, and in the upper halves by $c+\ell,\dots,c+n-1$, we get from (1) the inequality $\sum_{i=c+\ell}^{c+n-1} i \ge \sum_{i=c}^{c+s-1}$, i.e.

$$(2) \qquad (n+\ell+2c-1)(n-\ell) \ge s(s+2c-1), \quad \text{or} \quad (2n-\ell'+2c-1)\ell' \ge s(s+2c-1)$$

and this is equivalent to the fundamental inequality obtained in [1] (see also [2] or [9]). This inequality has been instrumental in establishing a number of important properties of PSDS; for details and other results on PSDS see e.g. [1], [2], [3], [5], [6], [7], [8], [10], [12], [13], [14].

In [2], the inequality (2) has been used to develop another inequality, easier to use, which will be useful in the proof of Theorem 3 below. Since [2] is not yet available in print we will repeat the main steps here.

Let us consider a PSDS with m components and let c_k denote the number of components of size k, $k \ge 2$. Then $m = \sum_{k\ge 2} c_k$ and we can write

$$(3) \quad \begin{cases} n = \frac{1}{2}\sum_{k\ge1} c_{2k}(4k^2+2k) + \frac{1}{2}\sum_{k\ge1} c_{2k+1}(4k^2+6k+2) \\ s = \frac{1}{2}\sum_{k\ge1} c_{2k}(3k^2+k) + \frac{1}{2}\sum_{k\ge1} c_{2k+1}(3k^2+3k) \\ \ell' = \frac{1}{2}\sum_{k\ge1} c_{2k}(k^2+k) + \frac{1}{2}\sum_{k\ge1} c_{2k+1}(k^2+k) \end{cases}$$

If we denote, for any positive integer p,

$$\varepsilon_p = \sum_{k\ge1} k^p c_{2k}, \qquad \omega_p = \sum_{k\ge1} k^p c_{2k+1}$$

then the second inequality in (2) yields

$$(\varepsilon_1+\varepsilon_2+\omega_1+\omega_2)(4c-2+3\varepsilon_1+7\varepsilon_2+4\omega_0+11\omega_1+7\omega_2) -$$
$$- (\varepsilon_1+3\varepsilon_2+3\omega_1+3\omega_2)(4c-2+\varepsilon_1+3\varepsilon_2+3\omega_1+3\omega_2) \ge 0$$

which implies

$$\varepsilon_1^2-\varepsilon_2^2+\omega_1^2-\omega_2^2+2\varepsilon_1\varepsilon_2+4\varepsilon_1\omega_1+2\varepsilon_1\omega_2-2\varepsilon_2\omega_2+2\omega_0(\varepsilon_1+\varepsilon_2+\omega_1+\omega_2) \ge$$
$$(4c-2)(\varepsilon_2+\omega_1+\omega_2) .$$

Adding $2\varepsilon_1^2$ to both sides we can transform the last inequality into

$$(\varepsilon_1+\varepsilon_2+\omega_1+\omega_2)(\varepsilon_1-\varepsilon_2+\omega_1-\omega_2) + 2\varepsilon_1(\varepsilon_1+\varepsilon_2+\omega_1+\omega_2) + 2\omega_0(\varepsilon_1+\varepsilon_2+\omega_1+\omega_2) \geq$$
$$2\varepsilon_1^2-(4c-2)\varepsilon_1+(4c-2)(\varepsilon_1+\varepsilon_2+\omega_1+\omega_2).$$

Dividing this last inequality by $\varepsilon_1+\varepsilon_2+\omega_1+\omega_2$ and denoting

$$\Delta_c = \frac{\varepsilon_1(\varepsilon_1-2c+1)}{\varepsilon_1+\varepsilon_2+\omega_1+\omega_2}$$

we get

$$3\varepsilon_1-\varepsilon_2+2\omega_0+\omega_1-\omega_2 \geq 4c-2+2\Delta_c.$$

Substituting from (3) we can transform the last inequality into

(4) $$c_2+c_3+c_4 \geq 2c-1 + \Delta_c + \sum_{k=4}^{\infty} \frac{1}{2}k(k-3)(c_{2k-1}+c_{2k}).$$

Furthermore,

$$2c-1+\Delta_c = \frac{(2c-1)(\varepsilon_2+\omega_1+\omega_2)+\varepsilon_1^2}{\varepsilon_1+\varepsilon_2+\omega_1+\omega_2} =$$
$$= \frac{(2c-2)(\varepsilon_2+\omega_1+\omega_2)}{\varepsilon_1+\varepsilon_2+\omega_1+\omega_2} + \frac{\varepsilon_1^2+\varepsilon_2+\omega_1+\omega_2}{\varepsilon_1+\varepsilon_2+\omega_1+\omega_2}$$

It is $\varepsilon_1 \leq \varepsilon_1^2$, $\varepsilon_1 \leq \varepsilon_2$, and therefore $\frac{\varepsilon_2+\omega_1+\omega_2}{\varepsilon_1+\varepsilon_2+\omega_1+\omega_2} \geq \frac{1}{2}$.

This implies $2c-1+\Delta_c \geq \frac{1}{2}(2c-2)+1 = c$, and (4) yields then

(5) $$c_2+c_3+c_4 \geq c + \frac{1}{2} \sum_{k\geq 4} k(k-3)(c_{2k-1}+c_{2k}).$$

Since (5) is weaker than (2) it holds as a strict inequality whenever (2) holds as a strict inequality.

Throughout this paper, we will use the symbols C_1, C_2, C_3, C_4 to denote the one-component perfect systems of difference sets represented by the following triangles:

$$C_1:\ \begin{matrix} & 3 & \\ 1 & & 2 \end{matrix} \qquad C_2:\ \begin{matrix} & 3 & \\ 2 & & 1 \end{matrix} \qquad C_3:\ \begin{matrix} & & 6 & & \\ & 4 & & 5 & \\ 1 & & 3 & & 2 \end{matrix} \qquad C_4:\ \begin{matrix} & & 6 & & \\ & 5 & & 4 & \\ 2 & & 3 & & 1 \end{matrix}.$$

A perfect system of difference sets will be called critical if its elements satisfy (2) as an equality. Equivalently, $\{D_1,\ldots,D_m\}$ is a critical PSDS if the elements of L,S,M correspond to the elements in the upper halves, lower halves, and middle rows of the difference triangles corresponding to the components of $\{D_1,\ldots,D_m\}$.

The reader will observe that a PSDS starting with c satisfying (5) as an equality satisfies also (2) as an equality and therefore is critical. A noncritical PSDS starting with c satisfies (5) as a strict inequality.

Only critical PSDS with c = 1 (the minimum start) will be considered in this paper. For $c > 1$ we would get different results. The reader will observe that the above PSDS C_1, C_2, C_3, C_4 are all critical (with $c = 1$).

Let us start our study of critical PSDS by investigating the possible position of 1. We get the following

Lemma 1. Let $\Delta = \{D_1, \ldots, D_m\}$ be a critical PSDS and let D_a be its component containing 1. Then D_a also contains s and s+1. Let $d_{ia}^1 = 1$. If the size p_a of D_a is odd then $i \neq r_a+1$ where $r_a = [p_a/2]$. If p_a is even then the sizes of all components of $\{D_1, \ldots, D_m\}$ are even numbers.

Proof. To simplify the notation we will drop the subscript a; hence e.g. d_i^r will stand for $d_{i,a}^{r_a}$ etc. Let $d_i^1 = 1$. If $i \leq r+\delta$ (where $\delta = p-2r$) we have $d_i^{r+1} = d_i^1 + d_{i+1}^r \geq s+1$ and $d_{i+1}^r \leq s$ and this implies $d_{i+1}^r = s$, $d_i^{r+1} = s+1$. If $i \geq r+1$ we get in a similar way $d_{i-r}^{r+1} = s$, $d_{i-r}^{r+1} = s+1$. If p_a is odd and $i = r+1$ we get $d_{r+2}^r = d_1^r = s$ and this is impossible. Furthermore, we see from these considerations that D_a contains s,s+1. If p_a is even then $s \in S$, $s+1 \in L$, hence $M = \emptyset$- there is no component of odd size.

Let $\Delta = \{D_1, \ldots, D_m\}$ be a PSDS and let $n = d_{hg}^1$, where D_g is one of the components. Then we will say that n is represented as the difference v-u in D_g if 1. $v-u = n$ 2. either $d_{h+1,g}^{r_g} = u$, $d_{hg}^{r_g+1} = v$ or $d_{h-r_g,g}^{r_g} = u$ and $d_{h-r_g,g}^{r_g+1} = v$.

Clearly, we must have $u \leq s$, $v \geq s+1$.

Lemma 2. Under the assumptions of Lemma 1, D_a contains 2 and at least one of the numbers s-1, s+2 (in addition to the numbers 1,s,s+1). The numbers 1,2 are neighbors in the first row of the triangle corresponding to D_a if and only if Δ is either C_1 or C_2. If 2 is represented in D_a as the difference of s+2 and s then Δ is either C_3 or C_4.

Proof. Let us denote by D_b the component of Δ containing 2; the subscript b will again be omitted. Let us define j $(1 \leq j \leq p_b)$ by $d_j^1 = 2$. Then $d_j^{r+1} = 2+d_{j+1}^r$ if $j \leq r+\delta$, $d_{j-r}^{r+1} = d_{j-r}^r+2$ if $j \geq r+1$. The two cases are similar (they can be transformed into each other by writing the elements of each row in the triangle representing D_b in the opposite order), so only the first case will be discussed. Since $d_j^{r+1} \geq s+1$, $d_{j+1}^r \leq s$, we have either $d_j^{r+1} = s+2$, $d_{j+1}^r = s$ or $d_j^{r+1} = s+1$, $d_{j+1}^r = s-1$. This implies that $2 \in D_a$ (since, $s, s+1 \in D_a$) and, therefore, $D_b = D_a$. In the first case, $s+2 \in D_a$, in the second one, $s-1 \in D_a$.

Let us now distinguish the two cases (according to the way in which 2 is represented as the difference of two elements of D_a).

I. Let 2 be represented as the difference of s+2 and s. Let $d_j^1 = 2$, $d_j^{r+1} = s+2$, $d_{j+1}^r = s$ for some $j \le r+\delta$. Let $d_i^1 = 1$. If $i \le r+\delta$ we have $d_{i+1}^r = s$, hence $i = j$, which is impossible, hence $i \ge r+1$, and $d_{i-r}^r = s$; this implies $i = j+r+1$ and $p_a > 2$. Furthermore, $d_{i-r}^{r+1} = d_{j+1}^{r+1} = d_{j+1}^r + d_{j+r+1}^1 = s+1$ and $d_j^{r+2} = d_j^1 + d_{j+1}^{r+1} = s+3$, hence s+3 L and $M = \{s+1, s+2\}$, $p_a = 3$. Let us now denote $L' = L - \{s+3\}$, $S' = S - \{1,2,s\}$. If $L' \neq \emptyset \neq S'$ there exist a $y \in L'$ and an $x \in S'$ such that $y = x+3$ (since 3 must be in some component of our PSDS), and this is impossible, as $y \ge s+4$, $x \le s-1$. We conclude that $L' = \emptyset = S'$ and we have a PSDS with only one component of size 3, which must coincide with either C_3 or C_4.

II. Let 2 be represented as the difference of s+1 and s-1. Let $d_j^1 = 2$, $d_j^{r+1} = s+1$, $d_{j+1}^r = s-1$ for some $j \le r+\delta$. If $d_i^1 = 1$ we can see as in case I that $i \ge r+1$ and $d_{i-r}^r = s$, $d_{i-r}^{r+1} = s+1$ which implies $i = j+r$. If 1,2 were neighbors we would have $r = 1$, $p = 2$ or 3. If $p = 2$, $3 \in L$ and we get C_1 or C_2. If $p = 3$, $3 \in M$ and the first row of the triangle corresponding to D_a must contain a number > 3 which belongs to S- and this is impossible.

<u>Lemma 3.</u> Let $\Delta = \{D_1, \ldots, D_m\}$ be a critical perfect system of difference sets different from each of C_1, C_2, C_3, C_4. Then $3 \in D_a$ (i.e. 3 is in the same component as 1 and 2). Let k be defined by $d_{ka}^1 = 3$. Then either 3 is represented in D_a as the difference of s+3 and s and $k \le r_a + {}_a$ or 3 is represented in D_a as the difference of s+2 and s-1 and $k \ge r_a+1$.

Proof. Let us denote by D_α the component containing 3; the subscript α will again be omitted. Let k be defined by $d_k^1 = 3$ (in D_α); now we have the following three possibilities.

A. $d_{k+1}^r = s$, $d_k^{r+1} = s+3$ (if $k \le r+\delta$)

or $d_{k-r}^r = s$, $d_{k-r}^{r+1} = s+3$ (if $k \ge r+1$)

B. $d_{k+1}^r = s-1$, $d_k^{r+1} = s+2$ (if $k \le r+\delta$)

or $d_{k-r}^r = s-1$, $d_{k-r}^{r+1} = s+2$ (if $k \ge r+1$)

C. $d_{k+1}^r = s-2$, $d_k^{r+1} = s+1$ (if $k \le r+\delta$)

or $d_{k-r}^r = s-2$, $d_{k-r}^{r+1} = s+1$ (if $k \ge r+1$)

Under our assumptions, if D_a contains 1,2, it contains s, s+1, s-1; this means that, in all 3 cases considered here, $3 \in D_a$, hence $D_\alpha = D_a$.

We will now show that the above case C is impossible. In case C, if $k \le r+\delta$, we have $d^r_{k+1} = s-2$, $d^{r+1}_k = s+1$. If $d^1_j = 2$, we have $d^{r+1}_j = s+1$ and we see that $j = k$ which is impossible. If $k \ge r+1$ we have $d^{r+1}_{k-r} = s+1$, hence $k = r+j$. According to the proof of Lemma 1 (Case I), we have also $i = r+j$ where $d^1_i = 1$, hence $i = k$ and this is again impossible.

Let us now consider case A. If $k \ge r+1$, we have $d^r_{k-r} = s$. If i is defined by $d^1_i = 1$ we have by the proof of Lemma 2 (Case I) $i = r+j$, $d^r_{i-r} = s$, hence $i = k$ and this is a contradiction.

In case B, if $k \le r+\delta$, we have $d^r_{k+1} = s-1$; we also have $d^r_{j+1} = s-1$ where $d^1_j = 2$, and this is a contradiction again. This completes the proof of Lemma 2.

<u>Lemma 4</u>. Let Δ satisfy the assumptions of Lemma 3. Let j,k be defined again by $d^1_j = 2$, $d^1_k = 3$ $(j \le k+\delta)$. If 3 is represented as the difference of $s+3$ and s then $k = j-1$. If 3 is represented as the difference of $s+2$ and $s-1$ then $k = j+r+1$.

Proof. We have $d^1_{j+r} = 1$ (see Lemma 2) and $d^r_j = s$, and this implies $j = k+1$ in the first case. In the second case, we have $d^r_{j+1} = s-1 = d^r_{k-r}$ and this implies $k = j+r+1$.

<u>Theorem 1</u>. Let Δ satisfy the assumptions of Lemma 3. Then Δ cannot have any component of odd size.

Proof. Let us assume the opposite. Then, according to Lemma 1, D_a is of odd size. Let us consider two separate cases.

A. 3 is represented as the difference of $s+3$ and s.

Then $d^1_{j-1} = 3$, $d^1_j = 2$, $d^1_{j+r} = 1$, $d^r_j = s$, $d^{r+1}_j = s+1$, $d^{r+1}_{j-1} = s+3$, and $d^{r+2}_{j-1} = d^1_{j-1}+d^{r+1}_j = s+4$. This implies $s+4 \in L$, $M = \{s+1,s+2,s+3\}$, $p_a = 5$. There are two possibilities for the position of $s+2$ in the triangle representing D_a: either $s+2 = d^{r+1}_{j+1}$ or $s+2 = d^{r+1}_{j-2}$. In the first case, $s+2 = d^{r+1}_{j+1} = d^r_{j+1}+d^1_{j+r+1} = s-1+d^1_{j+r+1}$, hence $d^1_{j+r+1} = d^1_{j-1} = 3$ which is impossible. In the second case, $j = 3$, $d^1_2 = 3$, $d^1_3 = 2$, $d^1_5 = 1$, $d^r_j = d^2_3 = s$, $d^3_1 = s+2 = d^1_1+3+2$, i.e. $d^3_1 = s-3$, and $d^2_1 = s-3+3 = s = d^2_3$ which is a contradiction.

B. 3 is represented as the difference of $s+2$ and $s-1$.

Then $d^1_j = 2$, $d^1_{j+r} = 1$, $d^1_{j+r+1} = 3$ (see Lemma 4). We have $d^{r+2}_j = d^1_j+d^{r+1}_{j+1} = 2+s+2 = s+4 \in L$, so again $M = \{s+1,s+2,s+3\}$ with $s+1 = d^{r+1}_j$,

$s+2 = d^{r+1}_{j+1}$. We have two possibilities for the position of $s+3$ this time: either $s+3 = d^{r+1}_{j+2}$ or $s+3 = d^{r+1}_{j-1}$. We observe that $p_a = 5$, $r_a = 2$ again. In the first case, $d^1_j = 2$, $d^1_{j+2} = 1$, $d^1_{j+3} = 3$, and $s+3 = d^3_{j+2} = d^1_{j+2}+d^1_{j+3}+d^1_{j+4} = 4+d^1_{j+4}$ which implies $d^1_{j+4} = s-1 = d^2_{j+1}$ and we have a contradiction. In the second case, $s+3 = d^3_{j-1} = d^1_{j-1}+d^1_j+d^1_{j+1}$. Since $s = d^2_j = d^1_j+d^1_{j+1}$ we have $d^1_{j+1} = s-2$, and we conclude that $d^1_{j-1} = 3 = d^1_{j+3}$ and we have a contradiction again.

Remark. It has been shown in [2] that the average number of differences in the components of any PSDS cannot exceed 21. However, the largest average number of differences ever achieved — to the best of our knowledge — is ten (see [11] and [12]). It was hoped that this can be improved by constructing a (critical) PSDS with one component of size 3 and several components of size 5 [6]. Using a computer, P.J. Laufer attempted to construct such PSDS with up to six components of size 5. All results were negative; Theorem 1 shows why.

To investigate in more detail the last remaining case (when all components are of even size), let us substitute into (2) for n,s,ℓ and write the result as an equality (we consider critical PSDS). We get

$$\Big(\sum_{a=1}^{m} r_a^2\Big)^2 - 2\Big(\sum_{a=1}^{m} r_a^2\Big)\Big(\sum_{a=1}^{m} r_a\Big) - \Big(\sum_{a=1}^{m} r_a\Big)^2 + 2\sum_{a=1}^{m} r_a^2 = 0 .$$

Let us put $x = \sum_{a=1}^{m} r_a$, $y = \sum_{a=1}^{m} r_a^2$. Then the last equation becomes $y^2-2xy-x^2+2y = 0$. Solving for y in terms of x we obtain (since $y \geq 0$)

$$y = x-1 + \sqrt{(x-1)^2+x^2} .$$

Since x,y are positive integers, x must be such that $(x-1)^2+x^2$ is a perfect square. Our search (for $x \leq 120$) provided the following pairs x,y:

$x = 1, \quad y = 1$

$x = 4, \quad y = 8$

$x = 21, \quad y = 49$

$x = 120, \quad y = 169$.

If $x = 1$, $y = 1$, we get either C_1 or C_2. If $x = 4$, $y = 8$, then, necessarily, $m = 2$, and $r_1 = r_2 = 2$. However, no PSDS with 2 components of size 4 can exist (see [7]), Theorem 4.2). For $x = 21$, $y = 49$, our complete search revealed the following possible candidates for critical PSDS (by c_k we denote the number of components of size k, by m the total number of components):

1. $c_2 = 2,\quad c_4 = 5,\quad c_6 = 3,\quad m = 10$

2. $c_2 = 5,\quad c_4 = 2,\quad c_6 = 4,\quad m = 11$

3. $c_2 = 9,\quad c_4 = 2,\quad c_6 = 0,\quad c_8 = 2,\quad m = 13$

4. $c_2 = 7,\quad c_4 = 2,\quad c_6 = 2,\quad c_3 = 1,\quad m = 12$

5. $c_2 = 11,\quad c_4 = 1,\quad c_6 = 1,\quad c_8 = 0,\quad c_{10} = 1,\quad m = 14$

6. $c_2 = 8,\quad c_4 = 4,\quad c_6 = 0,\quad c_8 = 0,\quad c_{10} = 1,\quad m = 13.$

The question is still open whether any of the above 6 possibilities really yields a critical PSDS. At this moment, the problem seems to be too difficult to decide even when using a computer.

The authors are indebted to the referee for pointing out that the above search for x can be replaced by using the Pell equation. The equation $x^2+(x-1)^2 = u^2$ is equivalent to $1 = u^2-2x^2+2x$ or $2 = 2u^2-4x^2+4x$ or $(2x-1)^2-2u^2 = 1$ which is a Pell equation. The Pell equation $M^2-2u^2 = -1$ has the general solution $M = A_k$, $u = B_k$ where $A_k+B_k\sqrt{2} = (1+\sqrt{2})^{2k+1}$. Using this formula we can obtain the same values of x as above.

To summarize, we have:

<u>Theorem 2</u>. There exists no critical PSDS, different from C_1,C_2,C_3,C_4, with the sum of half-sizes ≤ 20. There exists no critical PSDS with the sum of half-sizes equal to 21 and fewer than 10 components. There exist no critical PSDS with the sum of half-sizes between 22 and 119.

We would like to conclude this study by conjecturing that the only critical PSDS with start one are C_1,C_2,C_3,C_4.

The above results yield immediately the following extension of earlier results about the number of small components.

<u>Theorem 3</u>. Every PSDS with start one, which is different from C_1,C_2,C_3,C_4, and such that the sum of the half-sizes of its components is less than or equal to 119, has at least two small components (i.e. components of size 2,3, or 4).

Proof. If our PSDS has c_k components of size k, $k \geq 2$, then it satisfies (5) with c = 1:

$$c_2+c_3+c_4 \geq 1 + \frac{1}{2}\sum_{k\geq 4} k(k-3)(c_{2k-1}+c_{2k}) . \tag{6}$$

If the PSDS in question is noncritical, (6) holds as a strict inequality, hence $c_2+c_3+c_4 \geq 2$. If it is critical, then it must be one of the above-mentioned six possible candidates. However, each of them has at least seven small components.

Acknowledgments

The research was sponsored by NSERC grants No. A-7329 and A-9232 and by an Ontario-Quebec Exchange Program grant. The authors benefited from the referee's comments made on an earlier version of this paper.

References

[1] J. Abrham, "Bounds for the sizes of components in perfect systems of difference sets." Ann. Discrete Math. 12 (1982) (to appear).

[2] J. Abrham and A. Kotzig, "Inequalities for perfect systems of difference sets." Working paper No. 81-030, November 1981, Department of Industrial Engineering, University of Toronto.

[3] J.-C. Bermond, A.E. Brouver, and A. Germa, "Systèmes de triplets et différences associées." Proc. Colloque C.N.R.S. - Problèmes combinatoires et théorie des graphes, Orsay 1976, pp. 35-38 (1976).

[4] J.-C. Bermond, A. Kotzig and J. Turgeon, "On a combinatorial problem of antennas in radio-astronomy." Proc. 18th Hungarian Combinatorial Colloquium, North-Holland 1976, 135-149 (1976).

[5] F. Biraud, E.J. Blum, and J.C. Ribes, "Some new possibilities of optimum synthetic linear arrays for radio-astronomy." Astronomy and Astrophysics 41, 409-413 (1975).

[6] A. Kotzig, "Modern applications of elementary mathematics." Invited address, Third Annual Atlantic Math. Days, Halifax 1979.

[7] A. Kotzig and J. Turgeon, "Regular perfect systems of difference sets." Discrete Math. 20, 249-254 (1977).

[8] A. Kotzig and J. Turgeon, "Perfect systems of difference sets and additive sequences of permutations." Proc. 10th S.-E. Conf. Combinatorics, Graph Theory, and Computing, Boca Raton 1979, Vol. II, Congr. Numerantium 24, 629-636 (1979).

[9] A. Kotzig and J. Turgeon, "Sur l'existence de petites composantes dans tout systeme parfait d'ensemble de differences." Ann. Discrete Math. 8, pp. 71-75 (1980).

[10] A. Kotzig and J. Turgeon, "On a conjecture of Paul Erdös for perfect systems of difference sets." Publ. C.R.M.A., Université de Montréal, No. 897, September 1979.

[11] P.J. Laufer, "Some new results on regular perfect systems." Abstracts of papers presented to the American Math. Society Vol. 1 (1980), No. 2, p. 214.

[12] P.J. Laufer, "Regular perfect systems of different sets of size 4." Ann. Discrete Math. 12 (1982) (to appear).

[13] D.G. Rogers, "Addition theorems for perfect systems of difference sets." J. London Math. Soc. (2), 23, 385-395 (1981).

[14] D.G. Rogers, "A multiplication theorem for perfect systems of difference sets." Discrete Math. (to appear).

Some Aspects of Coding Theory between Probability, Algebra, Combinatorics and Complexity Theory

Thomas Beth

Institut für Mathematische Maschinen und Datenverarbeitung I
Universität Erlangen-Nürnberg
8520 Erlangen, Federal Republic of Germany

1. Summary:

The aim of this survey report is to draw attention to some recent developments which seem to have changed the face of Coding Theory completely. While this area of applicable algebra -which has strongly been influenced [3] by hard problems of communications engineering- during the last two decades has become a main part of Combinatorics, reaching from Finite Geometries to Representation Theory [18], it has never been fully accepted as a part of Algebra itself - the reasons for this being manifold. On the one hand, Coding Theory can easily be mistaken for a part of Linear Algebra, while on the other hand a non-typical feature distinguishes it from the main concept of modern and classical algebra: The properties of codes are "basis-dependent" so that the many tools of "basis-free" algebra are not always helpful.

Due to some very recent publications ([13],[19],[29]) this situation may be changing very soon, as the interaction between these fields has provided new insights into both:

Results from Algebraic Geometry permit the construction of codes, which are better than those known before, while very well-known bounds on codes in turn improve Weil's bound for the number of points on a curve over a finite field.
On the other hand, a very recent paper [9] shows that the construction of extremely good codes is possible by rather elementary means.

The aim of this survey report is to introduce a general mathematical audience to the background, eventually leading to these new developments. As is usual for a survey the author has included results from many different fields, not just from his own one. Thus, this report hopefully is in the spirit of the classical understanding of research - providing a collection of material which is not even contained in the most recent book [30] on Coding Theory.

2. Introduction:

Coding Theory has been developed during the last four decades in order to improve or secure the quality of data transmission systems, where transmission can mean to transport data through space (e.g. telephone links, satellite communications etc.) or time

(e.g. magnetic tapes, optical discs etc.). Possible threats to the quality of such communication lines are additional noise on the line, erasures, distortions, bursts due to fading etc.
The general model for a transmission system embodying the feature of such error protection is shown in the following figure:

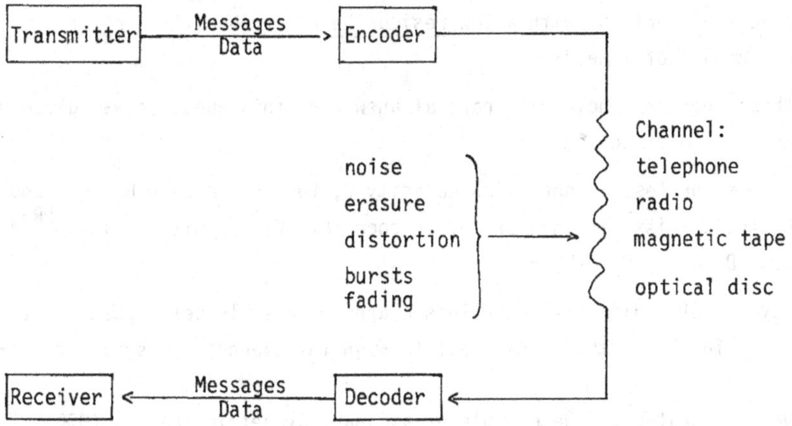

Figure 1.

At this point, some notation and definitions are necessary.

Henceforth, messages $\underline{m}$ are vectors of a fixed length k over a finite field $GF(q)$, i.e. $\underline{m} \in GF(q)^k$.

An encoder is an injective mapping

$$E : GF(q)^k \to GF(q)^n \quad ,$$

where $n(\geq k)$ is the length of the code $C = \text{im } E$, which consists of codewords $\underline{c} = (c_0,\ldots,c_{n-1}) \in C$.

The decoder is a surjective mapping

$$D : GF(q)^n \to GF(q)^k$$

performing a maximum likelihood decision [30] on the vectors coming out of the channel The channel is assumed to be discrete and memoryless [30] with known symbol-error distribution $p(x,y)$, $x,y \in GF(q)$, determining the channel capacity C. D maps the received word $\underline{w}$ to the message element most likely to have produced that result; for a symmetric discrete, memoryless channel this means the $\underline{m} \in GF(q)^k$ for which $E(\underline{m})$ differs from $\underline{w}$ in the fewest coordinates.

The practical problem now is

"How to find a good code?"

which more explicitly reads

"How to find an encoder/decoder pair (E,D) such that both can be computed in a small number of steps, such that $\mathcal{C} = \text{im } E$ has a high rate $R = \frac{k}{n}$ and the decoder D performs with a low residual error probability $P_{err}(D)$ using a small amount of time."

A beautiful - though not very helpful - partial answer to this question was given by Shannon [27], whose famous theorem reads

For a discrete memoryless channel with capacity C, for any R $(0 < R < C)$ and each $\varepsilon > 0$ there exists an integer n, a code $\mathcal{C} \subset GF(q)^n$ with $|\mathcal{C}| = q^{\lfloor Rn \rfloor}$ and a decoder D with $P_{err}(D) < \varepsilon$.

Here the capacity C of a discrete memoryless channel means, loosely speaking, the maximum amount of information that can be put through the channel per symbol transmitted.

This can best be illustrated by the example of a Binary Symmetric Channel (BSC) transmitting zeros and ones with the symbol error probability p of interchanging the two symbols. Its capacity $C_{BSC(p)}$ is easily computed [32] to be

$$C_{BSC(p)} = 1 + p\log_2 p + (1-p)\log_2(1-p).$$

Thus, for a noiseless channel $(p = 0)$ the capacity reaches its maximum of 100%, while in the jammed situation $(p = \frac{1}{2})$ the capacity is zero.

Proofs of Shannon's theorem for the simplest case of the Binary Symmetric Channel (BSC), giving the flavour of the proof technique, can be found in [30] or [32]. It should be remarked that this technique is based on a purely probabilistic argument so that Shannon's theorem gives a completely non-constructive answer to the question above in the following sense: A random code of length n and the appropriate size with the maximum likelihood decoding rule is very likely to have a very small residual error probability. For many reasons, though, it is desirable to be able to construct the code explicitly.

2. Linear Codes

An immediate approach to the problem mentioned above is to consider linear codes. A linear (n,k,q)-code is a k-dimensional subspace $\mathcal{C} = \text{im } E$, where $E \in \text{Hom}(GF(q)^k, GF(q)^n)$.

With respect to a suitable basis, the encoder E is given by the generating matrix G_E which maps the messages onto the codewords by $E : \underline{m} \to \underline{m}\, G_E$.

Hence C is the rowspace of G_E. Any basis $\underline{h}_1,\dots,\underline{h}_{n-k}$ of the dual code $C^\perp$ can be used to form the control matrix H of C, i.e. $H = [\underline{h}_1^t,\dots,\underline{h}_{n-k}^t]$. Thus an alternative description of the linear code C is, therefore, given by $C = \ker H$.

Using a probabilistic reasoning applied to control matrices, one can prove Shannon's theorem [32] in the class of linear codes over GF(q) as well. Yet, apart from the problem of finding an infinite class of linear codes with rates approaching channel capacity, the question as to how to decode linear codes efficiently is still not answered.

If the discrete memoryless channel is additive, i.e. $p(x,y) = \mu(x-y)$ with $\mu(o) = \rho$ and $\mu(x) = \frac{1-\rho}{q-1}$ for $x \in GF(q) \setminus \{o\}$, then the maximum likelihood decoding principle is equivalent to the method of minimum weight decoding, which reads:

For any $\underline{u} \in GF(q)^n$ define $D(\underline{u})$ to be the codeword $\underline{c} \in C$ with

$$\operatorname{wgt}(\underline{u}-\underline{c}) = \min_{\underline{x}\in C} \operatorname{wgt}(\underline{u}-\underline{x}),$$

where for $\underline{z} \in GF(q)^n$ the weight $\operatorname{wgt}(\underline{z})$ of $\underline{z}$ is given by $\operatorname{wgt}(\underline{z}) = |\{i \mid o \le i \le n-1,\ z_i \neq o\}|$.

Here again it should be pointed out that the concept of weight is heavily dependent on the choice of basis in the vector space.

Decoding by the minimum weight method means algebraically: For each coset $C+\underline{u}$ of C in $GF(q)^n$ the minimum weight vector $\underline{e}$, the coset leader, has to be computed. Comments on the complexity of this problem will be made later. Before doing so, however, the geometric meaning of this decoding procedure will be discussed more closely:

Defining the minimum weight d of C by

$$d = \min\{\operatorname{wgt}(\underline{c}) \mid \underline{c} \in C \setminus \{o\}\}$$

the code is said to be e-error-correcting if $2e+1 \leqq d$.

The geometric meaning behind this is clear, observing that a metric is defined on $GF(q)^n$ by the Hamming distance

$$d(\underline{x},\underline{y}) = \operatorname{wgt}(\underline{x}-\underline{y}) \quad \text{of} \quad \underline{x},\underline{y} \in GF(q)^n.$$

If C is e-error-correcting the spheres $S_e(\underline{c})$ of radius e around the codewords $\underline{c}$ are disjoint.

The following elementary estimate for deriving a lower bound for d from the structure of the control-matrix H of C will be used throughout the rest of this report.

Fact 1. If any $l-1$ rows of the control-matrix H of C are linearly independent, then $d \geq l$.

We now give an example.

Example 2. Let $r \geq 2$ be an integer, $q = 2$ and $n = 2^r-1$. Let H be the $(2^r-1)\mathrm{x}r$-matrix the rows of which are formed by the non-zero vectors of $GF(2)^r$. Then $C = \ker H$ is the r-th Hamming-code with $n = 2^r-1$, $k = 2^r-1-r$, $d = 3$ as is easily verified. These codes have especially attractive feature of being decodable nicely. Suppose the received vector is $\underline{u} \in GF(2)^n$. We assume that at most one error occured during transmission (recall $d = 3$). Then $\underline{u}$ has the form $\underline{u} = \underline{c} + \underline{e}$, where both the codeword $\underline{c}$ and the error-pattern $\underline{e}$, having at most one non-zero entry, are unknown. To determine these we compute the syndrome

$$\underline{u} \cdot H = \underline{c} \cdot H + \underline{e} \cdot H = \underline{o} + \underline{e} H$$

since $C = \ker H$. Now $\underline{e}H$ is the i-th row of H if $\underline{e}$ was of the form $\underline{e} = (o,..,o,1,o,...,o)$, where the entry 1 occurs in position i. Thus the error is in position i and can easily be corrected (confer [3],[6]).

Now it is time to come to the draw-backs of using linear codes. Decoding is generally a much harder problem - only Hamming codes are so nice. Recalling the remark about coset leaders we note that for an (n,k,q)-linear code a table of q^{n-k} coset-leaders has to be stored, if this simple decoding method is used. Even though more sophisticated methods can be applied in certain cases, the problem of decoding arbitrary linear (n,k,q)-codes is NP-complete (see [2]or[4]).
Finally, one further remark should be made here. Although Shannon's theorem holds for the class of linear codes over $GF(q)$, linear codes with given minimum distance will generally contain fewer codewords than unrestricted optimal codes. In other words, linear codes are subjected to quite restrictive bounds.

3. Bounds on the parameters of codes

An immediate restriction on the choise of the parameters of linear codes can easily be derived using Fact 1 from the previous section.

Observation 3. For a linear code of lenght n, dimension k and minimum weight d the inequality $k+d \leq n+1$ holds.

Codes satisfying this bound with equality are called Maximum Distance Separable (MDS)-Codes (confer [12],[18],[30]).

In a later section we shall return to this special class of codes.

Another bound that can easily be derived by standard counting arguments is given in the following:

Lemma 4. For a linear (n,k,q)-code C the equality $\sum_{c \in C} wgt(c) = nq^k - nq^{k-1}$ holds.

The proof of this lemma is by standard double counting (the non-zero entries in the matrix formed by the list of codewords). An easy but useful consequence of this lemma is the so-called Plotkin bound:

$$(q^k-1)d \leq nq^k(\frac{q-1}{q}) .$$

For given q let A(n,d) denote the maximum possible size of a linear code $C < GF(q)^n$ of length n and minimum distance at least d.

Using this notation two important bounds are the following:

Lemma 5 (Plotkin bound). If $d > n\frac{q-1}{q}$ then

$$A(n,d) \cdot (d-\Theta n) \leq d, \quad \text{where } \Theta = \frac{q-1}{q}.$$

Lemma 6 (Sphere-packing bound).

$$A(n,d) \cdot \sum_{i=0}^{\lfloor\frac{d-1}{2}\rfloor} \binom{n}{i}(q-1)^i \leq q^n.$$

The proof is again by double counting [30]. Equality in this bound is achieved by the so-called Perfect Codes (see [18],[28],[30]).
Whereas the above bounds are upper bounds, a powerful lower bound has been given by Gilbert and Varshamov [18].

Lemma 7 (Gilbert-Varshamov bound).

$$\text{If} \quad \sum_{i=0}^{d-2} \binom{n-1}{i}(q-1)^i < q^{n-k} \quad \text{then} \quad A(n,d) > q^k.$$

Again, the proof makes use of Fact 1. One can construct a control matrix for such a code by iteration. If i rows have been chosen such that any d-1 are linearly independent and if, furthermore,

$$\sum_{j=1}^{d-2} (q-1)^j\binom{i}{j} < q^{n-k}-1, \quad \text{then add a new row to H.}$$

Important asymptotic versions of bounds are derived immediately from these finite versions. For a linear code of length n, dimension k and minimum distance d let $R = \frac{k}{n}$ denote the rate of the code and the ratio $\delta = \frac{d}{n}$ the relative distance.

If the length n tends to infinity (as Shannon's theorem proposes), the asymptotic behaviour is of considerable interest. Observation 3 and Lemma 4 yield

(3') $R + \delta \lesssim 1$

(4') $\delta \lesssim \frac{q-1}{q}$ for linear codes over $GF(q)$ of sufficiently large rate.

Here $f \lesssim g$ means $f(n) \le g(n)(1+o(n))$ as $n \to \infty$.

In the same manner the asymptotic Plotkin bound may be derived.

Theorem 8: For linear codes over $GF(q)$

$$R \lesssim \begin{cases} 1-\frac{q}{q-1}\delta & \text{for } \delta \in [o, \frac{q-1}{q}] \\ 0 & \text{for } \delta \in [\frac{q-1}{q}, 1] \end{cases}$$

Proof: Take any "long" (n,k,q)-code with minimum weight d. Shorten it, i.e. take all codewords having zeros in fixed positions and omit these coordinates, to a length n fulfilling the condition

$$d > n'\frac{q-1}{q} \quad \text{of Lemma 5.}$$

Then this shortened code still has minimum weight d. Plotkin's bound, however, gives $A(n',d) \le \frac{d}{d-\frac{q-1}{q}n'} < d$. This in turn implies $q^k < d \cdot q^{n-n'}$. With $n \to \infty$ and $d = \delta n$, we have $\frac{n'}{n} \to \delta\frac{q}{q-1}$ and $R \le 1 - \delta\frac{q}{q-1}$. □

The most interesting conclusion can be derived from Lemma 7 giving the Gilbert-Varshamov asymptotic lower bound.

Theorem 9. Let $0 \le \delta < \frac{q-1}{q}$. For any q there exist linear codes C_i over $GF(q)$ with rates R_i and relative minimal distance $\delta_i > \delta$ satisfying

$$R_i \gtrsim 1 - H_q(\delta_i),$$

where $H_q(x) = x\log_q(q-1) - x\log_q x - (1-x)\log_q(1-x)$.

Proof: We may assume that $q = 2$.

One can estimate partial binomial sums as follows:

Let $\frac{1}{2} < \lambda < 1$. Then

$$2^{x\cdot\lambda\cdot n}\sum_{i=\lambda n}^{n}\binom{n}{i} \le \sum_{i=o}^{n} 2^{xi}\binom{n}{i} = (1+2^x)^n .$$

Thus $\sum_{i=\lambda n}^{n}\binom{n}{i} \le (2^{-x\lambda}+2^{+x(1-\lambda)})^n$. For $x = \log_2(\frac{\lambda}{1-\lambda})$ this inequality becomes

$$\sum_{i=\lambda n}^{n}\binom{n}{i} \le 2^{n H_2(\lambda)} .$$

Symmetry of binomial coefficients and Sterling's formula then give

$$\frac{2^{n H_2(\delta)}}{\sqrt{8n\,\delta(1-\delta)}} \le \sum_{i=o}^{\delta n}\binom{n}{i} \le 2^{n H_2(\delta)} .$$

Applying this to Lemma 7 for suitable values of n_i, k_i, d_i yields the assertion. □

The asymptotic form of the sphere-packing bound gives an upper bound for R using the same technique [18].

<u>Theorem 10.</u> $R \lesssim 1 - H_q(\frac{\delta}{2})$.

Using tools from Representation Theory, Linear Programming and the theory of Special Functions this bound can be sharpened to the McEliece-Rodemich-Rumsey-Welch-bound [11] for the case $q=2$. This will be of importance later:

<u>Theorem 11:</u> For $q=2$ we have $R \lesssim 1 - H_2(\frac{1}{2} - \sqrt{\delta(1-\delta)}\,)$.

Before going to the next sections, a rough sketch should provide some intuitive understanding of these bounds. (The true upper bound of the (hatched) region of asymptotically admissible codes will not be drawn).

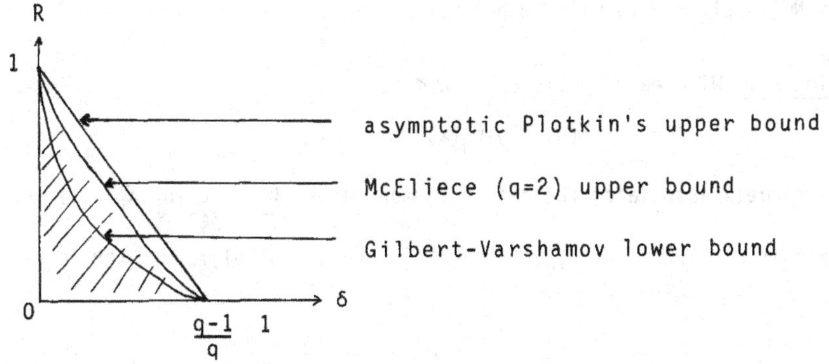

Codes which are on or above the Gilbert-Varshamov bound for large n are called "good" codes. They exist by Theorem 9. By Shannon's theorem, however, there are much "better" codes with respect to R and P_{err}. These, of course, cannot have a uniformly high minimum weight. However, there still remains the major problem of constructing good codes which are well decodable.

4. Codes which are well decodable: Cyclic Codes

Definition. Let $n \in \mathbb{N}$, $(n,q) = 1$ and

$$GF(q)[\mathbb{Z}_n] \simeq GF(q)[x]/(x^n-1)$$

be the group ring R_n. An ideal $C \triangleleft R_n$ is called a cyclic code.

How can one determine the rate and the minimum distance of such a code? An analysis of the structure of R_n will help to answer this question. Since $(n,q) = 1$, R_n is a semisimple ring:

$$R_n \simeq \bigoplus_{i=1}^{s} GF(q)[x]/(p_i(x)),$$

where the $p_i(x)$ are the irreducible factors of $x^n-1 = \prod_{i=1}^{s} p_i(x)$. Thus a cyclic code C is always a partial sum

$$C \simeq \bigoplus_{i \in I} GF(q)[x]/(p_i(x))$$

for some subset $I \subseteq \{1,\ldots,s\}$. Hence $c(x) \in R_n$ is a codeword $c \in C$

iff $\quad c(x) \equiv 0 \mod p_j(x)$ for each $j \in \{1,\ldots,s\} \setminus I$,

iff $\quad c(x_j) = 0$ for each root α_j of $p_j(x)$, $j \notin I$.

Using these basic facts almost all properties of cyclic codes can easily be derived. In this context, a simple observation, which may be considered as a finite version of Hilbert's Nullstellensatz, will be helpful.

Observation 12. With each cyclic code $C \triangleleft R_n$

$$C = \bigoplus_{i \in I} GF(q)[x]/p_i(x)$$

there is uniquely associated the set of common zeros $V_C = \bigcup_{j \notin I} \{\alpha_j, \alpha_j^q \ldots\}$ of the code polynomials in C. Then $\dim_{GF(q)} C = n - |V_C|$ and a control matrix for C is given by

$$H = \begin{pmatrix} 1 \\ \alpha_j \\ \cdot \\ \cdots\cdots\cdot\cdots\cdots \\ \cdot \\ \alpha_j^{n-1} \end{pmatrix}_{j \notin I} .$$

This matrix shows the close connection to the technique of Discrete Fourier Transforms (see, for instance, [6],[18],[21]) which immediately yields fast and efficient decoding algorithm (see, for instance, [3],[6] [32]). A rewording of Fact 1 gives an estimate for the minimum distance of a cyclic code.

Corollary 13 (BCH-bound). Let l be the length of the longest arithmetic progression of exponents of elements of V_C with a stepwidth relatively prime to n. Then the minimum weight d of C satisfies the inequality $d \geq l + 1$.

Before discussing the consequences we give an example to illustrate the usefulness of these observations.

Example 14. Consider the Hamming Code H_3 [5]. We observe that its control matrix consists of all points of PG(2,2) and rearrange the rows according to the Singer group of PG(2,2). Then the set of rearranged codewords is an ideal

$$H_3' \triangleleft GF(2)[x]/(x^7-1).$$

Over $GF(2)[x]$

$$x^7-1 = (x-1)(x^3+x+1)(x^3+x^2+1).$$

Since $x^3+x+1 = p(x)$ is the indicator function of the difference set $\{0,1,3\} \bmod 7$, H_3' is the ideal generated by $p(x)$. So let α be a root of p. Then $V_{H_3'} = \{\alpha,\alpha^2,\alpha^4\}$. Hence $\dim H_3' = 7-3 = 4$ and $d \geq 3$ as $\ell = 2$.

The corresponding control matrix is given by

$$H' = \begin{pmatrix} 1 \\ \alpha \\ \alpha^2 \\ \alpha^3 \\ \alpha^4 \\ \alpha^5 \\ \alpha^6 \end{pmatrix} = \begin{pmatrix} 1&0&0 \\ 0&1&0 \\ 0&0&1 \\ 0&1&1 \\ 1&1&0 \\ 1&1&1 \\ 1&0&1 \end{pmatrix}$$

Decoding this code is even easier than described in Example 2 and mathematically more attractive (confer [3] or [6]).

This short description of the properties of cyclic codes shows that they are

- well described

- easily designed

and, moreover,

- very practical,

as there are decoding procedures (see [3],[20],[22]) for cyclic codes of length n requiring $O(n \log^2 n)$ steps.

After these good properties, one has to mention the sad ones. Although some improvements on the BCH-bound are known (see [7] or [23]), it is bad asymtotically as the following theorem shows [18].

Theorem 15. There does not exist an infinite sequence of primitive BCH-codes of length n over $GF(q)$, where primitive means $n = q^m - 1$, with δ and R approaching non-zero limits.

Though cyclic codes may be very interesting for specific applications (e.g. the (23,12)-Golay code has $\delta = \frac{7}{23}$, $R = \frac{12}{23}$, so that (δ, R) lies far above the asymptotic Plotkin bound). Theorem 15 shows that in terms of asymptotic behaviour they do not meet any of the bounds discussed.

Recently other constructions were proposed which do not have those disadvantages. The first one is due to Ahlswede and Dueck [1] using random arguments. They show that

for $n \to \infty$ there exists an $\underline{x} \in GF(2)^n$ and $\pi_1, \ldots, \pi_{\lfloor R \cdot n \rfloor} \in S_n$ such that

$$C = \{\underline{x}, \underline{x}^{\pi_1}, \underline{x}^{\pi_2}, \underline{x}^{\pi_1 \pi_2}, \ldots\}$$

is a code of rate R (close to capacity) with $P_{err} \to 0$.

The disadvantage of this approach is clear. The proof is non-constructive and does not give any computationally feasible decoding rule.

The second one is just being published by Delsarte and Piret [9] who give an algebraic construction of codes with a feasible encoding/decoding algorithm which simultaneously reach channel capacity and have probability of erroneous decoding tending to zero.

This problem was a major challenge during the last few years. It may briefly be mentioned that the construction of these codes makes use of the idea of concatenated codes which are defined as follows:

First consider a so-called outer code, (for instance, a cyclic code over a field $GF(q^m)$) consisting of codewords $\underline{c} = (c_0, \ldots, c_{n-1})$. Here each c_i is a vector of length m over $GF(q)$ and thus a codeword of the so-called inner code. Delsarte and Piret succeeded in designing suitable outer and inner codes to achieve their result.

It should, however, be mentioned that this is only possible because they drop the requirement of a least d.

On the other hand, this condition was not omitted in the investigations going back to Goppa (see [13],[15],[19],[29]) whose results will be discussed in the sequel.

5. Codes which hopefully are better: Goppa-Codes

For the construction of Goppa-Codes one considers an extension field $GF(q^m)$ of $GF(q)$. Let $G \in GF(q^m)[z]$ be a polynomial of degree r. Let

$$P = \{\alpha_1,\dots,\alpha_n\} \subset GF(q^m)$$

be a set of places in $GF(q^m)$ such that $P \cap V_G = \emptyset$, i.e. P does not contain a zero of G.

The Goppa-Code $\Gamma(P,G)$ is the subspace of vectors $\underline{c} \in GF(q)^n$ fulfilling the equation

$$(*) \qquad \sum_{i=0}^{n-1} \frac{c_i}{z-\alpha_i} = 0 \quad \text{in} \quad GF(q^m)[z]/(G).$$

The properties of Goppa-Codes can easily be derived from this construction.

Theorem 16. The Goppa-Code $\Gamma(P,G)$ has dimension $k \geq n - mr$ and minimum weight $d \geq r+1$.

Proof. Consider the equation (*). By Lagrange interpolation a vector $\underline{c} \in GF(q)^n$ is a codeword iff

$$\sum \frac{-c_i}{G(\alpha_i)^{-1}} \cdot \frac{G(z)-G(\alpha_i)}{z-\alpha_i} = 0 \quad \text{in} \quad GF(q^m)[z] .$$

Thus the coefficients of $z^0,\dots,z^{r-1}$ have to satisfy a system of linear equations of rank at most r over $GF(q^m)$ and thus rank at most $r \cdot m$ over $GF(q)$. Hence $\dim \Gamma(P,G) \geq n - mr$. Equation (*) also shows that at least $r+1$ of the c_i's have to be non-zero in order to give a rational function the enumerator of which is divisible by G. □

The latter argument shows implicitly that Goppa-Codes are well decoded by the Berlekamp-Massey-Patterson algorithm (confer [3],[20],[22]), using the approximation by rational functions, with complexity $O(n \log^2 n)$.

Another combinatorial observation leads to an estimation of the asymptotic behaviour of Goppa-Codes.

Observation 17. ([18],[30]) Let $I_{q^m}(r)$ be the number of irreducible polynomials of degree r over $GF(q^m)$ and assume

$$(**)\qquad \sum_{w=r+1}^{d-1} \lfloor \frac{w-1}{r} \rfloor (q-1)^w \binom{q^m}{w} < I_{q^m}(r).$$

Then there exists a Goppa-Code $\Gamma(P,G)$ over $GF(q)$ with an irreducible polynomial $G \in GF(q^m)[z]$ of degree r admitting $GF(q^m)$ and having parameters

$$n = q^m, \quad k \geq n - rm \quad \text{and minimum weight } d.$$

Proof. By the same reasoning as in the proof of the minimum weight bound, it is clear that the denominator $D(z)$ of the expression (*) for a codeword of weight w is of degree $w-1$. Thus $D(z)$ can be divided by $\lfloor \frac{w-1}{r} \rfloor$ irreducible polynomials of degree r. □

We can now prove the asymptotic result on Goppa-codes [18].

Theorem 18. For $n \to \infty$ there exist Goppa-codes asymptotically meeting the Gilbert-Varshamow bound.

Proof. Observe that $I_{q^m}(r) \geq \frac{1}{r}(q^m - (r-1)q^{\frac{m}{2}})$ (see [3] or [18]). Substitution of this slightly sharper bound in condition (**) yields, using Lemma 7, the statement for $n \to \infty$.

For many years this result has been believed to be best possible.

The last section of this report gives a short description of very recent results which show that the final word has not been said, yet.

6. Generalized Goppa-Codes are better

During the years 1979-81 Goppa [13] pointed out that the construction of Goppa-codes is possible over other extensionfields (e.g. function fields) and that this might lead to better codes if one uses tools from algebraic geometry. This idea has recently been used by Tsfasman, Vladut and Zink who give a very compressed plan of attack in [29].

Goppa's idea is the following (see [13],[19]).

Let X be a smooth projective curve of genus g over $GF(q)$. Let $\{P_0, P_1 \ldots, P_n\}$ be the set of rational points of X over $GF(q)$.

Let $D = \sum_{i=1}^{n} P_i$ and $G = \Sigma m_Q Q$ be two effective divisors [31], where G is assumed to be rational (i.e. $m_Q = m_{Q^\sigma}$ for all conjugates Q^σ of Q) and of degree a. Furthermore, the supports of the divisors D and G should be disjoint. Let $\Omega(D-G)$ denote the GF(q)-vector space of differentials ω, whose divisors (ω) dominate G-D, (i.e. $(\omega) \geq G-D$).

Let $2g-2 < a \leq n+g-1$. The linear map

$$res_D : \Omega(D-G) \to \mathbb{F}_q^{\,n}, \text{ given by}$$
$$\omega \to res_D(\omega) = (res_{P_1}(\omega),\dots,res_{P_n}(\omega))$$

is injective [19], since $a > 2g-2$.
The <u>generalized Goppa-code</u> is the subspace $\overline{\Gamma}(D,G) = im(res_D)$ of $\mathbb{F}_q^{\,n}$. By Riemann-Roch's theorem (see [31],[24]), $\overline{\Gamma}(D,G)$ has the dimension $k = n+g-1-a$.
The weight $wgt(\omega) = wgt(res_D(\omega))$ in a natural way is given by $wgt(\omega) = |\{i \mid res_{P_i}(\omega) \neq 0\}|$ being the number of poles of ω, which is equal to the number of zeros of (ω) reduced by the degree of (ω). Thus we obtain the following inequality for the minimum weight d: $d \geq a-2g-2$.

<u>Remark 19.</u> The close relation between the generalized Goppa-codes and those of section 5 can be worked out without too much effort. The similarity of the concepts becomes clear by the residue map res which, in the case of the definition of the elementary Goppa-codes of section 5, is just the mapping

$$\sum_{i=1}^{n} \frac{c_i}{z-\alpha_i} \to (c_1,\dots,c_n),$$

while the "disjointness" of the divisors G and D corresponds to the condition that G(z) should have no zeros in the set $\{\alpha_1,\dots,\alpha_n\}$.

To consider the asymptotic behaviour of generalized Goppa-codes we restrict our attention to a slightly less general situation.

Let $\{P_0,\dots,P_n\}$ be the set of all rational points on X, $D = \sum_{i=1}^{n} P_i$ and $G = a P_0$.
Then the code $\overline{\Gamma}(D,G)$ has the rate $R = \frac{n+g-1-a}{n}$ and relative distance $\delta \geq \frac{a+2-2g}{n}$.
For the sequel of a closer look at the asymptotic behaviour of the quotient $\frac{g-1}{n}$ is is necessary. By Weil's theorem [14], the number $n(X)+1$ of rational points (i.e. GF(q)-points on a curve X of genus g(X)) is estimated by the following inequality [19]: $|n(X)-q| \leq 2g(X)\sqrt{q}$.

If we define $\gamma_q(X) = \frac{g(X)-1}{n(X)}$ then (see [19],[29])

$$\gamma_q = \liminf_X \gamma_q(X) \geq \frac{1}{2\sqrt{q}} .$$

A consequence of this result was observed by Tsfasman [19].

Theorem 20. The line $R+\delta = 1-\gamma_q$ with $0 \leq R,\delta \leq 1-\gamma$ lies completely in the region of asymptotically admissible codes over $GF(q)$.

Proof. Recall that a is allowed to run over the interval $2g-2 < a \leqq n+g-1$. Taking the limit over all curves X and observing that d is bounded from below by $a-2g-2$, the statement becomes obvious. □

This immediately yields two consequences. The first one is an improvement of Weil's inequality for the cases $q=2$ and $q=3$ [19].

Corollary 21. For $q=2,3$ the asymptotic Plotkin bound gives $\gamma_2 \geq \frac{1}{2}$ and $\gamma_3 \geq \frac{1}{3}$.

Remark 22. This improves the classical Weil bounds $\gamma_2 > \frac{1}{2\sqrt{2}}$ and $\gamma_3 > \frac{1}{2\sqrt{3}}$.

Yet another observation implies a further improvement of this estimate.

Theorem 23. $\gamma_2 \geq 0.525$.

Proof. By Theorem 20, the line $R+\delta = 1-\gamma_2$ has to lie under McEliece's bound (see Theorem 11). Thus γ_2 has to be large enough that the line $R+1 = 1-\gamma_2$ is a tangent of McEliece's function $H_2(\frac{1}{2}-\sqrt{\delta(1-\delta)}\,)$. Differentiation gives a tangent for $\gamma_2 = 0.525\ldots$. □

For $q>3$ no such simple corollaries of theorems in Coding Theory are known to yield improvements in this area of Algebraic Geometry.

On the other hand, the tools provided so far will give a contribution to Coding Theory improving the Gilbert-Varshamov lower bound. These results are due to Zink and Vladut [29] and Thara ([15],[16]).

Theorem 24. For $q = p^2$, P a prime, there are families of curves X over $GF(q)$ satisfying $\lim_X \gamma_q(X) = \frac{1}{\sqrt{q}-1}$.

Corollary 25. Let $S_1 < S_2$ be the solutions of the equation $H_q(S) - S = \frac{1}{\sqrt{q}-1}$. For $q = p^2$, $p \geq 7$, there are generalized Goppa-codes over GF(q) which asymptotically "lie" on the line-segment connecting $(S_1, H(S_1))$ and $(S_2, H(S_2))$. Thus they are clearly above the Gilbert-Varshamov bound.

Proof. Observe that for $q \geq 7^2$ the straight line $R + S = 1 - \frac{1}{\sqrt{q}-1}$ lies above the Gilbert-Varshamov bound, for all $S \in (S_1, S_2)$.

7. Conclusion

The results from the last section show that generalized Goppa-codes are better than all other known families of codes: The reader should note that they only exist for relatively large alphabets. No easy approach for the interesting case q=2, which in most practical application is of major interest, seems to be known [19]. Besides, for generalized Goppa-codes it is necessary to compute the genus of a given curve as well as a basis of the space of differentials. In principle this can be performed by Coates algorithm [8]. Nevertheless, the question of efficient decoding algorithms is not settled [19]. Although Goppa [13] refers to Mahler's p-adic approximation algorithm [17], a general method has not been developed yet.

Finally it should be pointed out that the Delsarte-Piret approach [9], cf.Sect.5, seems to be far more promising, as they construct codes achieving capacity (which is far beyond all bounds discussed so far) by dropping the requirement of a uniform minimum weight.

It should be a challenge to all coding theorists to use this approach to construct codes lying above the Gilbert-Varshamov bound by elementary means.

Acknowledgement:

The author wants to thank Professor R.Ahlswede (Bielefeld) for bringing the key paper [29] to his attention. The author is also grateful to Professor W.D.Geyer (Erlangen) for some helpful remarks in connection with the papers [29] and [19] and - last but not least - to Dr. P.J.Cameron (Oxford) for reading this paper and making several comments and suggestions.

References:

[1] R.Ahlswede, G.Dueck: The best known codes are Highly Probable and can be produced by a few permutations, Universität Bielefeld, Materialien XXIII, 1982

[2] A.V.Aho, J.E.Hopcroft, J.D.Ullmann: The Design and Analysis of Computer Algorithms, Addison-Wesley, 1974

[3] E.R.Berlekamp: Algebraic Coding Theory, McGraw-Hill, 1968

[4] E.R.Berlekamp, R.McEliece, H.V.Tilburg: On the inherent intractability of certain coding problems, IEEE Trans.Inf.Th. 24, 384-386, 1978

[5] T.Beth, V.Strehl: Materialien zur Codierungstheorie, Berichte IMMD, Erlangen, 11 Heft 14, 1978

[6] T.Beth et al.: Materialien zur Codierungstheorie II, Berichte IMMD, Erlangen, 12 Heft 10, 1979

[7] P.Camion: Abelian Codes, Math.Res.Center, Univ. of Wisconsin, Rept.1059, 1970

[8] J.H.Davenport: On the Integration of Algebraic Functions, LNCS 102, Springer 1981

[9] P.Delsarte, P.Piret: Algebraic Constructions of Shannon Codes for Regular Channels, IEEE Trans.Inf.Th. 28, 593-599, 1982

[10] M.Eichler: Introduction to the Theory of Algebraic Numbers and Functions, Acad.Press, 1966

[11] R.McEliece, E.R.Rodemich et al.: New upper bounds on the rate of a code via the Delsarte-MacWilliam inequalities, IEEE Trans.Inf.Th. 23, 157-166, 1977

[12] G.Falkner, W.Heise et al.: On the existence of Cyclic Optimal Codes, Atti Sem. Mat.Fis., Univ.Modena, XXVIII, 326-341, 1979

[13] V.D.Goppa: Codes on Algebraic Curves, Soviet Math.Dokl. 24, 170-172, 1981

[14] J.Hirschfeld: Projective Geometries over Finite Fields, Oxford University 1979

[15] Y.Ihara: Some remarks on the number of rational points of algebraic curves over finite fields, J.Fac.Science, Tokyo, 28, 721-724, 1981

[16] Y.Ihara: On modular curves over finite fields, in: Discrete subgroups of Lie groups and applications to moduli, Oxford Univ.Press 1973

[17] K.Ireland, M.Rosen: A Classical Introduction to Modern Number Theory, Springer GTM 84, 1982

[18] F.J.MacWilliams, N.J.A.Sloane: The Theory of Error-Correcting codes, North-Holland, 1978

[19] Y.I.Manin: What is the maximum number of points on a curve over $\mathbb{F}_2$? Fac.Science, Tokyo, 28, 715-720, 1981

[20] J.L.Massey: Step by Step decoding of the BCH-codes, IEEE Trans.Inf.Th. 11, 580-585, 1965

[21] H.Nussbaumer: The Fast Fourier Transform and Convolution Algorithms, Springer, Wien, 1980

[22] N.J.Patterson: The algebraic decoding of Goppa-Codes, IEEE Trans.Inf.Th.,21, 203-207, 1975

[23] C.Roos: A result on the minimum distance of a linear code with applications to cyclic codes, manuscript, 1982

[24] G.Schellenberger: Ein schneller Weg zum Riemann-Roch'schen Theorem, Studienarbeit, Erlangen, 1981

[25] G.Schellenberger: Codes on Curves, to appear in: Proceedings Colloquium Geometry and its Applications, Passa di Mendola, 1982

[26] W.M.Schmidt: Equations over Finite Fields - An Elementary Approach, Springer LNM 536, 1970

[27] C.E.Shannon: Communication in the presence of noise, Proc.IEEE, 37, 10-21, 1949

[28] A.Tietäväinen: On the nonexistence of perfect codes over finite fields, SIAM J.Appl.Math 24, 88-96, 1973

[29] M.A.Tsfassman, S.G.Vladut, T.Zink: Modular curves, Skimura curves and Goppa codes, better than Gilbert-Varshamov bound, Manuscript1982

[30] J.H.van Lint: Introduction to Coding Theory, Springer GTM 86, 1982

[31] B.L.van der Waerden: Algebra II, Springer HTB 23, 1966

[32] I. Zech: Methoden der fehlerkorrigierenden Codierung, Studienarbeit, Erlangen 1982.

GENERALIZED SCHUR NUMBERS

Albrecht Beutelspacher and Walter Brestovansky

Fachbereich Mathematik der Universität,
Saarstr. 21, D-6500 Mainz, West Germany

An *n-partition* of a set X is a set $\pi = \{A_1,\ldots,A_n\}$ of subsets of X such that any element of X is contained in exactly one element of π. The n-partition $\pi = \{A_1,\ldots,A_n\}$ of a set X of integers is said to be *m-sum free*, if in no *component* A_i of π there are m (not necessarily distinct) integers $a_1,\ldots,a_m$ with the property that $a_1+\ldots+a_{m-1} = a_m$ $(i \in \{1,\ldots,n\})$.

Issai SCHUR [4] proved in 1916 that, for a given n, there is an integer v such that no n-partition of $\{1,\ldots,v\}$ is 3-sum free. More generally, using RAMSEY's theorem, one can show that this is true for an arbitrary integer $m \geq 2$. (See for instance HALDER and HEISE [3], p. 142.) The smallest number v such that no n-partition of $\{1,\ldots,v\}$ is m-sum free will be denoted by $v = \sigma(m,n)$. These numbers $\sigma(m,n)$ are called *SCHUR numbers*. It is easy to check that for any number $v' \geq \sigma(m,n)$, there is no m-sum free n-partition of $\{1,\ldots,v'\}$.

SCHUR numbers have been thoroughly investigated (see e. g. [5]). In this paper we shall first determine the SCHUR numbers $\sigma(m,2)$. Moreover, in Section 2, we shall give new lower bounds for the numbers $\sigma(3,6)$ and $\sigma(3,7)$.

Afterwards, we define *generalized SCHUR numbers* for arithmetic progressions in an obvious way. Here, an interesting phenomenon occurs: There are generalized SCHUR numbers which are not finite. Therefore, in Section 3, we look for conditions which assure that these numbers are finite (or, infinite, respectively).

Finally, in Section 4, we shall give an explicit formula for the generalized SCHUR numbers with $n = 2$.

1. SCHUR numbers

In this Section we are concerned with the ordinary SCHUR numbers $\sigma(m,n)$. Clearly, $\sigma(m,1) = m-1$ and $\sigma(2,n) = 1$. Therefore, we shall always suppose $m \geq 3$. Our first tool is to determine $\sigma(m,2)$. In order to do this, the following general lemma is useful.

1.1 LEMMA. $\sigma(m,n) \geq m\cdot\sigma(m,n-1) - 1$.

Proof. It is to show that there exists an m-sum free n-partition of the set $\{1,\ldots,m\cdot\sigma(m,n-1)-2\}$.

By the definition of $\sigma = \sigma(m,n-1)$, there is an m-sum free (n-1)-partition $\{A_1,\ldots,A_{n-1}\}$ of $\{1,\ldots,\sigma-1\}$. Let us define the following sets:

$$A_n = \{\sigma,\sigma+1,\ldots,(m-1)\sigma - 1\}, \quad Y = \{(m-1)\sigma,\ldots,m\sigma - 2\}.$$

Moreover, the sets $B_1,\ldots,B_{n-1}$ are defined as follows:

$$(m-1)\sigma - 1 + s \in B_j \Leftrightarrow s \in A_j \quad (j \in \{1,\ldots,n-1\}).$$

We claim that $\pi = \{A_1 \cup B_1,\ldots,A_{n-1} \cup B_{n-1},A_n\}$ is an m-sum free n-partition of $X = \{1,\ldots,m\sigma-2\}$.

For: Obviously, π is an n-partition of X. In order to show that π is m-sum free we distinguish the following cases:

Case 1. Let $a_1,\ldots,a_{m-1}$ be elements of A_j for a $j \in \{1,\ldots,n-1\}$. Since A_j is m-sum free, $a_1+\ldots+a_{m-1} \notin A_j$. Moreover, since $a_1+\ldots+a_{m-1} \leq (m-1)(\sigma-1)$, $a_1+\ldots+a_{m-1}$ is no element of Y.

Case 2. Suppose $b_1 \in B_j$ and $a_2,\ldots,a_{m-1} \in A_j$ for a $j \in \{1,\ldots,n-1\}$. Then there exists an element a_1 in A_j with $b_1 = (m-1)\sigma - 1 + a_1$. It follows

$$b_1 + a_2+\ldots+a_{m-1} = a_1+a_2+\ldots+a_{m-1} + (m-1)\sigma - 1.$$

Since A_j is sum free, $a_1+\ldots+a_{m-1} \notin A_j$; consequently $b_1+a_2+\ldots+a_{m-1} \notin B_j$.

Case 3. Let b_1 and b_2 be elements of B_j and $c_3,\ldots,c_{m-1}$ be elements of $A_j \cup B_j$ $(j \in \{1,\ldots,n-1\})$. Since $m \geq 3$, it follows

$$b_1 + b_2 + c_3+\ldots+c_{m-1} \geq b_1 + b_2 \geq 2(m-1)\sigma > m\sigma,$$

hence $b_1+b_2+c_3+\ldots+c_{m-1} \notin A_j \cup B_j$.

Case 4. If $a_1,\ldots,a_{m-1}$ are elements of A_n, then

$$a_1+\ldots+a_{m-1} \geq (m-1)\sigma,$$

so $a_1+\ldots+a_{m-1} \notin A_n$.

Together we have shown that π is m-sum free. □

Since $\sigma(m,1) = m-1$, the above Lemma implies immediately

<u>1.2 COROLLARY</u>. $\sigma(m,n) \geq m^n - (m^{n-1}+\ldots+m+1)$. □

<u>1.3 THEOREM</u>. $\sigma(m,2) = m^2 - m - 1$.

<u>Proof</u>. In view of 1.2 we have only to show that $\sigma(m,2) \leq m^2-m-1$ holds. *Assume on the contrary* that there exists an m-sum free 2-partition $\{A_1,A_2\}$ of $\{1,\ldots,m^2-m-1\}$.

Without loss in generality we can suppose $1 \in A_1$. This implies $(m-1)\cdot 1 \notin A_1$, so $m-1 \in A_2$. Therefore $(m-1)\cdot(m-1) \notin A_2$, hence $(m-1)^2 \in A_1$. Now,

$$m^2-m-1 = (m-2)\cdot 1 + 1\cdot(m-1)^2 \notin A_1;$$

consequently $m^2-m-1 \in A_2$.

Consider now the number m. If $m \in A_1$, then

$$(m-1)^2 = (m-2)\cdot m + 1\cdot 1 \notin A_1;$$

if $m \in A_2$, then

$$m^2-m-1 = (m-2)\cdot m + 1\cdot(m-1) \notin A_2:$$

in both cases a contradiction. □

2. The SCHUR numbers $\sigma(3,6)$ and $\sigma(3,7)$

On the SCHUR numbers $\sigma(3,n)$, the following is known (see 1.3, [1] and [2]):

$$\sigma(3,2) = 5,$$
$$\sigma(3,3) = 14,$$
$$\sigma(3,4) = 45,$$
$$\sigma(3,5) \geq 158.$$

With the aid of a computer we could show the following

2.1 THEOREM. $\sigma(3,6) \geq 476$ and $\sigma(3,7) \geq 1430$.

(*Note* that 1.1 implies

$$\sigma(3,6) \geq 3\cdot\sigma(3,5) - 1 \geq 473, \text{ and}$$
$$\sigma(3,7) \geq 3\cdot\sigma(3,6) - 1 \geq 1427.)$$

The above Theorem implies in particular (cf. [5]):

2.2 COROLLARY. $\sigma(3,n) \geq \dfrac{2859\cdot 3^{n-7}+1}{2}$ for $n \geq 7$. □

Proof of Theorem 2.1.

(a) The following 3-sum free 6-partition $\{A_1,\ldots,A_6\}$ of $\{1,\ldots,475\}$ shows $\sigma(3,6) \geq 476$. For each integer $a \leq 475$ and any $i \in \{1,\ldots,6\}$ it holds

$$a \in A_i \Leftrightarrow 476-a \in A_i.$$

So, we list only the integers less than 239.

A_1: 1, 4, 10, 16, 21, 23, 28, 34, 40, 43, 45, 48, 54, 60, 98, 104, 110, 113, 115, 118, 124, 130, 135, 137, 142, 148, 154, 157, 159, 181, 203, 227, 232, 238.

A_2: 2, 3, 8, 9, 14, 19, 20, 24, 25, 30, 31, 37, 42, 47, 52, 65, 70, 88, 93, 106, 111, 116, 121, 127, 128, 133, 134, 138, 139, 144, 149, 150, 155, 156, 161, 167, 184, 195, 218, 224, 230, 235.

A_3: 5, 11, 12, 13, 15, 29, 32, 33, 35, 36, 39, 53, 55, 56, 57, 59, 77, 79, 81, 99, 101, 102, 103, 105, 119, 122, 123, 125, 126, 129, 143, 145, 146, 147, 153, 163, 169, 171, 173, 193, 213, 215, 217, 233, 237.

A_4: 6, 7, 17, 18, 22, 26, 27, 38, 41, 46, 50, 51, 75, 83, 107, 108, 112, 117, 120, 131, 132, 136, 140, 141, 151, 152, 160, 221, 222, 231.

A_5: 44, 49, 58, 61, 62, 63, 64, 66, 67, 68, 69, 71, 72, 73, 74, 76, 78, 80, 82, 84, 85, 86, 87, 89, 90, 91, 92, 94, 95, 96, 97, 100, 109, 114, 212, 219, 225, 226, 229, 234, 236.

A_6: 158, 162, 164, 165, 166, 168, 170, 172, 174, 175, 176, 177, 178, 179, 180, 182, 183, 185, 186, 187, 188, 189, 190, 191, 192, 194, 196, 197, 198, 199, 200, 201, 202, 204, 205, 206, 207, 208, 209, 210, 211, 214, 216, 220, 223, 228.

(b) The following 3-sum free 7-partition $\{B_1,\dots,B_7\}$ of $\{1,\dots,1429\}$ shows $\sigma(3,7) \geq 1430$. For $i \in \{1,\dots,6\}$ it holds $A_i \subseteq B_i$. Moreover, for $i \in \{1,\dots,7\}$ and any integer $a \leq 1429$ we have

$$a \in B_i \Leftrightarrow 1430-a \in B_i.$$

Therefore, we list only the integers a with $476 \leq a \leq 715$.

B_1: 477, 499, 521, 545, 550, 556, 562, 567, 591, 613, 635, 637, 640, 646, 652, 657, 659, 664, 670, 676, 679, 681, 684, 690, 696.

B_2: 479, 485, 502, 513, 536, 542, 548, 553, 559, 564, 570, 571, 576, 599, 610, 627, 633, 638, 639, 644, 645, 650, 655, 656, 660, 661, 666, 667, 673, 678, 683, 688, 701, 706.

B_3: 481, 487, 489, 491, 511, 531, 533, 535, 551, 555, 557, 561, 577, 579, 581, 601, 621, 623, 625, 631, 641, 647, 648, 649, 651, 665, 668, 669, 671, 672, 675, 689, 691, 692, 693, 695, 713, 715.

B_4: 478, 539, 540, 549, 563, 572, 573, 634, 642, 643, 653, 654, 658, 662, 663, 674, 677, 682, 686, 687, 711.

B_5: 530, 537, 543, 544, 547, 552, 554, 558, 560, 565, 568, 569, 575, 578, 582, 680, 685, 694, 697, 698, 699, 700, 702, 703, 704, 705, 707, 708, 709, 710, 712, 714.

B_6:

B_7: 476, 480, 482, 483, 484, 486, 488, 490, 492, 493, 494, 495, 496, 497, 498, 500, 501, 503, 504, 505, 506, 507, 508, 509, 510, 512, 514, 515, 516, 517, 518, 519, 520, 522, 523, 524, 525, 526, 527, 528, 529, 532, 534, 538, 541, 546, 566, 574, 580, 583, 584, 585, 586, 587, 588, 589, 590, 592, 593, 594, 595, 596, 597, 598, 600, 602, 603, 604, 605, 606, 607, 608, 609, 611, 612, 614, 615, 616, 617, 618, 619, 620, 622, 624, 626, 628, 629, 630, 632, 636. □

3. SCHUR numbers of arithmetic progressions: Finiteness conditions

Let a and d be non-negative integers; by $N_{a,d}$ we denote the *arithmetic progression*

$$N_{a,d} = \{a, a+d, a+2d, \ldots\}.$$

If there exists an integer $v \in N_{a,d}$ such that no n-partition of $\{a,a+d,\ldots,v\}$ is m-sum free, then we denote the smallest such integer by $\sigma_{a,d}(m,n)$. If there is no such integer, we put $\sigma_{a,d}(m,n) = \infty$. In other words: $\sigma_{a,d}(m,n) = \infty$, if there is an m-sum free n-partition of $N_{a,d}$.

Clearly, $\sigma_{1,1}(m,n) = \sigma(m,n)$. Moreover, it is easy to check that for any $v \in N_{a,d}$ with $v \geq \sigma_{a,d}(m,n)$ it holds that no n-partition of $\{a,a+d,\ldots,v\}$ is m-sum free.

In this Section we shall deal with the question, whether $\sigma_{a,d}(m,n) = \infty$ holds or not.

3.1 LEMMA. Let $a_1,\ldots,a_{m-1}$ be elements of $N_{a,d}$. Then $a_1+\ldots+a_{m-1} \in N_{a,d}$ if and only if d divides $(m-2)a$.

Proof. By definition of $N_{a,d}$, there exist non-negative integers $s_1,\ldots,s_{m-1}$ with

$$a_i = a + ds_i \quad (i \in \{1,\ldots,m-1\}).$$

It follows

$$a_1+\ldots+a_{m-1} = a + (m-2)a + d(s_1+\ldots+s_{m-1}).$$

Thus, $a_1+\ldots+a_{m-1}$ is in $N_{a,d}$ if and only if $(m-2)a$ is a multiple of d. □

3.2 COROLLARY. If $d \nmid (m-2)a$, then $\sigma_{a,d}(m,n) = \infty$. □

The following Lemma shows that it is sufficient to consider arithmetic progressions $N_{a,d}$ with $\gcd(a,d) = 1$.

3.3 LEMMA. Let g be a common divisor of a and d. Then

$$\sigma_{a,d}(m,n) = g\cdot\sigma_{\frac{a}{g},\frac{d}{g}}(m,n).$$

Proof. For a subset A of $N_{a,d}$ we define $\frac{A}{g} = \{\frac{x}{g} \mid x \in A\}$. Clearly, $\pi = \{A_1,\ldots,A_n\}$ is an n-partition of $\{a,a+d,\ldots,a+td\}$ if and only if $\frac{\pi}{g} = \{\frac{A_1}{g},\ldots,\frac{A_n}{g}\}$ is an n-partition of $\{\frac{a}{g},\frac{a+d}{g},\ldots,\frac{a+td}{g}\}$. Moreover, π is m-sum free if and only if $\frac{\pi}{g}$ is m-sum free. □

Remark. For an integer g we define $\infty\cdot g = g\cdot\infty = \infty$ and $\infty + g = g + \infty = \infty$.

From now on, we suppose always that a, d, m and n are positive integers with $n \geq 2$, $m \geq 3$ and $d \mid (m-2)a$. Denote by k the positive integer with $dk = (m-2)a$.

The proof of the following Theorem is analoguous to 1.1 and will be omitted here.

3.4 THEOREM. $\sigma_{a,d}(m,n) \geq m\cdot\sigma_{a,d}(m,n-1) - a$.□

3.5 PROPOSITION. If m is even and k is odd, then $\sigma_{a,d}(m,n) = \infty$.

Proof. In view of 3.4 it is sufficient to prove the assertion for $n = 2$. Define

$$A_1 = \{a + sd \mid s \text{ odd}\} \text{ and } A_2 = \{a + sd \mid s \text{ even}\}.$$

Then $\{A_1,A_2\}$ is an m-sum free 2-partition of $N_{a,d}$.

Namely: If $a+s_1\cdot d,\dots,a+s_{m-1}\cdot d \in A_1$, then

$$a+s_1\cdot d + \dots + a+s_{m-1}\cdot d = a + (m-2)a + (s_1+\dots+s_{m-1})d$$

$$= a + (k + s_1+\dots+s_{m-1})d.$$

Now, $s_1+\dots+s_{m-1}$ is the sum of an odd number of odd summands; since k is supposed to be odd, $k + s_1+\dots+s_{m-1}$ is even. Thus $a + (k + s_1+\dots+s_{m-1})d \notin A_1$.

On the other hand, any sum of $m-1$ elements of A_2 is in A_1.

This shows $\sigma_{a,d}(m,n) = \infty$.□

3.6 THEOREM. Suppose

(i) a is not a multiple of d (in particular, $d \neq 1$) and

(ii) $n \geq m-2$.

Then $\sigma_{a,d}(m,n) = \infty$.

Proof. Again, by 3.4 it is sufficient to show $\sigma_{a,d}(m,m-2) = \infty$. In order to do this, we define for $j \in \{1,\dots,m-2\}$:

$$A_j = \{a + sd \mid s \equiv j \pmod{m-2}\} = \{a + sd \mid s \in N_{j,m-2}\}.$$

We claim that $\{A_1,\dots,A_{m-2}\}$ is an m-sum free (m-2)-partition of $N_{a,d}$.

Namely: Let $a_1,\dots,a_{m-1}$ be elements of A_j ($j \in \{1,\dots,m-2\}$). Then there exist non-negative integers $t_1,\dots,t_{m-1}$ with

$$a_i = a + (j + t_i(m-2))d \quad (i \in \{1,\dots,m-1\}).$$

It follows

$$a_1 + \dots + a_{m-1}$$

$$= a + (m-2)a + [(m-1)j + (t_1+\dots+t_{m-1})(m-2)]d.$$

$$= a + [k + (m-1)j + (t_1+\dots+t_{m-1})(m-2)]d.$$

If this element were in A_j as well, it would follow

$$k+j \equiv k + (m-1)j + (t_1+\dots+t_{m-1})(m-2) \equiv j \pmod{m-2}, \text{ or}$$

$$k \equiv 0 \pmod{m-2}.$$

Thus $m-2$ would be a divisor of $k = \frac{(m-2)a}{d}$, which forces a to be a multiple of d:

a contradiction to the assumptions of our Theorem.□

3.7 THEOREM. $\sigma_{a,d}(m,n) \leq a\cdot\sigma_{1,d}(m,n)$.

Proof. Without loss in generality, we can suppose that $\sigma_{1,d}(m,n)$ is finite.

Denote by $\pi = \{A_1,\ldots,A_n\}$ an arbitrary n-partition of $\{a,a+d,\ldots,a\cdot\sigma_{1,d}(m,n)\}$ (*Note* that $\sigma_{1,d}(m,n) \in N_{1,d}$, so $a\cdot\sigma_{1,d}(m,n) \in N_{a,d}$.) It is to show that π is not m-sum free. For this purpose, consider the following n-partition $\pi_o = \{B_1,\ldots,B_n\}$ of $\{1,1+d,\ldots,\sigma_{1,d}(m,n)\}$:

$$B_j = \{1+sd \mid a + as\cdot d \in A_j\} \qquad (j \in \{1,\ldots,n\}).$$

By the definition of $\sigma_{1,d}(m,n)$, in at least one component B_h of π_o there are elements $b_1,\ldots,b_{m-1}$ with $b_1+\ldots+b_{m-1} \in B_h$. If $b_i = 1 + s_i\cdot d$, then

$$a + as_i\cdot d \in A_h \qquad (i \in \{1,\ldots,m-1\})$$

and

$$a+as_1d + \ldots + a+as_{m-1}d = a(b_1+\ldots+b_{m-1}) \in A_h.$$

Thus π is not m-sum free. Consequently, $\sigma_{a,d}(m,n) \leq a\cdot\sigma_{1,d}(m,n)$.□

Remark. The above Theorem says among other things: If $\sigma_{1,d}$ is finite, then $\sigma_{a,d}$ is finite as well. However, the converse is not true: 3.3 and 1.3 imply

$$\sigma_{2,2}(4,2) = 2\cdot\sigma_{1,1}(4,2) = 2\cdot\sigma(4,2) = 22.$$

On the other hand, by 3.6, $\sigma_{1,2}(4,2) = \infty$.

4. The SCHUR numbers $\sigma_{a,d}(m,2)$

The aim of this Section is to prove the following Theorem:

4.1 THEOREM. Denote by a, d and m positive integers with $d \mid (m-2)a$ and $m \geq 3$. Then

$$\sigma_{a,d}(m,2) = \begin{cases} \infty, \text{ if } m \text{ is even and } k = \dfrac{(m-2)a}{d} \text{ is odd} \\ a(m^2-m-1) \text{ otherwise.} \end{cases}$$

This Theorem will be proved by a series of Lemmas.

4.2 LEMMA. $\sigma_{a,d}(m,2) > a(m^2-m-1) - d$.

Proof. It is to show that there exists an m-sum free 2-partition of the set $X = \{a,a+d,\ldots,a(m^2-m-1)-d\}$. Define

$$B_1 = \{a,a+d,\ldots,(m-1)a-d\},$$

$$B_2 = \{(m-1)a,(m-1)a+d,\ldots,(m-1)^2a-d\},$$

$$B_3 = \{(m-1)^2a,(m-1)^2a+d,\ldots,(m^2-m-1)a-d\}.$$

Since $(m-2)a$ is divisible by d, the sets X, B_1, B_2 and B_3 are subsets of $N_{a,d}$. As in the proof of 1.1 one can show that $\{B_1 \cup B_3, B_2\}$ is an m-sum free 2-partition of X.□

4.3 LEMMA. Suppose that one of the following conditions holds:

(i) m is odd, or

(ii) m and $k = \frac{(m-2)a}{d}$ are even.

Then $\sigma_{a,d}(m,2) \leq a(m^2-m-1)$.

Proof. Assume that there exists an m-sum free 2-partition $\{A_1, A_2\}$ of $X = \{a, a+d, \ldots, a(m^2-m-1)\}$. We can suppose $a \in A_1$. Then $(m-1)a \in A_2$, so $(m-1)^2 a \in A_1$. Consequently,

$$(m^2-m-1)a = 1\cdot(m-1)^2 a + (m-2)\cdot a \in A_2.$$

Since $m \geq 3$, the integer $z = a + 2(m-2)a = a + 2kd$ is an element of X. We claim that z is in A_1. (*Otherwise*, we would have

$$a(m^2-m-1) = (m-2)\cdot(m-1)a + 1\cdot z \notin A_2:$$

a contradiction.)

(i) Since $m-1$ is even, it would follow

$$(m-1)^2 a = \frac{m-1}{2}\cdot a + \frac{m-1}{2}\cdot z \notin A_1,$$

a contradiction.

(ii) Since k is even, the integer $y = a + \frac{k}{2}\cdot d$ is in X. Moreover,

$$(m-3)\cdot a + 2\cdot y = a + 2kd = z \in A_1$$

implies that $y \in A_2$.

Consider now the element $x = (m^2-3m+3)a = a + (m-1)kd$ of X. On the one hand, we have

$$x = (m-3)\cdot(m-1)a + 2\cdot(a + \frac{k}{2}\cdot d) \notin A_2.$$

But, using that m is even, we get from

$$x = 1\cdot a + \frac{m-2}{2}\cdot a + \frac{m-2}{2}\cdot(a+2kd) \notin A_1$$

that x is no element of A_1 either.□

By 3.5, 4.2 and 4.3, Theorem 4.1 is proved.

We conclude with the following Corollary to 4.1:

4.4 COROLLARY. Suppose $m \geq 3$. If d is odd, then

$$\sigma_{a,d}(m,2) = (m^2-m-1)a.$$

In particular, $\sigma_{a,1}(m,2) = (m^2-m-1)a$.

Proof. Suppose that m is even. Then from $dk = (m-2)a$ it follows that dk is even as well. So, if d is odd, k has to be divisible by 2. Now, 4.1 implies that $\sigma_{a,d}(m,2)$ is finite.□

References

1. Abbott, H.L. and Moser, L.: Sum-free sets of integers. Acta arith. XI (1966), 393-396.

2. Fredricksen, H.: Schur Numbers and the Ramsey Numbers N(3,3,...,3;2). J. Combinat. Theory (A) 27 (1979), 376-377.

3. Halder, H.-R. and Heise, W.: Einführung in die Kombinatorik. München - Wien, Hanser 1976.

4. Schur, I.: Über die Kongruenz $x^m+y^m \equiv z^m \pmod{p}$. Jahresber. Deutsch. Math. Verein. 25 (1916), 114-117.

5. Wallis, W.D., Street, A.P. and Wallis, J.S.: Room squares, sum-free sets, Hadamard matrices. Springer lecture notes series, vol. 292, 1972.

Description of Spherically Invariant Random Processes by Means of G-Functions

H. Brehm
Lehrstuhl für Nachrichtentechnik
Universität Erlangen - Nürnberg
Cauerstrasse 7, D-8520 Erlangen

Spherically invariant random processes are generalizations of the well known Gaussian process. Their joint probability densities are functions of a non-negative definite quadratic form, but there is no exponential dependence from the argument as in the Gaussian case. Though some special relations between these densities themselves and the characteristic function of the process are known, in most cases explicite notations in terms of familiar functions are not available.
The use of G-functions, which form a class of higher-transcendental functions, yields comprehensive explicite notations. Thus, quantitative solutions of problems, where spherically invariant random processes are involved, can be achieved easily.

1. Introduction

A random process $\xi(t)$ is completely characterized, if the joint probability density function (PDF) is known for each random vector, taken from the process by sampling its amplitudes at arbitrary values of the parameter t. Because the number of the elements of such a random vector should not be limited, one has to know all higher-order PDFs.
An equivalent characterization is given by means of the characteristic function (CF), defined as the Fourier-transform of the PDF, whenever all higher-order CFs are available.
As an example we refer to the well known Gaussian process with normally-distributed amplitudes. In this case all the nth-order PDFs as well as the CFs are given by the exponential function $\exp(-q^2/2)$ of an argument q^2 , that is a non-negative definite quadratic form of n variables.

Spherically invariant random processes (SIRPs) were introduced as generalizations of the Gaussian process or under equivalent points of view. Their PDFs and CFs are only functions of such a quadratic form, too, but there exists no exponential dependence as in the Gaussian case.
The earliest work in this area was done by Lord (1954) [1] and Kingman (1963) [2] in connection with the classical "random-flight-problem" and by Vershik (1964) [3], who discovered that an ergodic SIRP necessarily has a normal distribution. In consequence of this result Blake and Thomas (1968) [4] uttered some doubt on the physical significance of non-normal SIRPs.
Independently McGraw and Wagner (1968) [5] looked at random processes with concentric ellipses as contour-lines of their second-order PDFs. This feature reveals a necessary but not sufficient condition, that the process under consideration is a SIRP.
The work of Picinbono (1970) [6], Kingman (1972) [7], and Yao (1973) [8] yielded representation theorems asserting, that each SIRP is equivalent to a univariate randomization of a Gaussian process. The randomization has to be performed over a variable, that multiplies the covariance function of the normally distributed process. These theorems are based on properties of completely monotone functions, earlier (1931 - 1938) evaluated by Bochner [9-12], Schoenberg [13-15], and Widder [16].
So far it had become obvious, that a SIRP is completely characterized by its mean value, its covariance function, and (in addition to the Gaussian case) likewise by its univariate PDF or CF. Higher-order PDFs

are Hankel-transforms of the CF, they also can be obtained by differentiating functions, that are closely related to the univariate or bivariate PDF, respectively.
Further work on SIRPs, concerned to detection problems, was done by Picinbono and Vezzosi (1970-72) [17, 18] and by Goldman (1974-76) [19, 20] starting with the assumption, that higher-order PDFs should be known in principle. A similar understatement was made by Leung and Cambanis (1978) [21], who gave an expression to calculate the Shannon Lower Bound for Rate Distortion Functions.

Recently (1978) some new results were obtained [22], stimulated by experimental studies on speech signals, performed by Wolf and Brehm (1973) [23]. These signals were found to be realizations of a SIRP [24 - 26] under a certain constraint, that reveals to be not restrictive in most cases of practical interest. Motivated by this important result, the work [22] discovers, that one has to become acquainted with G-functions in order to achieve a mathematically treatable description of SIRPs. This statement stems from the fact, that in most cases higher-order PDFs cannot be expressed by commonly used functions despite a new and likewise simple relation between their densities, given in terms of Laplace-transforms.

Now we will view at the essential parts of the work reported here. In the following section we shall deal with the properties of SIRPs, referring especially to Yao [8] and some additional results, that are given in [22]. We will start with the assumption, that the first-order PDF explicitely is prescribed or known from experimental data. Furthermore it should be certified, that the concerned process is a SIRP, as may be deduced from measured second-order PDFs under certain further assumptions. At first we will see, that multivariate PDFs of an odd order can easily be obtained from the first-order PDF only by means of differentiations. Because of the fact, that PDFs have to be non-negative valued functions of their variables, we then will find an unique solution of an integral equation, using a representation-theorem for completely monotone functions due to Widder [16]. This solution yields a new relation between the first-order PDF of a SIRP and its higher-order ones, expressed in terms of Laplace-transforms. Though this transformation is commonly considered to be very familiar, results are not easily obtained, as will be illustrated by examples. In the case of a Gamma-distribution, which is of great importance in the fields of speech processing, multivariate PDFs of even order can explicitely be given only by means of G-functions.

As may be expected, the properties of these higher-transcendental functions are not well known. Therefore we should look at them in some more detail. Consequently, in the first part of the third section we shall introduce the G-function by its definition as a Mellin-Barnes-integral and list up those properties, that are referred to in the following. As will be seen, the special interest in these functions is due to the fact, that they form a set, closed under the operations of differentiation, integration, and commonly used integral-transformations. The values of each G-function, dependent on its argument and some parameters, can be calculated by means of an algorithm given in [22]. In the second part of this section we will show, how to express all quantities characterizing a SIRP, e.g. CF or especially higher-order PDFs in terms of G-functions.

Finally, in the fourth section there are given some illustrations, how to apply the obtained results in cases of practical interest. Referring to experimental data, received from bandlimited speech signals, we will describe their first-order PDF in terms of a G-function. Then we will present some higher-order PDFs, calculated by means of the algorithm mentioned above. Furthermore we shall present and evaluate an explicite expression for the Shannon Lower Bound of Rate Distortion Functions, which are of great interest in the fields of information theory.

2. Spherically Invariant Random Processes

At first we will agree upon some notations and definitions, that are used in the following. We shall deal with random vectors $\underline{\xi} = \mathrm{col}(\xi_1, \ldots, \xi_n)$ whose elements $\xi_\nu = \xi(t_\nu)$ are random variables, taken from a stationary random process $\xi(t)$ by sampling its amplitudes at arbitrarily chosen instances t_ν, $\nu = 1, \ldots, n$. Ordering the same elements in a row constitutes the transposed vector $\underline{\xi}^T$. We will interpret $\underline{x} = (x_1, \ldots, x_n)$ as a vector, too, or only as a collection of n ordinary real variables. $\underline{\underline{A}} = (a_{ik})$ is a quadratic matrix with real elements a_{ik} and $\det(\underline{\underline{A}}) = A$.

It is well known, that the matrix $\underline{\underline{A}}$ is non-negative definite iff the inequality concerning the quadratic form of the variables $\underline{x}^T \underline{\underline{A}}\, \underline{x} = \Sigma\, a_{ik}\, x_i\, x_k \geq 0$ holds for each choice of the vector $\underline{x}$.

Thus, we will use a shorthand notation $p_{\underline{\xi}}(\underline{x})$ for the nth-order probability density function (PDF) $p_{\xi_1, \ldots, \xi_n}(x_1, \ldots, x_n)$ of the random vector $\underline{\xi}$. An expectation value will likewise be noted by $\int dx_1 \ldots \int dx_n\, g(x_1, \ldots, x_n) = \int d\underline{x}\, g(\underline{x}) = \langle g(\underline{\xi})\rangle$. Finally an equation like $p_{\underline{\xi}}(\underline{x}) = f(\underline{x}^T \underline{\underline{A}}\, \underline{x})$ will indicate, that the PDF is given by a function $f(\cdot)$, the argument of which is solely a quadratic form of the variables collected in $\underline{x}$.

Now, starting with a definition of a spherically invariant random process (SIRP) we will outline those properties of SIRPs that are of common interest, especially in the fields of communication engineering. Without any loss of generality we will assume, that a given stationary random process $\xi(t)$ has zero mean and unit variance. For convenience we further assume, that each pair of different random variables, taken from the process by sampling its amplitudes, is uncorrelated, i.e. $\langle\xi_i\, \xi_k\rangle = \delta_{ik}$ with Kronecker's delta. Thus, $\xi(t)$ constitutes a SIRP iff all its higher-order PDFs are spherically invariant functions of their variables. Consequently, these PDFs can be given in the form

$$p_{\underline{\xi}}(\underline{x}) = \pi^{-n/2}\, f_n(\underline{x}^T\, \underline{x}) \equiv \pi^{-n/2}\, f(\underline{x}^T\, \underline{x};\, n)\,. \qquad (2\text{-}1)$$

The type of the functions introduced here by the notation $f_n(\cdot) = f(\cdot;\, n)$ depends on n, the order of the PDF. These functions have to be properly determined for each prescribed univariate PDF.

Obviously the well known Gaussian process is a SIRP according to the relation

$$p_{\underline{\xi}}(\underline{x}) = (2\pi)^{-n/2} \exp(-\underline{x}^T \underline{x}/2) = \pi^{-n/2} \cdot 2^{-n/2} \exp(-\underline{x}^T \underline{x}/2),$$

i.e.
$$f(s;\, n) = 2^{-n/2} \exp(-s/2). \tag{2-2}$$

In this special case the type of the functions $f(s;\, n)$ is exponential and affected by n, only due to the simple term $2^{-n/2}$.

The assumptions made above concerning zero mean, unit variance, and decorrelation are not restrictive at all, because the PDF of a random vector, which is generated by linear mapping $\underline{\eta} = \underline{\underline{A}}\, \underline{\xi} + \hat{\underline{y}}$ is easily obtained [31]

$$p_{\underline{\eta}}(\underline{y}) = M^{-1/2} \cdot \pi^{-n/2} f(q^2;\, n) \text{ with } q^2 = (\underline{y} - \hat{\underline{y}})^T \underline{\underline{M}}^{-1} (\underline{y} - \hat{\underline{y}}) \tag{2-3}$$

from (2-1). The mean value of the new variables is $\hat{\underline{y}}$ and their covariance matrix is given by $\underline{\underline{M}} = \underline{\underline{A}}\, \underline{\underline{A}}^T$. In the case of a second-order PDF contour-lines, i.e. lines of equal height, are concentric ellipses.

Now, the problem is to calculate higher-order PDFs from the univariate PDF, which may be given in terms of a mathematically treatable function, i.e. we assume to know $f(s;\, 1)$ according to rel. (2-1). In general, this problem has no unique solution. On the other hand it is well known, that the order of a PDF can be reduced by integration, which results in

$$\pi^{-n/2} f(r^2;n) = \pi^{-(n+2)/2} \int_{-\infty}^{\infty} dy_1 \int_{-\infty}^{\infty} dy_2\; f(r^2+y_1^2+y_2^2;n+2) \tag{2-4}$$

an expression, that shows how to calculate $f(\cdot;\, n)$ from $f(\cdot;\, n+2)$. Introducing spherical coordinates $(\rho,\, \phi)$ and substituting $r^2 = s$, $s + \rho^2 = x$ we obtain

$$f(s;\, n) = \int_s^{\infty} dx\; f(x;\, n+2) \text{ for } 0 < s < \infty \tag{2-5}$$

and after differentiation with respect to s

$$\frac{-d}{ds} f(s;\, n) = f(s;\, n+2)\ . \tag{2-6}$$

Thus, all multivariate PDFs of an odd order can easily be calculated

$$f(s;\ 2m+1) = (-1)^m \frac{d^m}{ds^m} f(s;\ 1) \tag{2-7}$$

by differentiating the function f(s; 1) closely related to the univaria te PDF. Likewise all multivariate PDFs of an even order can be calculated from

$$f(s;\ 2m) = (-1)^m \frac{d^m}{ds^m} f(s;\ 0) \tag{2-8}$$

if the formally introduced function f(s; 0) is known.
This function may be determined in accordance to rel. (2-4). But a straight-forward handling is prevented by the fact, that a resultant integral equation cannot be solved uniquely without some proper assumptions. Because we look for PDFs, all functions f(s; n) in eqs. (2-7,8) should have non-negative values. This statement is equivalent with the restriction, that both the functions f(s; 1) as well as f(s; 0) must be completely monotone. Consequently [16] there exists a unique representation of f(s; 1)

$$f(s;1) = \int_0^\infty dt\ e^{-st}\ \phi(t) = \mathsf{L}(\phi(t);s); \quad s \geq 0 \tag{2-9}$$

as the Laplace-transform of a non-negative valued function $\phi(t)$. This expression is closely related to "Yao's representation theorem". Based on relation (2-9) we can proceed to calculate f(s; 0) by integrating f(s; 1) in accordance to (2-4). Starting with

$$f(r^2;0) = \pi^{-1/2} \int_{-\infty}^{\infty} dy\ f(r^2+y^2;1) \tag{2-10}$$

we obtain

$$\begin{aligned} f(s;0) &= \pi^{-1/2} \int_{-\infty}^{\infty} dy \int_0^\infty dt\ e^{-(s+y^2)t}\ \phi(t) \\ &= \pi^{-1/2} \int_0^\infty dt\ e^{-st}\ \phi(t) \int_{-\infty}^{\infty} dy\ e^{-y^2 t} \\ &= \int_0^\infty dt\ e^{-st}\ [t^{-1/2}\ \phi(t)] \end{aligned} \tag{2-11}$$

$$f(s;0) = \mathsf{L}(t^{-1/2}\ \phi(t);s)\ .$$

Because of the fact that f(0; 0) = 1 according to a normalized univariate PDF, all integrals do exist. Furthermore we conclude that f(s; 0) is completely monotone as Laplace-transform of the non-negative valued function $t^{-1/2}\,\phi(t)$. The ultimate result is, that all higher-order PDFs of a SIRP may be calculated

$$\begin{aligned} f(s;n) &= L(t^{(n-1)/2}L^{-1}(f(p;1);t);s) \\ p_{\underline{\xi}}(\underline{x}) &= \pi^{-n/2}\, f(r^2;n); \quad r^2 = \sum_{j=1}^{n} x_j^{\,2} \end{aligned} \tag{2-12}$$

from the univariate PDF only by means of the familiar Laplace-transformation.

The procedure to be performed may be illustrated by an example. The (two-sided) Laplace-distribution is defined by its first-order PDF

$$p_\xi(x) = 2^{-1/2}\exp(-2^{-1/2}|x|), \quad -\infty \le x \le \infty. \tag{2-13}$$

Thus we have

$$f(s;\ 1) = (\tfrac{\pi}{2})^{1/2}\exp(-2^{1/2}\, s^{1/2}), \quad 0 \le s \le \infty\ , \tag{2-14}$$

and the inverse Laplace-transform is found [32] to be

$$\phi(t) = 2^{-1}\, t^{-3/2}\exp[-1/(2t)], \tag{2-15}$$

a non-negative valued function. From this we conclude, that a univariate Laplace-distribution is consistent with the assumption, that the random process in question is a SIRP. We now refer to [32] again and find

$$\begin{aligned} f(s;n) &= L(\tfrac{1}{2}\, t^{n/2-2}\, e^{-1/(2t)};s) \\ &= (2s)^{-(n/2-1)/2}\, K_{\frac{n}{2}-1}(2^{1/2}\, s^{1/2}) \end{aligned} \tag{2-16}$$

an expression in terms of modified Bessel-functions K_ν, which guarantees a comprehensive formulation of all higher-order PDFs.

Another distribution described by a special Gamma-PDF

$$p_\xi(x) = \frac{1}{4}\,(\frac{3}{\pi})^{1/2}\, s^{-1/2}\, e^{-s} \qquad \text{with } s = 3^{1/2}\,|x|/2\ , \tag{2-17}$$

which is a consistent first-order PDF of a SIRP, too, is of outstanding interest in the fields of speech processing. In this case a comprehensive formulation of higher-order PDFs only may be given in terms of G-functions. Therefore we conclude that higher-order PDFs may be calculated by means of the very familiar Laplace-transformation, but we have to keep aware of the fact, that in available tables results are either not listed or only given in terms of higher transcendental functions, whose properties are commonly unknown.

At the first glance one may argue, that each univariate PDF, found by an experiment, requires the acquaintance with a new class of higher transcendental functions. Fortunately, as will be seen in the following sections, this is not true, because the use of G-functions will allow to describe a large variety of random processes. This statement should stimulate the interest in the properties of G-functions, that will be treated in the following section.

Before doing so, we will look at another definition of SIRPs related to the univariate CF of the process under consideration. By definition we have for the nth-order CF

$$C_{\underline{\xi}}(\underline{v}) = \pi^{-n/2} \int_{-\infty}^{\infty} \Big(\prod_{j=1}^{n} dx_j \Big)\, e^{i\underline{v}^T\underline{x}}\, f(t^2;n) \qquad \text{with } t^2 = \sum_{j=1}^{n} x_j^2 \,. \tag{2-18}$$

The complexity of this formula considerably reduces because of the spherical symmetry [1] of $p_{\underline{\xi}}(\underline{x})$. For convenience we assume again, that the process $\xi(t)$ is uncorrelated with zero mean and unit variance. At first we find, that the CFs are not affected by their order n. They depend only on the sum of the squares of their variables

$$C_{\underline{\xi}}(\underline{v}) = g(\underline{v}^T\,\underline{v}) \,. \tag{2-19}$$

From this result we may deduce another definition of a SIRP, sometimes used in the literature. Obviously a random process $\xi(t)$ is a SIRP iff the type of all higher-order CFs does not depend on the order n. This definition seems to be more straight-forward. However it is of lower practical interest, because no facilities are introduced to measure the CFs of a random process. Nevertheless, we can calculate higher-order PDFs from the CF by means of the Hankel-transformation as it is known from [1] in cases of spherical symmetry. With $C_\xi(v) = g(v^2)$ we obtain an expression

$$\{2^{n/2}\, r^{(n-1)/2}\, f(r^2;n)\} = \int_0^\infty ds\{s^{(n-1)/2}\, g(s^2)\}\, (rs)^{1/2}\, J_{\frac{n-2}{2}}(rs) \tag{2-20}$$

from which we see, that

$$s^{(n-1)/2}\, g(s^2) \text{ and } 2^{n/2}\, r^{(n-1)/2}\, f(r^2;\, n) \tag{2-21}$$

form a pair of Hankel-transforms.

Though a lot of formulas are available [34], we should not be surprised about the fact, that comprehensive solutions avoiding the G-functions cannot be found in the case of the Gamma-distribution.

Another interesting feature of SIRPs is, that the random variables $\rho_n, \phi_1, \ldots, \phi_{n-1}$, defined by

$$\xi_1 = \rho_n \prod_{j=1}^{n-1} \sin\Phi_j \;, \quad \xi_k = \rho_n \cos\Phi_{n+1-k} \prod_{j=1}^{n-k} \sin\Phi_j \quad \text{for } k\varepsilon(2,n-1) \tag{2-22}$$

$$\xi_n = \rho_n \cos\Phi_1 \;,$$

are mutually statistically independent. The functions $f(r^2;\, n)$, characterizing a SIRP, do not affect the PDFs of the new variables $\phi_1, \ldots, \phi_{n-1}$. Therefore a SIRP is completely described only by the univariate PDFs

$$p_{\rho_n}(r) = \frac{2}{\Gamma(n/2)}\, r^{n-1}\, f(r^2;\, n),\; 0 \leq r \leq \infty \tag{2-23}$$

depending on n.

Sometimes, as we will see later on, it is more convenient to look at the PDFs of other random variables, defined by $\sigma_n = n^{-1/2}\,\rho_n$, which can be interpreted as estimation-values of the standard deviation of $\xi(t)$. These PDFs are given by

$$p_{\sigma_n}(r) = n^{1/2}\, p_{\rho_n}(n^{1/2}\, r),\; 0 \leq r \leq \infty \tag{2-24}$$

an expression, which should tend to $\delta(r-1)$, i.e. to Dirac's delta-function centered about $r = 1$ in the limit $n \to \infty$, if the process under consideration has to be ergodic. However, the limiting PDF

$$p_\sigma(r) = \lim_{n\to\infty} p_{\sigma_n}(r) \tag{2-25}$$

differs from the delta function for all non-Gaussian SIRPs, as will be seen in the following section.

3. G-Functions

3.1 DEFINITIONS AND PROPERTIES

The G-functions, which were introduced by Meijer in 1936, are generalizations of the hypergeometric functions. Here, we will restrict ourselves to listing definitions and those properties of the functions, that are needed to describe SIRPs later on. For detailed information the reader should consult original research documents by Meijer [27], Erdélyi et. al. [28], Luke [30], and Brehm [22]. In special cases G-functions can be expressed in terms of other higher transcendental functions, which might be more familiar to the reader. A listing of such correspondences is given by Luke [30].
In the literature G-functions are named in different, but equivalent forms

$$G_{pq}^{mn}(z) = G_{pq}^{mn}(z|{}^{a_p}_{b_q}) = G_{pq}^{mn}(z|{}^{a_1,\dots,a_p}_{b_1,\dots,b_q}) \ . \qquad (3\text{-}1)$$

They are functions of the complex variable z and depend furthermore on two sets of complex parameters $a_p \equiv (a_1, \dots, a_p)$ and $b_q \equiv (b_1,\dots,b_q)$, which are ordered into four groups

$$G_{pq}^{mn}\left(z \,\middle|\, \frac{a_1,\dots,a_n}{b_1,\dots,b_m} \,\middle|\, \frac{a_{n+1},\dots,a_p}{b_{m+1},\dots,b_q} \right) . \qquad (3\text{-}2)$$

The non-negative integers m, n, p, q obviously refer to the number of elements in the different groups. Within any group the elements may be interchanged, as becomes evident from the definition of the G-function in terms of a Mellin-Barnes-integral

$$G_{pq}^{mn}(z|{}^{a_p}_{b_q}) = (2\pi i)^{-1} \int_C ds\ z^s \frac{\prod_{j=1}^{m}\Gamma(b_j-s) \prod_{j=1}^{n}\Gamma(1-a_j+s)}{\prod_{j=m+1}^{q}\Gamma(1-b_j+s) \prod_{j=n+1}^{p}\Gamma(a_j-s)} \ . \qquad (3\text{-}3)$$

If a group contains no elements, the corresponding empty product in the integrand is interpreted as unity. Because of the factor $z^s \equiv \exp\,(s \log z)$ the G-function generaly is a multivalued function of the complex variable z. The well known Gamma-function $\Gamma(s)$ is analytic in the whole s-plane, except for the points $s = 0, -1, -2, \dots,$ and $s = \infty$.

Therefore, each parameter $b_j \in (b_1, \dots, b_m)$ gives rise to an infi-

nite number of simple poles at the points, where $b_j - s$ is a negative integer or equal to zero. The same holds for each parameter $a_k \in (a_1, \ldots, a_n)$ at those points, where $1 - a_k + s$ equals a negative integer or zero.
Poles resulting from any b_j must not coincide with those resulting from any a_k , but there may occur poles of higher order, if at least two elements b_{j1}, b_{j2} differ by an integer or zero.
The path of integration goes from $\sigma - i\infty$ to $\sigma + i\infty$ so that all poles of $\Gamma(b_j - s)$, $j = 1, \ldots, m$, lie to the right of the path, whereas all poles of $\Gamma(1 - a_k + s)$, $k = 1, \ldots, n$, lie to the left. The convergence of the integral depends on several relations between the numbers m, n, p, q and on other constraints, that may be omitted here.
If $q \geq 1$ and either $p < q$ or $p = q$ and $|z| < 1$, the path may be bent to a loop beginning and ending at $+\infty$, as illustrated below for three parameters a_1, b_1 and b_2.

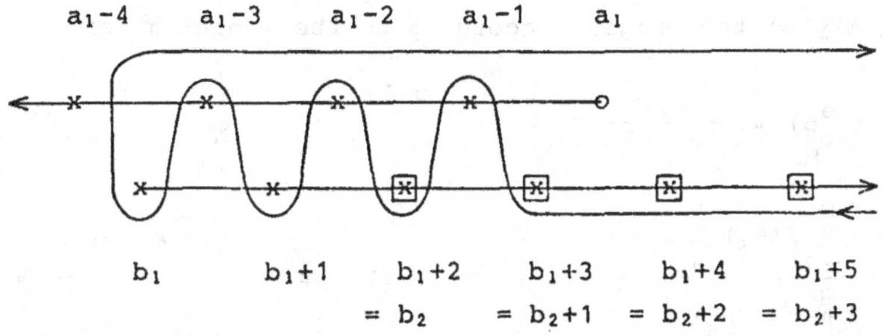

The integral can be evaluated as a sum of residues. Under the restriction, that all poles are simple, the result is a weighted sum of hypergeometric series, as given by Luke [30]. In the case of second-order poles, one may proceed using L'Hospital's theorem, as done by Luke [30], but the resultant formulas exhibit a considerable increase in complexity.
A rigorous approach to the evaluation of the integral as a sum of residues in the case of higher-order poles is involved with some formal difficulties, because there have to be made differentiations of the integrand, that is represented as a product of functions, divided by another product of functions, all of them depending on the variable s. A general concept for an evaluation based on the logarithm of the integrand has been given by Brehm [22]. It has been transferred into an algorithm, designed to be implemented on a digital computer. Thus, values of the G-functions can be computed, using sums of series even in the case of higher-order poles.

Limiting forms of the G-functions for small arguments can be derived from those series representations. The dominant part of these forms is destined by that b_τ, which shows the minimum real part of all elements $b_1, \ldots, b_m$. The result is

$$G^{mn}_{pq}\left(z\,\middle|\,{a_p \atop b_q}\right) \sim z^{b_\tau} \frac{\prod\limits_{1}^{m}{}^{*}\,\Gamma(\beta_j)\prod\limits_{1}^{n}\Gamma(1-\alpha_j)}{\prod\limits_{m+1}^{q}\Gamma(1-\beta_j)\prod\limits_{n+1}^{p}\Gamma(\alpha_j)} \quad \text{for } z \to 0 \qquad (3\text{-}4)$$

with $R(b_\tau) = \min R(b_j),\ j\varepsilon(1,m);\quad \alpha_j = a_j-b_\tau;\ \beta_j = b_j-b_\tau,$
$b_j-b_k \neq \text{integer} \quad , \ j\neq k;\quad j,k\varepsilon(1,m).$

If there are more elements with minimum real part, one has to distinguish whether some of them are equal or not. In the case of real parameters and $b_1 = \ldots = b_\mu = \min\,(b_1, \ldots, b_m)$, there exists a logarithmic singularity at the origin, according to the limiting form

$$G^{mn}_{pq}\left(z\,\middle|\,{a_p \atop b_q}\right) \sim \frac{-(-1)^{\mu^*}}{(\mu-1)!}\, z^{b_\mu}(\log z)^{\mu-1} \,.$$

$$\cdot \frac{\prod\limits_{\mu+1}^{m}\Gamma(\beta_j)\prod\limits_{1}^{n}\Gamma(1-\alpha_j)}{\prod\limits_{m+1}^{q}\Gamma(1-\beta_j)\prod\limits_{n+1}^{p}\Gamma(\alpha_j)} \quad \text{for} \quad z \to 0 \qquad (3\text{-}5)$$

with $\mu^* = \mu - \sum\limits_{1}^{\mu-1}\beta_j$

$\alpha_j = a_j-b_\mu$

$\beta_j = b_j-b_\mu$

$R(b_\mu) = \min R(b_j), \quad j\varepsilon(1,m).$

The asymptotic behaviour of the G-functions is either exponential, logarithmic or algebraic, depending on the given grouping of the parameters. In the special case $n = 0$ and $m = q$, occurring in section 4, we have

$$G^{q\,0}_{p\,q}\left(z\,\middle|\,{a_p \atop b_q}\right) \sim H_{p,q}(z) = \frac{(2\pi)^{(\sigma-1)/2}}{\sigma^{1/2}} \exp[-\sigma z^{1/\sigma}]\,z^{\Theta} \sum_{k=0}^{\infty} M_k\, z^{-k/\sigma}$$

for $|z| \to \infty$, $|\arg z| \leq (\sigma+\varepsilon)\pi-\delta$, $\delta > 0$. (3-6)
σ, Θ, M_k are constants not depending on z.

For a detailed information, especially in the remaining cases, the reader may consult Luke [30] or Brehm [22].

Before we look at some further properties of the G-functions, we first have to agree upon notations concerning the sets of parameters. With

$$\left|\begin{matrix}\alpha, a_p \\ b_q, \alpha\end{matrix}\right) \equiv \left|\frac{\alpha, a_1, \ldots, a_n \,|\, a_{n+1}, \ldots, a_p}{b_1, \ldots, b_m \,|\, b_{m+1}, \ldots, b_q, \alpha}\right)$$
$$\left|\begin{matrix}a_p, \alpha \\ \alpha, b_q\end{matrix}\right) \equiv \left|\frac{a_1, \ldots, a_n \,|\, a_{n+1}, \ldots, a_p, \alpha}{\alpha, b_1, \ldots, b_m \,|\, b_{m+1}, \ldots, b_q}\right) \tag{3-7}$$

it is indicated, how to adjoin a new parameter α to two positions of the four parameter groups. Likewise

$$\left|\begin{matrix}a_p+\alpha \\ b_q-\beta\end{matrix}\right) \equiv \left|\begin{matrix}(a_1+\alpha), \ldots, (a_p+\alpha) \\ (b_1-\beta), \ldots, (b_q-\beta)\end{matrix}\right) \tag{3-8}$$

means, that the value α has to be added to all parameters a_p and the value β has to be subtracted from all parameters b_q.
Now, there exist two rules to change the order of G-functions (i.e. to diminish or enlargen the values of m, n, p, q) by deleting or adjoining common parameters in corresponding groups

$$G^{m\ \ n+1}_{p+1\ q+1}\left(z\left|\begin{matrix}\alpha, a_p \\ b_q, \alpha\end{matrix}\right.\right) = G^{mn}_{pq}\left(z\left|\begin{matrix}a_p \\ b_q\end{matrix}\right.\right)$$
$$G^{m+1\ \ n}_{p+1\ q+1}\left(z\left|\begin{matrix}a_p, \alpha \\ \alpha, b_q\end{matrix}\right.\right) = G^{mn}_{pq}\left(z\left|\begin{matrix}a_p \\ b_q\end{matrix}\right.\right) . \tag{3-9}$$

A power of the argument may be extracted from or included in the G-function itself according to

$$z^{\sigma}\, G^{mn}_{pq}\left(z\left|\begin{matrix}a_p \\ b_q\end{matrix}\right.\right) = G^{mn}_{pq}\left(z\left|\begin{matrix}a_p+\sigma \\ b_q+\sigma\end{matrix}\right.\right) . \tag{3-10}$$

In the discussion of the G-functions, we can without loss of generality suppose that $p \leq q$ in view of the important relation

$$G^{mn}_{pq}\left(z\left|\begin{matrix}a_p \\ b_q\end{matrix}\right.\right) = G^{nm}_{qp}\left(z^{-1}\left|\begin{matrix}1-b_q \\ 1-a_p\end{matrix}\right.\right) , \quad \arg(z^{-1}) = -\arg(z) . \tag{3-11}$$

So far, we have gathered some interesting properties of the G-functions, which can be derived from the definition (3-3). The outstanding importance of these functions for describing SIRPs, however, stems from the fact, that differentiations, integrations, and especially integral-transformations can be performed within the class of G-functions.

In the following we will give a listing of and some comments to those relations.

Differentation

$$z^k \frac{d^k}{dz^k}\left[G^{mn}_{pq}\left(z\middle|{}^{a_p}_{b_q}\right)\right] = G^{m\ \ n+1}_{p+1\ q+1}\left(z\middle|{}^{0,a_p}_{b_q,k}\right) \tag{3-12}$$

$$z^k \frac{d^k}{dz^k}\left[G^{mn}_{pq}\left(z^{-1}\middle|{}^{a_p}_{b_q}\right)\right] = (-1)^k\, G^{m\ \ n+1}_{p+1\ q+1}\left(z^{-1}\middle|{}^{1-k,a_p}_{b_q,1}\right) \tag{3-13}$$

$$\frac{d}{dz}\left[z^{-b_1}\, G^{mn}_{pq}\left(z\middle|{}^{a_p}_{b_q}\right)\right] = -z^{-(1+b_1)}\, G^{mn}_{pq}\left(z\middle|{}^{a_p}_{1+b_1,b_2,\ldots,b_q}\right) \ . \tag{3-14}$$

Here we see, that differentiation can easily be done by formal alterations in the set of the parameters.

Integration

The indefinte integral

$$\begin{aligned} F(x) &= \int_0^x dy\ \ G^{mn}_{pq}\left(\lambda y^\gamma\middle|{}^{a_p}_{b_q}\right) \\ &= F(\infty) - \int_x^\infty dy\ G^{mn}_{pq}\left(\lambda y^\gamma\middle|{}^{a_p}_{b_q}\right) \end{aligned} \tag{3-15}$$

is solved within the class of G-functions by

$$F(x) = F(\infty) - \gamma^{-1}\lambda^{-1/\gamma}\ G^{m+1\ \ n}_{p+1\ q+1}\left(\lambda x^\gamma\middle|{}^{(a_p+\frac{1}{\gamma}),1}_{0,(b_q+\frac{1}{\gamma})}\right) \ . \tag{3-16}$$

Thus, if it is possible to express either the cumulative distribution function or the corresponding density function in terms of G-functions, then it is possible to do this for both of them.

Integral-transformations are based upon the so called "master formula", already given by Meijer.

$$\int_0^\infty dx\, G^{mn}_{pq}\left(\lambda x\Big|{a_p \atop b_q}\right) \cdot G^{\mu\nu}_{\sigma\tau}\left(\omega x\Big|{c_\sigma \atop d_\tau}\right) =$$

$$= \frac{1}{\lambda}\, G^{n+\mu\ m+\nu}_{q+\sigma\ p+\tau}\left(\frac{\omega}{\lambda}\ \Big|\ {-b_m, c_\nu;\ c_{\sigma-\nu}, -b_{q-m} \atop -a_n, d_\mu;\ d_{\tau-\mu}, -a_{p-n}}\right) \tag{3-17}$$

$$= \frac{1}{\omega}\, G^{m+\nu\ n+\mu}_{p+\tau\ q+\sigma}\left(\frac{\lambda}{\omega}\ \Big|\ {a_n, -d_\mu;\ -d_{\tau-\mu}, a_{p-n} \atop b_m, -c_\nu;\ -c_{\sigma-\nu}, b_{q-m}}\right) .$$

Specializing one of the two G-functions in the integrand yields the fol lowing transforms.

Laplace Transforms

With the familiar definition

$$L(f(x);s) = \int_0^\infty dx\, f(x) e^{-sx} \tag{3-18}$$

we have

$$L\left(G^{mn}_{pq}\left(\lambda x\Big|{a_p \atop b_q}\right);s\right) = \frac{1}{\lambda}\, G^{n+1\ m}_{q\ \ p+1}\left(\frac{s}{\lambda}\Big|{-b_q \atop 0, -a_p}\right)$$

$$= \frac{1}{s}\, G^{m\ \ n+1}_{p+1\ q}\left(\frac{\lambda}{s}\Big|{0, a_p \atop b_q}\right) \tag{3-19}$$

a relation, that also may be read in the reverse direction yielding

$$L^{-1}\left(G^{mn}_{pq}\left(\lambda s\Big|{a_p \atop b_q}\right);x\right) = \frac{1}{\lambda}\, G^{n\ \ m}_{q\ p+1}\left(\frac{x}{\lambda}\Big|{-b_q \atop -a_p, 0}\right) \tag{3-20}$$

the formula for the inverse Laplace transformation.

Fourier Transforms

The Fourier-transformation may be splitted into the Fourier-Cosine- and the Fourier-Sine-transformation

$$F_C(f(x);\ y) = \int_0^\infty dx\ f(x)\ \cos(xy)$$

$$F_S(f(x);\ y) = \int_0^\infty dx\ f(x)\ \sin(xy) . \tag{3-21}$$

In the case of G-functions we have

$$F_c(G_{pq}^{mn}(\lambda x^2|{}^{a_p}_{b_q});y) = \pi^{1/2}y^{-1}G_{p+2\ q}^{m\ n+1}(\frac{4\lambda}{y^2}|{}^{1/2,a_p,0}_{b_q})$$

$$= \frac{\pi^{1/2}}{2\lambda^{1/2}}G_{q\ p+2}^{n+1\ m}(\frac{y^2}{4\lambda}|{}^{1/2-b_q}_{0,1/2-a_p,1/2}) \quad (3\text{-}22)$$

$$F_s(G_{pq}^{mn}(\lambda x^2|{}^{a_p}_{b_q});y) = \pi^{1/2}y^{-1}G_{p+2\ q}^{m\ n+1}(\frac{4\lambda}{y^2}|{}^{0,a_p,1/2}_{b_q})$$

$$= \frac{\pi^{1/2}}{2\lambda^{1/2}}G_{q\ p+2}^{n+1\ m}(\frac{y^2}{4\lambda}|{}^{1/2-b_q}_{1/2,1/2-a_p,0}). \quad (3\text{-}23)$$

Because a SIRP has a first-order PDF, that is an even function of its argument, the corresponding CF is twice the F_c-transform of the PDF.

Hankel Transform

With the definition

$$J_K(f(x);y,k) = \int_0^\infty dx\ (xy)^{1/2}\ J_k(xy)\ f(x) \quad (3\text{-}24)$$

we get

$$J_K(G_{pq}^{mn}(\lambda x^2|{}^{a_p}_{b_q});y,k) = (2\lambda)^{-\frac{1}{2}}\ G_{q\ p+2}^{n+1\ m}(\frac{y^2}{4\lambda}\Big|{}^{1/2\ -\ b_q}_{\frac{1+2k}{4},\frac{1}{2}-a_p,\frac{1-2k}{4}}) . \quad (3\text{-}25)$$

Thus, it is possible, starting with a first-order PDF expressed in terms of a G-function, to calculate the corresponding CF and then to determine higher order PDFs as Hankel transforms.

Mellin Transform

This transformation, defined by

$$M(f(x);z) = \int_0^\infty dx\ x^{z-1}\ f(x) \quad (3\text{-}26)$$

results in terms of Γ-functions

$$M(G_{pq}^{mn}(x|{}_{b_q}^{a_p});z) = \frac{\prod_1^m \Gamma(b_j+z) \prod_1^n \Gamma[1-(a_j+z)]}{\prod_{m+1}^q \Gamma[1-(b_j+z)] \prod_{n+1}^p \Gamma(a_j+z)} . \qquad (3\text{-}27)$$

It is a powerful tool, to determine all the moments of a prescribed distribution.

In all cases there might exist some constraints concerning the number or values of the parameters. Details are listed in Luke [30] or Brehm [22].

3.2 COMPLETE CHARACTERIZATION OF SIRPs BY MEANS OF G-FUNCTIONS

In this section we will aim at a complete description of a SIRP using MEIJER's G-function. Therefore we start under the assumption, that the first-order PDF of the process is prescribed or well fitted by

$$p_\xi(x) = A\, G_{pq}^{mn}(\lambda x^2|{}_{b_q}^{a_p}) \qquad \text{for } -\infty \le x \le \infty \qquad (3\text{-}28)$$

i.e. by a G-function, whose parameters are properly chosen. The special dependence on the square of the variable x is not necessary but recommended, because the PDF is an even function. Naturally we have to certify, that G(s) is completely monotone in order that the representation (3-28) is consistent with the properties of a SIRP. This may be done in several ways, that are discussed in detail in [22]. Likewise the normalizing factor

$$A = \lambda^{1/2} \frac{\prod_{m+1}^q \Gamma(\frac{1}{2}-b_j) \prod_{n+1}^p \Gamma(\frac{1}{2}+a_j)}{\prod_1^m \Gamma(\frac{1}{2}+b_j) \prod_1^n \Gamma(\frac{1}{2}-a_j)} \qquad (3\text{-}29)$$

must be finite. In this expression the constant λ yields unit variance, if it is chosen equal to

$$\lambda = (-1)^\varepsilon \frac{\prod_{j=1}^q (\frac{1}{2}+b_j)}{\prod_{j=1}^p (\frac{1}{2}+a_j)} \qquad (3\text{-}30)$$

$$\varepsilon = n - (q-m) .$$

In cases of infinite variance (e.g. Cauchy-distribution) we may choose $\lambda = 1$. Obviously both results have been obtained by means of Mellin-transformation according to (3-27), which also yields all *moments of the distribution*

$$\langle \xi^{2k+1} \rangle = 0$$

$$\langle \xi^{2k} \rangle = ((-1)^{\varepsilon} \lambda^{-1})^k \cdot \frac{\prod_1^q (\frac{1}{2}+b_j)_k}{\prod_1^p (\frac{1}{2}+a_j)_k}, \quad \text{with} \quad \varepsilon = n-(q-m), \quad (\alpha)_k \equiv \frac{\Gamma(\alpha+k)}{\Gamma(\alpha)} \qquad (3\text{-}31)$$

and k = 0, 1, 2, ...

Looking at (3-15, 16) we find, that the
Cummulative distribution function

$$P_\xi(x) = 1 - \frac{A}{2\lambda^{1/2}} G^{m+1\ n}_{p+1\ q+1}\left(\lambda x^2 \middle| {a_p+1/2,1 \atop 0,b_q+1/2}\right)$$

$$P_\xi(-x) = 1 - P_\xi(x) \qquad \text{both for } x \geq 0 \qquad (3\text{-}32)$$

is given by another G-function with slightly changed and two additional parameters. Because the PDF is an even function, F_C-transformation (3-22) leads to the
characteristic function

$$C_\xi(v) = A\,\lambda^{-1/2}\,\pi^{1/2}\,G^{n+1\ m}_{q\ p+2}\left(\frac{v^2}{4\lambda} \middle| {1/2-b_q \atop 0,1/2-a_p,1/2}\right) \qquad (3\text{-}33)$$

and likewise by means of the rules (3-19, 20) concerning Laplace-transformation all
higher-order probability density functions (PDFs)

$$p_{\underline{\xi}}(\underline{x}) = \pi^{-\nu/2} f(s;\nu)\ ; \qquad s = \sum_{j=1}^{\nu} x_j^2$$

$$\text{with} \quad f(s;\nu) = \pi^{1/2} A\, s^{\frac{1-\nu}{2}} G^{m+1\ n}_{p+1\ q+1}\left(\lambda s \middle| {a_p,0 \atop (\nu-1)/2,b_q}\right) \qquad (3\text{-}34)$$

are obtained without any further calculations. Finally, and this completes the characterization of the SIRP, we will list the
PDFs of the random variables ρ_ν *and* σ*, defined by (2-23, 25)*

$$p_{\rho_\nu}(r) = 2A \frac{\pi^{1/2}}{\Gamma(\frac{\nu}{2})} G^{m+1\ n}_{p+1\ q+1}\left(\lambda r^2 \middle| {a_p,0 \atop \frac{\nu-1}{2},b_q}\right); \ 0 \leq r \leq \infty \qquad (3\text{-}35)$$

$$p_\sigma(r) = 2(2\pi)^{1/2} A\ G^{m\ n}_{m+1\ q}\left(2\lambda r^2 \middle| {a_p,0 \atop b_q}\right),\ 0 \leq r \leq \infty\ . \qquad (3\text{-}36)$$

Referring to these results, comprehensive notations are available, that may be the basis of further calculations.

4. Applications

4.1 SPEECH AS A REALIZATION OF A SIRP

As mentioned above, an application of the theory outlined in the preceding sections must be based on results of experimental studies. Firstly, one has to find a proper fit to the first-order PDF of the random process in question, and secondly, one has to verify that this process is a SIRP.

Therefore, experimental studies with speech signals were performed [23]. This was done in the time domain, i.e. the signals were sampled and the relative frequency of the occurrence of quantized amplitude values was measured. Thus, it had been assumed, that the random process, whose realizations were examined, is stationary as well as ergodic. At the first glance, this assumption might appear to be very restrictive. However, for most applications it is not, as recently has been stated again by Abut et. al. [35].

Measurements were done with speech signals, bandlimited to the frequency range from 300 to 3400 Hz, according to the requirements of telephone channels. The number of evaluated samples was 10^6 or 10^7 in the first- or second-order case, respectively. In consequence statistical variances are small enough to guarantee, that the measured relative frequencies are fairly good approximations to the corresponding PDFs. In a logarithmic scale fig. 4.1 shows experimental points of the first-order PDF versus amplitude values, normalized to unit variance.

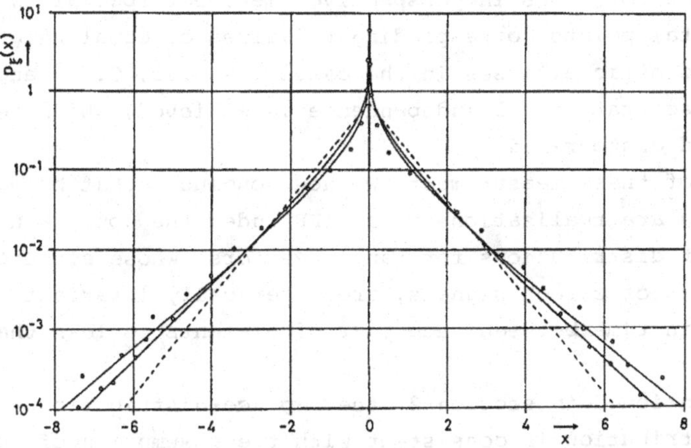

Fig. 4.1: Experimental points of the first-order PDF, fitted by (------) Laplace-, (———) K_0-, and (———) Gamma-distribution

Together with the measured points there are drawn three curves representing the following PDFs

$$\text{Laplace} \qquad p_\xi(x) = 2^{-1/2} \exp(-2^{1/2}|x|) \qquad (4\text{-}1)$$

$$K_o \qquad p_\xi(x) = \pi^{-1} K_o(|x|) \qquad (4\text{-}2)$$

$$\text{Gamma} \qquad p_\xi(x) = \frac{3^{1/2}}{4\pi^{1/2}} \left[\frac{3^{1/2}}{2}|x|\right]^{-1/2} \exp\left(\frac{-3^{1/2}}{2}|x|\right). \qquad (4\text{-}3)$$

It can be seen, that the best fit is achieved by the Gamma-PDF. Nevertheless there even may exist another PDF, that yields a significantly better fit. This question will be discussed later on.

One may ask, whether it is recommended to look for an optimum fit, because the PDF might vary for different speech signals considerably. However, measurements have certified, that the first-order PDF of bandlimited speech is almost unaffected by the personal characteristics of the speaker as well as the used language.

For the following it is of great importance, that measurements have proved, that the type of the distribution remains unchanged by linear filtering, a typical feature of SIRPs.

The shape of the second-order PDF depends on τ, the distance in time between the relevant random variables $\xi(t)$ and $\xi(t+\tau)$, i.e. on the statistical dependencies between these variables, which are essentially affected by the personal characteristics of the speaker. Experimental results are illustrated in fig. 4.2, where four maps of contour-lines are given, referring to the four PDFs in perspective view, for special values of τ. The experimental points corresponding to values of equal height are well fitted by concentric ellipses in the cases τ = 0.25, 0.75, and 1.0 msec. For τ = 50 msec statistical independence is achieved, which results in diamond-shaped contour-lines.

On the basis of these measurements we now conclude, that bandlimited speech signals are realizations of a SIRP under the constraint $\tau < 5$ msec, i.e. the joint distributions for random vectors, whose elements are sampled amplitudes of speech signals, are spherically invariant whenever the distance in time between each pair of elements is less than 5 msec.

As already mentioned in section 2, the Laplace-distribution as well as the Gamma-distribution is consistent with the assumption of spherical invariance. Following our general concept we now have to look for a representation in terms of G-functions. With the relation

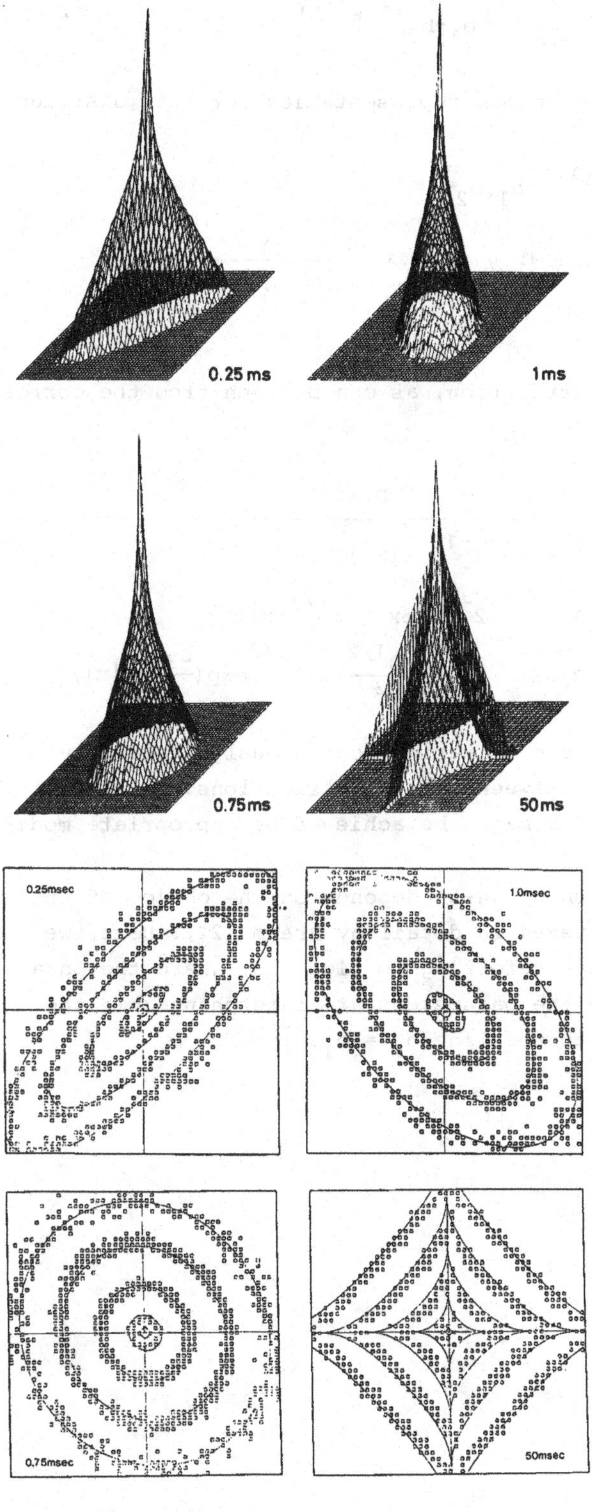

Fig. 4.2: Second-order PDFs of speech signals, given in perspective views and contour-lines with τ = 0.25, 0.75, 1.0, and 50 msec.

$$G_{02}^{20}(z|b_1,b_2) = 2\,z^{(b_1+b_2)/2}\,K_{b_1-b_2}(2\,z^{1/2}) \qquad (4\text{-}4)$$

given in Luke [30] we have the common representation for both distributions

$$p_\xi(x) = A\,G_{02}^{20}(\lambda x^2|b_1,b_2)$$

with

$$\lambda = (\tfrac{1}{2}+b_1)(\tfrac{1}{2}+b_2) \quad \text{and} \quad A = \frac{\lambda^{1/2}}{\Gamma(\frac{1}{2}+b_1)\Gamma(\frac{1}{2}+b_2)} \qquad (4\text{-}5)$$

which even includes the K_o-distribution, as can be seen from the correspondences listed below.

b_1	b_2	A	λ	$p_\xi(x)$
0	0	$(2\pi)^{-1}$	1/4	$\pi^{-1}K_o(\|x\|)$
1/2	0	$(2\pi)^{-1/2}$	1/2	$2^{-1/2}\exp(-2^{1/2}\|x\|)$
1/4	-1/4	$(3/2)^{1/2}(4\pi)^{-1}$	3/16	$\frac{3^{1/2}}{4\pi^{1/2}}(\frac{3^{1/2}}{2}\|x\|)^{-1/2}\exp(\frac{-3^{1/2}}{2}\|x\|)$

Because the parameters b_1, b_2 can be varied continuously, there may exist continuous transitions between those distributions. Consequently an even better fit of the PDF might be achieved by appropriate modifications of b_1 and b_2.

The type of the functions given by (4-5) depends on the choice of the parameters b_1 and b_2, as discussed in detail by Brehm [22]. Here, we only will give the main results referring to fig. 4.3, where the parameter plane is drawn. Because the parameters are interchangeable the figure is symmetric relative to the line $b_2 = b_1$.

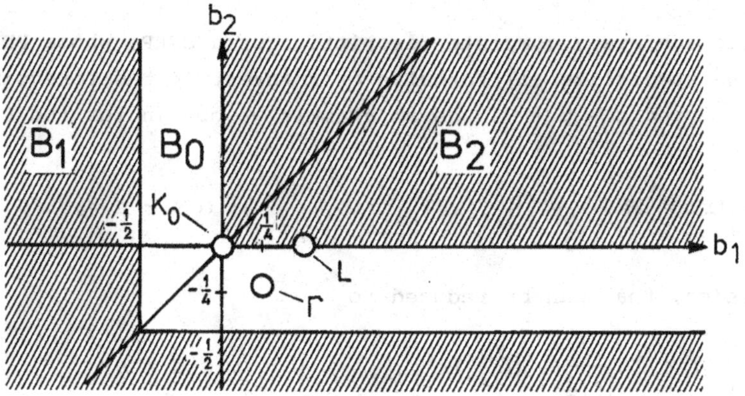

Fig. 4.3: Unshaded Area indicates possible combinations of b_1, b_2 for G_{02}^{20}-PDF

Though the functions $G_{02}^{20}(x^2)$ show an exponential decay for large arguments according to (3-6) they cannot be normalized, if $b_1 \leq -1/2$ or $b_2 \leq -1/2$. This stems from singularities in the origin and may be deduced from the identity.

$$G_{02}^{20}(x^2|_{b_1,b_2}) = x^{2b_2} G_{02}^{20}(x^2|_{(b_1-b_2),0}) \tag{4-6}$$

and the limiting forms (3-4, 5) for small arguments. On the other hand, $G_{02}^{20}(s)$ is not completely monotone, if $b_1 > 0$ and $b_2 > 0$. Therefore first-order PDFs for SIRPs are only given in the nonshaded area. Here we find the generalized Laplace-distributions for $b_2 = 0$ and $-1/2 < b_1$, which tend to the Gaussian-distribution in the limit $b_1 \to \infty$. Another well known class of distributions, whose members are the generalized Gamma-distributions, is found along the line $b_2 = b_1 - 1/2$, from which only a subset may be identified with first-order PDFs of a SIRP.

It should be mentioned here, that representations of first-order PDFs of SIRPs are not restricted to the subclass G_{02}^{20} of the G-functions. An unlimited number of further subclasses, from which G_{01}^{10}, G_{12}^{20}, G_{13}^{30} are only a few examples, are found following a strategy given by Brehm [22].

In order to achieve a complete description of a SIRP, whose PDF is given by (4-5) we specialize all formulas of interest in the foregoing section to m = q = 2 and n = p = 0. As an example we obtain for the PDF of the radius

$$p_{\rho_\nu}(r) = 2A \frac{\pi^{1/2}}{\Gamma(\frac{\nu}{2})} \; G^{30}_{13}\left(\lambda r^2 \,\middle|\, {0 \atop \frac{\nu-1}{2},\, b_1,\, b_2}\right); \; 0 \le r \le \infty, \tag{4-7}$$

an expression, that can be reduced to

$$p_{\rho_\nu}(r) = 2A \frac{\pi^{1/2}}{\Gamma(\frac{\nu}{2})} \; G^{20}_{02}\left(\lambda r^2 \,\middle|\, \frac{\nu-1}{2},\, b_1\right) \quad ; \; 0 \le r \le \infty \tag{4-8}$$

i.e. to modified Bessel-functions, under consideration of eq. (3-9), in the case of generalized Laplace-distributions. Not only for the sake of a more convenient scaling of the drawings it is recommended to look at the equivalent PDF $p_{\sigma_n}(r)$, where now in the notation the subscript n is chosen instead of ν. In fig. 4.4 results are shown for the three distributions of interest here with values n = 1, 2, 4, 8, and n = ∞.

At first we recognize, that in each case the limiting PDF $p_\sigma(r)$ do not agree at all with Dirac's delta-function $\delta(r-1)$ centered at the value r = 1, they even do not approximate it. This behaviour challenges the question, whether non-Gaussian SIRPs really might be proper models for processes, that are delt with in practice and for which ergodicity has to be assumed. A positive answer will be given in the following section based on the observed relatively fast convergence to the limiting PDF $p_\sigma(r)$ with increasing n.

At last we should pay attention to the fact, that, dependent on the type of the distribution, significant dissimilarities occur for small values of the argument. These dissimilarities will have considerable influence on the results to be presented later on.

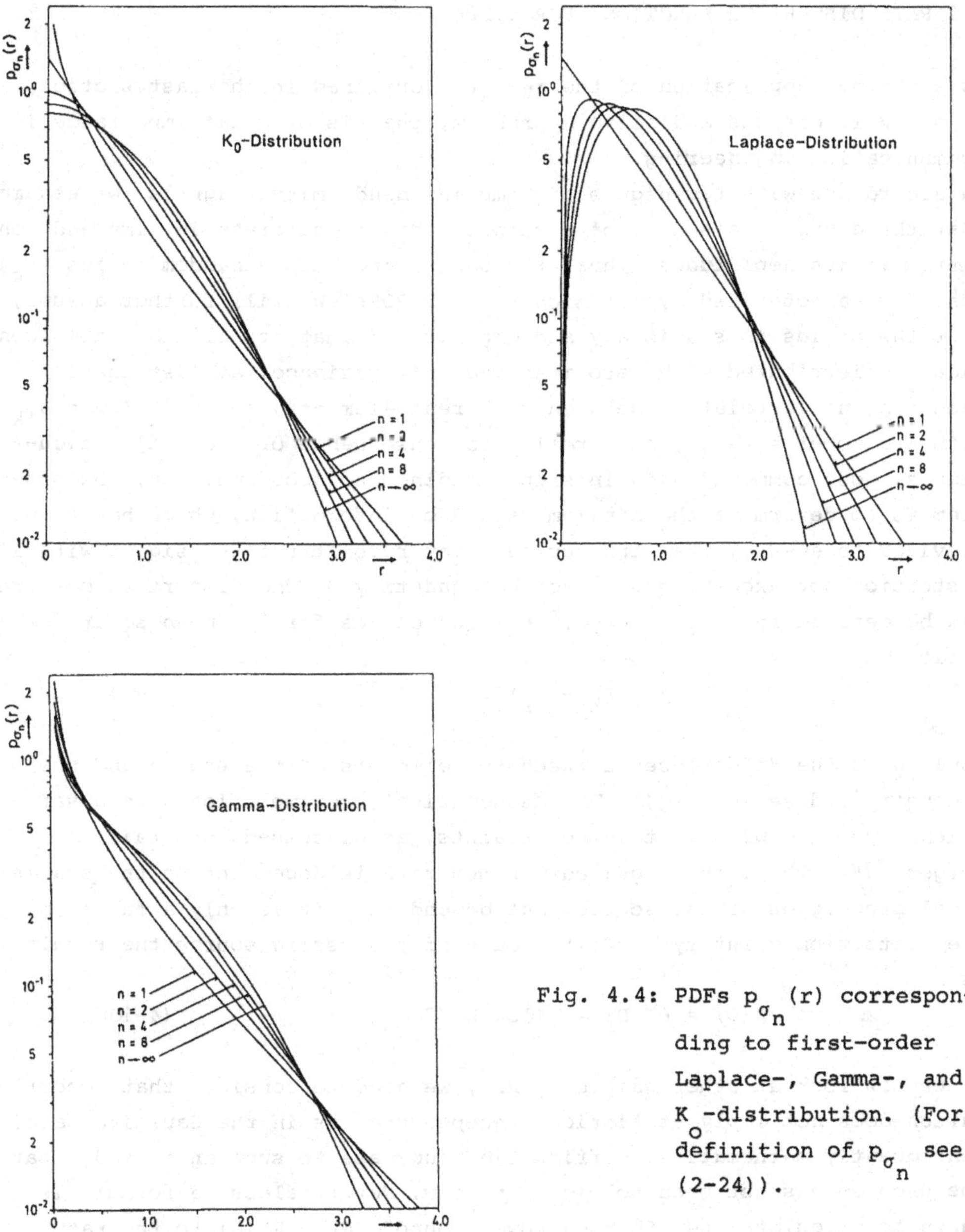

Fig. 4.4: PDFs $p_{\sigma_n}(r)$ corresponding to first-order Laplace-, Gamma-, and K_o-distribution. (For definition of p_{σ_n} see (2-24)).

4.2 RATE DISTORTION FUNCTIONS FOR SIRPs

As a further application of the results, outlined in the last section, we now will discuss and solve a problem, that is of great importance in communication engineering.
In accordance with techniques of sampling bandlimited signals, we assume that there exists a source of a signal, that is discrete in time and continuous in its amplitudes. Thus, the source realizes a random series $\{\xi_k\}$, that is characterized by its higher-order PDFs. We will further assume, that the series is stationary and ergodic and that its elements are identically distributed with zero mean and unit variance. At last there should be no correlation between different elements, i.e. $\langle\xi_i\ \xi_k\rangle = \delta_{ik}$ with Kronecker's delta, as a realistic consequence of prediction techniques, most commonly used in signal coding and decoding. Now, the problem is to determine the minimum rate R of information, which has to be provided to ensure, that the receiver may reconstruct the signal with a distortion not exceeding a prescribed quantity D. The distortion measure can be defined in several ways. Here, we decide for the mean square criterion

$$D = \langle(\xi_k - \eta_k)^2\rangle \qquad (4\text{-}9)$$

applied to the differences between the elements of the source and the reconstructed series $\{\eta_k\}$. The mathematical treatment leads to a variational problem with certain constraints, as discussed in detail in Berger [36]. Thus, the required minimum rate is dependent on the statistical properties of the source, but beyond that it is only a function of the distortion quantity D. In the case of a Gaussian source the result is

$$R(D) \mathrel{\hat{=}} R^G(D) = -(\log D)/2 \,. \qquad (4\text{-}10)$$

If we now look at other sampled SIRPs, we need to consider that decorrelation does not imply statistical independence as in the Gaussian case. Consequently mathematical difficulties increase to such an extend, that the problem has not been solved up to now. Nevertheless, a formula is known to calculate the "Shannon Lower Bound" (SLB) $R_L(D)$ to the rate distortion function R(D) in the case of SIRPs. According to the nomenclature the relation $R_L(D) \leq R(D)$ holds for each value of D and R_L has been found to be very tight to R, esp. for values $D \ll 1$.

The SLB is expressed in a limiting form

$$R_L(D) = R^G(D) - \lim_{n\to\infty} C(n) \tag{4-11}$$

$$C(n) = 2[n\,\Gamma(n/2)]^{-1} \int_0^\infty dr\; r^{n-1}\, f(r^2;\, n)\, \log[\pi^{-n/2} f(r^2;\, n)] \tag{4-12}$$

with a term $\lim_{n\to\infty} C(n)$ correcting the Gaussian case.

Now, we will show, how integration as well as the limiting process can be performed for all SIRPs which are characterized by G-functions. We introduce the PDF $p_{\sigma_n}(r)$ and obtain the following sum

$$C(n) = C_1(n) + C_2(n) + C_3(n) \qquad \text{with}$$

$$\begin{aligned} C_1(n) &= \tfrac{1}{2}\log(2\pi e) + \tfrac{1}{n}\log[\pi^{-n/2}\,\Gamma(n/2)\; n^{-n/2}/2] \\ C_2(n) &= \tfrac{1}{n}\int_0^\infty dr\; p_{\sigma_n}(r)\,\log[p_{\sigma_n}(r)] \\ C_3(n) &= \tfrac{1-n}{n}\int_0^\infty dr\; p_{\sigma_n}(r)\,\log r\,. \end{aligned} \tag{4-13}$$

Interchanging the order of integration and limiting process, we obtain $C_1(\infty) = 0$ from the asymptotic behaviour of the Γ-function and $C_2(\infty) = 0$ because the integral is bounded for each value of n. The remaining additive term

$$C_3(\infty) = -\int_0^\infty dr\; p_\sigma(r)\,\log r\,, \tag{4-14}$$

contains the PDF $p_\sigma(r)$ defined by eq. (2-25). Now integration is done by parts. Because of the fact, that we can express $p_\sigma(r)$ in terms of G-functions, the indefinite integrals

$$h_1(x) = \int^x dr\; p_\sigma(r) \qquad\qquad h_2(x) = \int^x dr\; [r^{-1}\, h_1(r)] \tag{4-15}$$

are principally known as well and the general result is

$$C(\infty) = \{-h_1(x)\,\log x + h_2(x)\}\Big|_{x=0}^{x=\infty}\,. \tag{4-16}$$

Specializing this result for $p_\xi(x) = A\, G^{20}_{02}(\lambda x^2 \,|\, b_1,\, b_2)$ leads to

$$C(\infty) = \tfrac{1}{2}[\chi(b + \tfrac{1}{2}) + \chi(b + \tfrac{1}{2}) - \chi(\tfrac{1}{2})] \tag{4-17}$$

under consideration of the behaviour of the G-functions for large and small arguments respectively. To achieve a simple notation the abbreviation $\chi(z) = \log z - \psi(z)$ has been introduced, where $\psi(z) \equiv d \log \Gamma(z)/dz$ is the digamma-function tabulated e.g. in [33].

Quantitative results are given in fig. 4.5. As could be expected, the rate decreases in the non-Gaussian cases on account of the remaining statistical dependencies after decorrelation. The significant differences between the curves emphasize, that it is an essential point, to have an optimum fit of the first-order PDF.

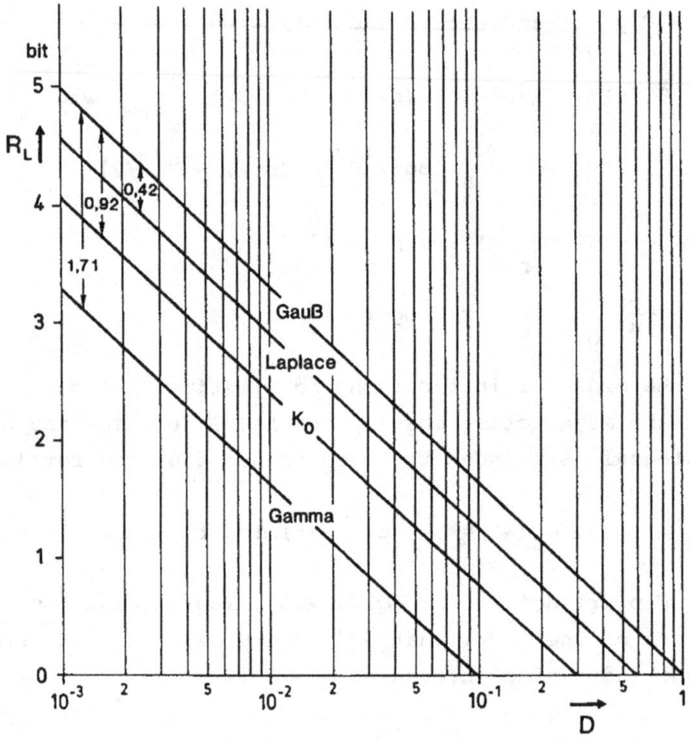

Fig. 4.5: Shannon Lower Bounds for Rate Distortion Functions of SIRPs with Gaussian-, K_o-, Laplace- and Gamma-distribution

One may argue that there have been obtained results, which cannot be applied to problems arising in the fields of speech processing, because the assumption of spherical invariance has to be restricted to a finite number of successive signal samples. Referring to pulse-code-modulation-(PCM)-systems, that are in use for the processing of bandlimited speech signals at a sampling rate of 8 kHz, we conclude that the restriction $\tau < 5$ msec is equivalent to the constraint $n < 40$. Thus, the results might become meaningless in the limit $n \to \infty$. On the other hand we have already recognized that the PDFs $p_{\sigma_n}(r)$ rapidly converge to $p_\sigma(r)$ with increasing n. In order to justify a quantitative statement the integral (4-12) has been evaluated numerically for finite values of n avoiding the limiting process. The results shown in fig. 4.6 certify, that for n = 40 the limiting values are reached up to about 90% in each of the three cases. Consequently we conclude that the assumption of invariance does not imply a severe restriction.

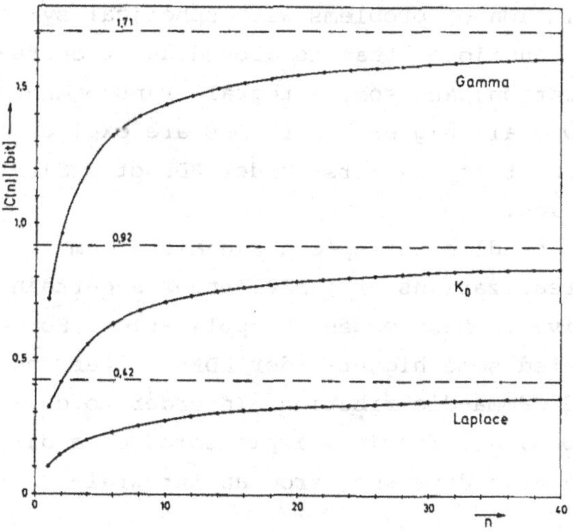

Fig. 4.6: Convergence of the correcting terms C(n) to the limiting values

Finally it should be emphasized, that the solution to this complex problem which never has been worked out so far, has been found now, because higher-order or equivalent PDFs could be given explicitely. We achieved this without detailed knowledge of the features of special G-functions used in intermediate steps of the calculation. We only needed a proper fit to the first-order PDF in terms of a G-function and then we have used some common relations between those functions, their integrals, and transforms. By further consideration of the behaviour of the G-functions for large and small arguments an expression for $C(\infty)$ has been obtained, the evaluation of which can be performed by means of a pocket calculator. Of course the computation of C(n), requires an algorithm to calculate values of several G-functions. If, however, the algorithm given in [22] once is implemented on a digital computer, there arises no problem at all. Altoghether, these facts should strengthen the conviction that G-functions are highly recommended to achieve new results of theoretical as well as practical interest in any field, where SIRPs are of importance as for instance in communication theory and speech processing.

5. Conclusions

SIRPs are generalizations of the very familiar Gaussian random process. They are of great importance, especially in the fields of communication engineering, because signals are treated there as realizations of random processes. For these SIRPs a complete characterization by means of higher-order PDFs has been obtained in an utmost comprehensive notation using Meijer's G-function.

Though these higher-transcendental functions are commonly not used, they are highly recommended for the solution of problems with spherical symmetry, because they form a set of functions, that is closed under operations like differentiation, integration, and some integral-transformations. Therefore, as has been shown, all higher-order PDFs are explicitely available under the assumption, that the first-order PDF of a SIRP is expressed in terms of a G-function.

These results have been applied to bandlimited speech signals, from which it is known, that they are realizations of a SIRP under a certain constraint, which is not restrictive in most cases of application. For convenience there have been evaluated some higher-order PDFs, referring to a univariate Laplace-, K_o-, and Gamma-distribution. In order to calculate the values of the G-functions, an algorithm implemented on a digital computer was used, which had been developed from an integral-representation of the G-function.

Finally, the solution of a problem of great interest, arising in the fields of information theory, has been found due to the fact, that explicite expressions for PDFs even of unlimited order are available now. The Shannon Lower Bound for the Rate Distortion Function of a decorrelated SIRP has been determined, referring to a limiting form of G-functions, only by means of a pocket calculator. By this result, the recommendation of the G-functions for a convenient description of SIRPs is emphasized essentially.

Acknowledgements

The author gratefully acknowledges the help of W. Stammler and K. Trottler in preparing this manuscript.

The work was supported by the Deutsche Forschungsgemeinschaft and most parts of it were done, whilst the author stayed with the Institute of Applied Physics of the University in Frankfurt.

References

[1] Lord, R. D.: The Use of the Hankel Transform in Statistics. Biometrika 41 (1954), pp. 44 - 55.

[2] Kingman, J.F.C.: Random Walks with Spherical Symmetry. Acta Math. (Stockholm) 109 (1963), pp. 11 - 53.

[3] Vershik, A.M.: Some Characteristic Properties of Gaussian Stochastic Processes. Theory of Probability and its Applications, IX (1964), pp. 353-356.

[4] Blake, I.F. and Thomas, J.B: On a Class of Processes Arising in Linear Estimation Theory. IEEE Trans. Information Theory, IT-14 (1968), pp. 12 - 16

[5] McGraw, D.K. and Wagner, J.F: Elliptically Symmetric Distributions. IEEE Trans. Information Theory, IT 14 (1968), pp. 110 - 120.

[6] Picinbono, B: Spherically Invariant and Compound Gaussian Processes. IEEE Trans. Information Theory, IT-16 (1970), pp. 77 - 79.

[7] Kingman, J.F.C.: On Random Sequences with Spherical Symmetry. Biometrika 59 (1972), pp. 492 - 494.

[8] Yao, K.: A Representation Theorem and its Applications to Spherically-Invariant Random Processes. IEEE Trans. Information Theory, IT-19 (1973), pp. 600 - 607.

[9] Bochner, S.: Monotone Funktionen, Stieltjessche Integrale und harmonische Analyse. Math. Annalen 108 (1933), pp. 399 - 408

[10] Bochner, S.: Completely Monotone Functions of the Laplace Operator for Torus and Sphere. Duke Math. J. 3 (1937), pp. 488 - 502.

[11] Bochner, S.: Stable Laws of Probability and Completely Monotone Functions. Duke Math. J. 3 (1937), pp. 726 - 728.

[12] Bochner, S.: Lectures on Fourier Integrals. Princeton University Press (1959)

[13] Schoenberg, I.J.: On Certain Metric Spaces Arising from Euclidian Spaces by a Change of Metric and their Imbedding in Hilbert Space. Annals of Mathematics 38 (1937), pp. 787 - 793.

[14] Schoenberg, I.J.: Metric Spaces and Completely Monotone Functions. Annals of Mathematics 39 (1938), pp. 811 - 841.

[15] Schoenberg, I.J.: Metric Spaces and Positive Definite Functions. Trans. Am. Math. Soc. 44 (1938), pp. 522 - 536.

[16] Widder, D.V.: Necessary and Sufficient Conditions for the Representation of a Function as a Laplace Integral. Trans. Am. Math. Soc. 33 (1931), pp. 851 - 892.

[17] Picinbono, B. and Vezzosi, G.: Détection d'un Signal Certain dans un Bruit non Stationnaire et non Gaussien. Ann. Télécommunic. 25 (1970) pp. 433 - 439.

[18] Vezzosi, G. and Picinbono, B.: Détection d'un Signal Certain dans un Bruit Sphériquement Invariant, Structure et Characteristiques des Récepteurs.
Ann. Télécommunic. 27 (1972), pp. 95 - 110.

[19] Goldmann, J.: Statistical Properties of a Sum of Sinusoids and Gaussian Noise and its Generalization to Higher Dimensions.
Bell Sys. Tech. J. 53 (1974), pp. 557 - 580.

[20] Goldmann, J.: Detection in the Presence of Spherically Symmetric Random Vectors.
IEEE Trans. Information Theory, IT-22 (1976), pp. 52 - 59.

[21] Leung, H.M. and Cambanis, S.: On the Rate Distortion Functions of Spherically Invariant Vectors and Sequences.
IEEE Trans. Information Theory, IT-24 (1978), pp. 367 - 373.

[22] Brehm, H.: Sphärisch Invariante Stochastische Prozesse.
Habilitationsschrift, Fachbereich Physik, Universität Frankfurt am Main (1978).

[23] Wolf, D. and Brehm, H.: Experimental Studies on one- and two-dimensional Amplitude Probability Densities of Speech Signals.
Proc. 1973 Int. Symp. on Inf. Theory, Ashkelon (Israel), IEEE-Cat. 73 CH 0753-4 IT, B4-6.

[24] Brehm, H. and Wolf D.: N^{th}-Order Joint Probability Densities of Non-Gaussian Stochastic Processes.
Proc. 1976 Int. Symp. on Inf. Theory, Ronneby (Schweden), IEEE-Cat. 76, CH 1095-9 IT, pp. 121 - 122.

[25] Wolf, D. and Brehm H.: Mathematical Treatment of Speech Signals.
Proc. 1977 Int. Symp. on Inf. Theory, Ithaca (USA), IEEE-Cat. 77, CH 1277-3 IT, E3, p. 105.

[26] Wolf, D.: Analytische Beschreibung von Sprachsignalen.
AEÜ 31 (1977), pp. 392 - 398.

[27] Meijer, C.S.: Nieuw. Arch. Wisk. 18 (1936), pp. 10-39,
Meijer, C.S.: Nederl. Akad. Wetensch. Proc. Ser. A 49 (1946), pp. 227-237, 344-356, 457-469, 632-641, 765-772, 936-943, 1063-1072 and 1165-1175.

[28] Erdelyi, A. (ed.): Higher Transcendental Functions.
Vol. I, II, III. McGraw-Hill Book Company, New York, Toronto, London, 1953.

[29] Luke, Y.: The Special Functions and their Approximations.
Vol. I, II. Academic Press New York, San Francisco, London, 1969.

[30] Luke, Y.: Mathematical Functions and their Approximations.
Academic Press, New York, San Francisco, London 1975.

[31] Papoulis, A.: Signal Analysis.
McGraw-Hill Book Company, New York, etc., 1977.

[32] Oberhettinger, F. and Badii, L.: Tables of Laplace Transforms.
Springer-Verlag, Berlin, Heidelberg, New York, 1973.

[33] Abramowitz, M. and Stegun I. (ed.): Handbook of Mathematical Functions.
Dover Publications, New York, 1965.

[34] Oberhettinger, F.: Tables of Bessel Transforms. Springer-Verlag, Berlin, Heidelberg, New York, 1972.

[35] Abut, H., Gray, R.M. and Rebolledo, G.: Vector Quantization of Speech and Speech-Like Waveforms. IEEE Trans. Acoust., Speech, Signal Processing, vol. ASSP-30, (1982), pp. 423 - 435.

[36] Berger, T.: Rate Distortion Theory. Prentice Hall, Englewood Cliffs, New Jersey, 1971.

GEOMETRIES FOR THE MATHIEU GROUP M_{12}

F.Buekenhout*
Département de Mathématique CP 216
Université Libre de Bruxelles
1050 Bruxelles - Belgium

1. Introduction.

In our search towards a geometric interpretation of the sporadic simple groups which would be close in all respects to the theory of buildings of Tits [13], [1], [14] it is necessary to introduce restrictions if we want to avoid the paralyzing effect of hundreds of objects. On the other hand, restrictions which would be made too soon could have a paralyzing effect as well or just mislead us. Therefore, it may be appropriate to explore in a fairly broad sense, a small group like the Mathieu group M_{12}. Our choice of M_{12} rather than M_{11} is motivated by several observations, in particular the presence of outer auto morphisms of order 2 to explain and so a potential existence of geometries admitting polarities.

There is no attempt towards formalization and a complete classification of all geometries satisfying given axioms. Nevertheless, our search is systematic enough to prepare such classifications.

2. Subgroup structure.

We shall work with the simple group M_{12} of order $95040=2^6.3^3.5.11$. For basic facts about this group such as the character table, the list of maximal subgroups and their structure, the action of these subgroups on the dodecads left invariant by M_{12} inside the larger group M_{24}, we refer to Lüneburg [9], Conway [4], Fischer-McKay [5], McKay [10] and their bibliography.
In the natural action on 24 points mentioned already, M_{12} has two orbits of 12 points, which is denoted by [12;12], while the automorphism group $M_{12}.2$ of M_{12} is transitive on the 24 points with two blocks of imprimitivity of size 12. This is summarized by $[12^2]$

There are 11 conjugacy classes of maximal subgroups in M_{12} of which six appear in pairs fused by $M_{12}.2$. Here is a description of

*This research was partially carried out at Ohio State University and at the Technische Universität Braunschweig.

these subgroups together with their index, their permutation character and the subdegrees of the corresponding permutation group, when available. The conventions are those of [4], [5].

M_{11}	$[1,11;12]$ $[12;1,11]$	12	$\pi=1+11_1$ $\pi=1+11_2$	$12=1+11$
$M_{10}.2$	$[2,10;6^2]$ $[6^2;2,10]$	66	$\pi=1+11_1+54$ $\pi=1+11_2+54$	$66=1+45+20$
$M_9.S_3$ $=3^2.2^{1+2}.S_3$	$[3,9;3^4]$ $[3^4;3,9]$	220	$\pi=1+11_1+54+55_3+99$ $\pi=1+11_2+54+55_3+99$	$220=1+12+27+72+108$
$M_8.S_4$ $=2^{1+4}.S_3=C(2A)$	$[4,8;4,8]$	495	$\pi=1+11_1+11_2+55_3+2.54+66+99+144$	$495=1+6+16+24+2.32+$ $+2.48+3.96$
$2\times S_5$ $=C(2B)$	$[6\times 2;6\times 2]$	396		
$2^2.2^3.S_3=N(2^2)$	$[4^3;4^3]$	495	$\pi=1+16_1+16_2+45+$ $+2.54+66+99+144$	$495=1+6+16+24+2.32+2.48$ $+3.96$
$A_4\times S_3$ $=N(2^2)=N(3_B)$	$[4\times 3;4\times 3]$	1320	$\pi=1+16_1+16_2+2.45+2.54$ $+55_3+2.66+2.99+2.120$ $+2.144+176$	
$L_2(11)$	$[12;12]$	144	$\pi=1+16_1+16_2+45+66$	$144=1+2.11+55+66$

Comments 1. The notation A.B describes a group with a normal subgroup A and a quotient group B. Notations such as $3^2,2^3$ are for elementary abelian groups and $2^{1+4},2^{1+2}$ for extraspecial groups of orders 32 and 8 respectively.

2. The subdegrees for $M_9.S_3$ given in [5] are not correct. The right numbers are easily obtained from the action of M_{12} on 3-sets, inside $[3,9;3^4]$.

3. In $2\times S_5$, 6×2 means that a grid consisting of 6 blocks of 2 points and 2 blocks of 6 points is invariant. The permutation character given in [5] does not seem to be correct. This holds also for the character of $L_2(11)$.

3. Geometric terminology.

We are using freely the terminology, notations and results of Buekenhout [2]. In particular, we shall denote by Γ a geometry of finite rank n, over some basic diagram Δ and we shall assume that M_{12} is a flag-transitive group of Δ-automorphisms of Γ. We assume that Γ has the properties (SC)(strong connectivity) and (IP)(intersection property). Since it turns out that we shall only have to consider linear diagrams, Γ can be seen in two dual ways, as a set of points together with distinguished subsets such that any intersection of some of them is still a member of their family.

Whenever we have to deal with a rank 2 geometry, we shall describe it by a picture of the following shape

```
0    g,d0,d1    1
o---------------o      B
s+1            t+1
 v              b
P0=[C].D       P1=[E].F
```

This is read as follows: 0 and 1 are labels given to the elements of the geometry and we call 0-elements (resp.1-elements) points (resp. lines); g is half of the girth in the incidence (bipartite) graph i.e. half of the elements in a shortest circuit; d_0 (resp.d_1) is the greatest distance which can be achieved in the incidence graph, from a point (resp.line); s+1 (resp.t+1) is the number of elements incident with a given line (resp.point); v (resp.b) is the number of points (resp.lines); P_0 (resp.P_1) denotes the stabilizer of a point p (resp. line) in M_{12} and C is the normal subgroup fixing each line incident with p; finally B denotes the Borel subgroup or stabilizer of a maximal flag in Γ which may also be identified with $P_0 \cap P_1$.

The most common of these pictures will be replaced by abbreviations inspired by those used in [1] such as : o——o for $o\overset{3,3,3}{\text{——}}o$, o═══o for $o\overset{4,4,4}{\text{——}}o$, o   o for $o\overset{2,2,2}{\text{——}}o$, $o\overset{C}{\underset{2}{\text{——}}}o$ for $o\overset{3,3,4}{\underset{2}{\text{——}}}o$ (complete graph), $\underset{n}{o}\overset{3,3,4}{\text{——}}\underset{n+1}{o}$ (affine plane of order n).

When it is available (and non trivial), we shall also give a picture of the incidence graph itself using a 0-element (or 1-element) as starting vertex and concentric circles with respect to it. This is inspired for instance from [3].

To describe a rank 3 geometry, we use similar conventions, for

instance

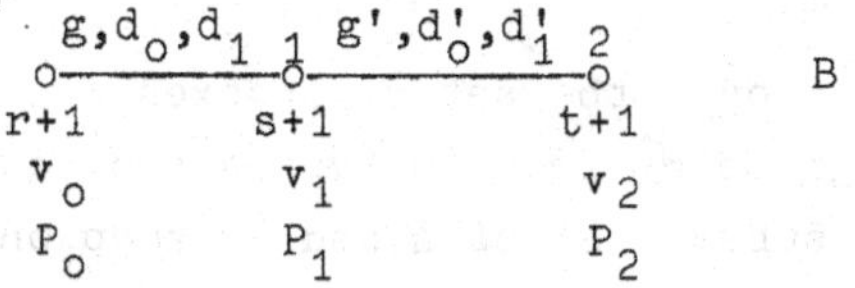

$$\underset{\substack{r+1 \\ v_0 \\ P_0}}{\overset{g,d_0,d_1}{\circ\!\!-\!\!-\!\!-\!\!-\!\!-\!\!\circ}} \underset{\substack{s+1 \\ v_1 \\ P_1}}{\overset{1}{}} \underset{\substack{t+1 \\ v_2 \\ P_2}}{\overset{g',d_0',d_1'}{\circ\!\!-\!\!-\!\!-\!\!-\!\!-\!\!\circ}} \overset{2}{} \qquad B$$

where P_0 is the stabilizer of a 0-element, v_0 is the number of 0-elements, the residue of a 2-elements is a rank 2 geometry with a diagram $\underset{\substack{r+1 \\ }}{\overset{g,d_0,d_1}{\circ\!\!-\!\!-\!\!-\!\!\circ}}\underset{s+1}{}_1$ and the residue of a 0-element is likewise related to the parameters g',d_0',d_1'.

4. Variations on the Steiner system $S(5,6,12)$.

When looking for a geometric interpretation of M_{12}, the most obvious idea is of course to consider the Steiner system $S(5,6,12)$ invariant by M_{12}, consisting of 12 points and 132 blocks of 6 points such that every set of 5 points is contained in a unique block.

From our viewpoint, it is natural to consider the blocks as hyperplanes of our geometry and to accept their intersections as elements of the geometry. This leads to a rank 5 geometry which is conveniently described by the following diagrams where we list only the most useful residues.

$$M_{12} \quad \underset{\substack{2 \\ 12 \\ M_{11}}}{\overset{0}{\circ}}\!\!-\!\!\underset{\substack{2 \\ 66 \\ M_{10}:2}}{\overset{1}{\circ}}\!\!-\!\!\underset{\substack{2 \\ 220 \\ M_9 \cdot S_3}}{\overset{2}{\circ}}\!\!\overset{C}{-\!\!-}\!\!\underset{\substack{3 \\ 495 \\ [2]4.S_4}}{\overset{3}{\circ}}\!\!\overset{Af}{-\!\!-}\!\!\underset{\substack{4 \\ 132 \\ S_6}}{\overset{4}{\circ}} \quad B=2$$

$$M_{11} \quad \underset{\substack{2 \\ 11 \\ M_{10}}}{\overset{1}{\circ}}\!\!-\!\!\underset{\substack{2 \\ 55 \\ M_9 \cdot 2}}{\overset{2}{\circ}}\!\!\overset{C}{-\!\!-}\!\!\underset{\substack{3 \\ 165 \\ [2]4.S_3}}{\overset{3}{\circ}}\!\!\overset{Af}{-\!\!-}\!\!\underset{\substack{4 \\ 66 \\ S_5}}{\overset{4}{\circ}} \qquad S_6 \quad \underset{\substack{6 \\ S_5}}{\overset{0}{\circ}}\!\!-\!\!\underset{\substack{15 \\ S_4.2}}{\overset{1}{\circ}}\!\!-\!\!\underset{\substack{20 \\ S_3 \times S_3}}{\overset{2}{\circ}}\!\!\overset{C}{-\!\!-}\!\!\underset{\substack{15 \\ [2]S_4}}{\overset{3}{\circ}} \qquad (1)$$

$$M_{10} \quad \underset{\substack{2 \\ 10 \\ M_9}}{\overset{2}{\circ}}\!\!\overset{C}{-\!\!-}\!\!\underset{\substack{3 \\ 45 \\ [2]4.2}}{\overset{3}{\circ}}\!\!\overset{Af}{-\!\!-}\!\!\underset{\substack{4 \\ 30 \\ S_4}}{\overset{4}{\circ}} \qquad M_9 \quad \underset{\substack{3 \\ 9 \\ [2]4}}{\overset{3}{\circ}}\!\!\overset{Af}{-\!\!-}\!\!\underset{\substack{4 \\ 12 \\ S_3}}{\overset{4}{\circ}}$$

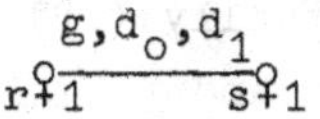

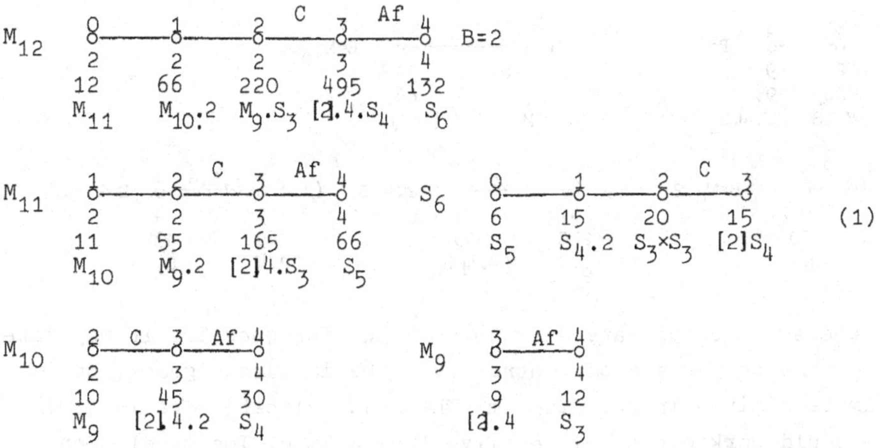

We observe that the parabolic subgroup $P_4 = S_6$ is the only one which is not maximal in M_{12}. It is of index 2 in an $M_{10}.2$ which is not conjugate to P_1.

5.Truncations of the Steiner system.

When Γ is a geometry of rank n over the set of indexes
$\{0,1,\ldots,n-1\}=\Delta$ we can get a new geometry from it by the procedure of
truncation. Therefore we choose a subset Δ' of Δ and we keep only
the elements of Γ whose type is in Δ'. This may look a little strange
since we left with the Steiner system which corresponds to $\Delta'=\{0,4\}$
in (1) and we wanted to avoid it in order to get (IP). Actually (IP)
may still hold in the truncation. For this it is necessary and
sufficient that Δ' be connected in the diagram structure on Δ. Taking
only the non trivial cases where $2 \leqslant |\Delta'| \leqslant 4$ this gives us 9 distinct
truncations to study.
Here they are.

The nine truncations of (1) having rank $\geqslant 2$ and (IP)

(2)
$$\underset{\substack{2\\12\\M_{11}}}{\overset{0}{\circ}}\quad\underset{\substack{2\\66\\M_{10}.2}}{\overset{1}{\circ}}\quad\underset{\substack{2\\220\\M_9.S_3}}{\overset{2}{\circ}}\overset{\subset}{}\underset{\substack{9\\495\\[2.4]S_4}}{\overset{3}{\circ}}\qquad B=2.4$$

(3)
$$\underset{\substack{3\\66\\M_{10}.2}}{\overset{1}{\circ}}\overset{\supset}{}\underset{\substack{2\\220\\M_9.S_3}}{\overset{2}{\circ}}\overset{\subset}{}\underset{\substack{3\\495\\[2]4S_4}}{\overset{3}{\circ}}\quad\underset{\substack{4\\132\\S_6}}{\overset{Af\;4}{\circ}}\qquad B=2^2$$

(4)
$$\underset{\substack{2\\12\\M_{11}}}{\overset{0}{\circ}}\quad\underset{\substack{2\\66\\M_{10}.2}}{\overset{1}{\circ}}\overset{\subset}{}\underset{\substack{10\\220\\[M_9].S_3}}{\overset{2}{\circ}}\qquad B=M_9$$

(5)
$$\underset{\substack{4\\220\\M_9.S_3}}{\overset{2\;\;3,6,6}{\circ}}\underset{\substack{3\\495\\[2]4S_4}}{\overset{3}{\circ}}\quad\underset{\substack{4\\132\\S_6}}{\overset{Af\;\;4}{\circ}}\qquad B=2.S_3$$

(6)
$$\underset{\substack{3\\66\\M_{10}.2}}{\overset{1}{\circ}}\overset{\supset}{}\underset{\substack{2\\220\\M_9.S_3}}{\overset{2}{\circ}}\overset{\subset}{}\underset{\substack{9\\495\\[2.4]S_4}}{\overset{3}{\circ}}\qquad B=2.4.2$$

(7)
$$\underset{\substack{2\\12\\M_{11}}}{\overset{0}{\circ}}\overset{\subset}{}\underset{\substack{11\\66\\[M_{10}]2}}{\overset{1}{\circ}}\qquad B=M_{10}$$

(8)
$$\underset{\substack{3\\66\\M_{10}.2}}{\overset{1\;\;3,5,6\;\;2}{\circ}}\underset{\substack{10\\220\\[M_9].S_3}}{\overset{}{\circ}}\qquad B=M_9.2$$

(9)
$$\underset{\substack{4\\220\\M_9.S_3}}{\overset{2\;\;4,7,8\;\;3}{\circ}}\underset{\substack{9\\495\\[2.4]S_4}}{\overset{}{\circ}}\qquad B=2.4.S_3$$

(10)
$$\underset{\substack{15\\495\\[2.4]S_4}}{\overset{3\;\;3,5,6\;\;4}{\circ}}\underset{\substack{4\\132\\S_6}}{\overset{}{\circ}}\qquad B=2.4.S_3$$

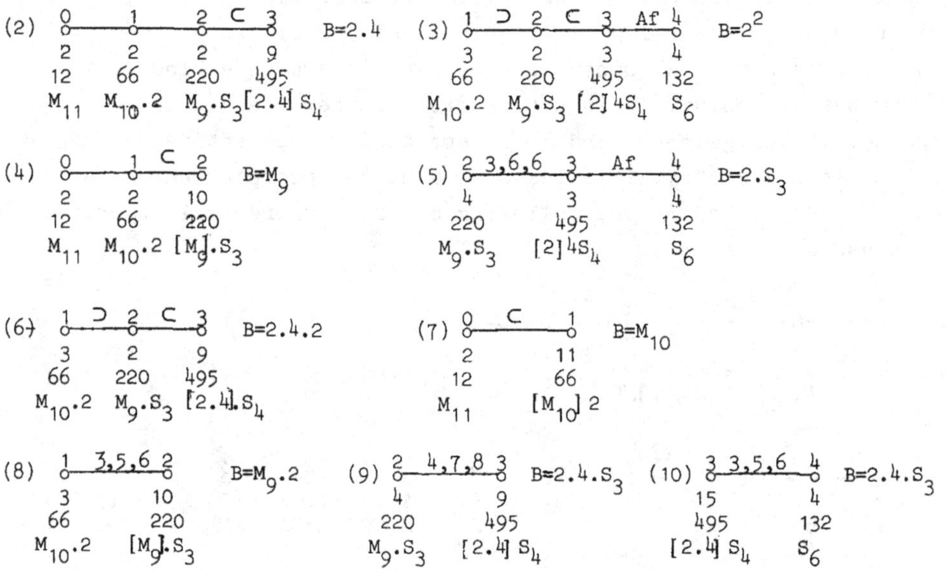

Most of these structures are easy to work out for oneself. It may take
some more time to get a good control over the incidence graph, beyond
the parameters given in the diagram. The most delicate here is (10)
which we could work out fairly easily, from a line. The development
from a point should be more interesting if we refer to our experience
based on the examples listed in [3].

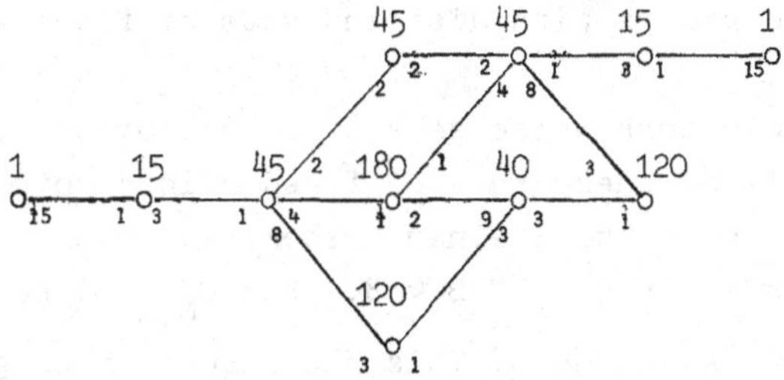

Among the geometries (2) to (10) a major role is played by (4) which appears as a residue of a less trivial geometry for the Conway group C_1 and which is a member of a family including other sporadic groups. In it, we observe that the characteristic 3 of the "ground field" appears clearly in all parameters (2,10 are equal to a power of 3 plus one) and in the structure of the local parabolic subgroups. This holds also in the residues of the geometry which we did not list. In this respect, (4) appears as optimal: it is the most complete (less truncated) geometry having such a good behaviour with respect to a prime number, namely 3.

6. Looking for quadrangles.

Let us observe all geometries obtained so far, in particular their rank 1 residues which are residues of flags F having r-1 elements, where r is the rank of the geometry. The stabilizer of F which we call a rank 1 parabolic subgroup acts transitively on the rank 1 residue of F. A non trivial observation is that in all but one cases, this group turns out to be doubly-transitive. The exception occurs in (10) with the action of S_6 on a line of 15 points. Here we remember that $S_6 \simeq Sp_4(2)$ which is acting on a generalized quadrangle of 15 points. Therefore, our lines of 15 points bear each a structure of generalized quadrangle with lines of 3 points or 3-lines. Let us look more closely at these 3-lines and their geometric structure.

Let Ω_{12} be the set of 12 0-elements of (1). Here the points of the geometry we want to study, are the 3-elements of (1) i.e.all subsets of 4 points of Ω_{12}. On the other hand, each such set is the set of fixed points of a unique involution of M_{12} which can be seen in the parabolic subgroup $2.4.S_4$. Such an involution is called central in M_{12}. Let us examine pairs of commuting central involutions whose product is central i.e. pure subgroups 2^2 of central type. These fall into two conjugacy classes namely:

$\mathcal{A}$ whose members have 3 involutions with disjoint sets of fixed points in Ω_{12};
$\mathcal{B}$ whose members have 3 involutions whose sets of fixed points intersect pairwise in 2 points and all of these sets of fixed points cover a 4-element of (1) i.e. a block of the Steiner system.
Let $A \in \mathcal{A}$. Then $C_{M_{12}}(A)$ has order 2^5. If $B \in \mathcal{B}$, then $C_{M_{12}}(B)$ has order 2^3. Therefore $M_{12}.2$ acts on $\mathcal{A}$ and on $\mathcal{B}$. This has intersecting geometric consequences.

Now, let p,q be central involutions such that $\langle p,q\rangle \in \mathcal{B}$. Then $C_{M_{12}}(D)$, $D=\langle p,q\rangle$, is elementary abelian of order 2^3 and it contains exclusively central involutions. Exactly one of these, say $p*q$ has the property that $\langle p,p*q\rangle$ and $\langle q,p*q\rangle$ are members of $\mathcal{B}$ while $p*q \neq pq$. Clearly $p*(p*q)=q$. Therefore $\langle p,q,p*q\rangle$ and its transforms under M_{12}, is a good choice for a line of 3 points on the set Ω_{495} of all central involutions. It can be checked that these are the lines of the generalized quadrangles of 15 points obtained from (10). Our analysis shows that $M_{12}.2$ acts on these lines of 3 points while it does not act on the quadrangles. This means that there are 264 quadrangles (instead of 132 so far) on which $M_{12}.2$ acts. Hence we get the following geometry for $M_{12}.2$.

(11) $M_{12}.2$

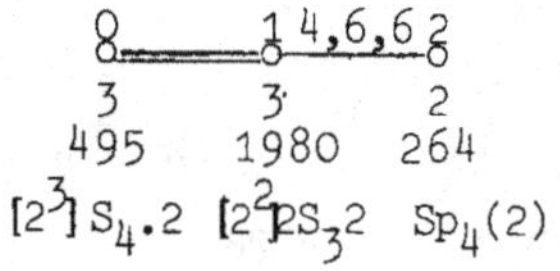

$B=2^3.2$

$Sp_4(2)$

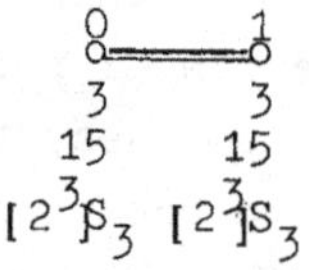

$[2]\ 4\ S_4$

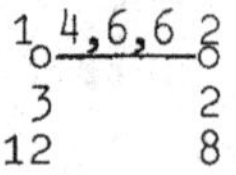

In the last residue we recognise a truncation of the cube namely

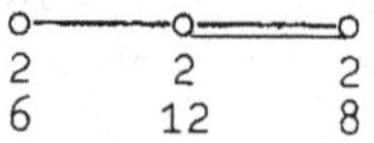

On this geometry (11), M_{12} has two orbits on the 2-elements and so we expect a phenomenon similar to that arising from the Dynkin diagram D_n
We indeed get

(12) M_{12}

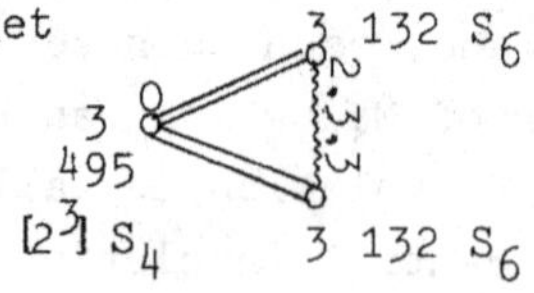

$B=2^3$

which admits polarities, but in which (IP) fails since a o—2,3,3—o (3, 4 under each node) contradicts (IP).

All polarities of (12) are conjugate. They admit 15 absolute points and their centralizer in M_{12} is $2 \times A_5$.

We did not analyze truncations of (11).

7. More on central involutions.

The following geometry is produced by Ronan and Stroth [12]; it arises implicitly in the work of Goldschmidt [6].

(13) 0 o—8,12,12—o 1 $\quad B = 2^{1+4}.2$

3   3

495   495

$[2^{1+4}]S_3$ $[2^2.2^3]S_3$

The incidence graph, seen from a 0-element, is as follows

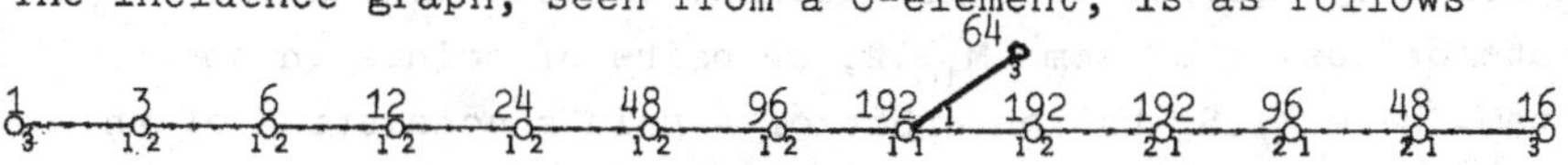

$M_{12}.2$ acts on this geometry which is close to a classical generalized polygon over $\mathbb{F}_2$, by the fact that B is a 2-Sylow subgroup of M_{12}.

8. Further with central involutions.

Let us consider as points now, the 132 blocks of S(5,6,12) and take as adjacency the fact that two blocks intersect in 4 points of S(5,6,12). Then we get a graph as follows

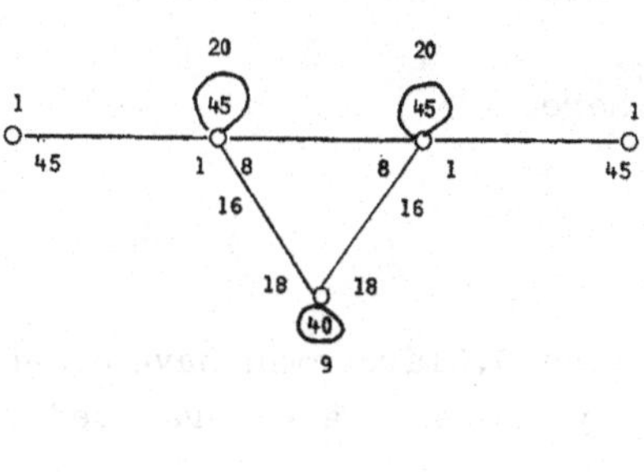

The group M_8S_4 has 2 orbits of 4 points in this graph and these orbits are cliques. Indeed, in S(5,6,12) M_8S_4 has an orbit of 4 points and one of 8 points. The first determines 4 blocks on it and on the second, there is an invariant pairing such that any 3 pairs in it constitute a block. This gives our two orbits of 4 points of $M_8.S_4$ in Ω_{132}. Taking these two families of 4-cliques as elements of a geometry, of different types, we obtain

(14) 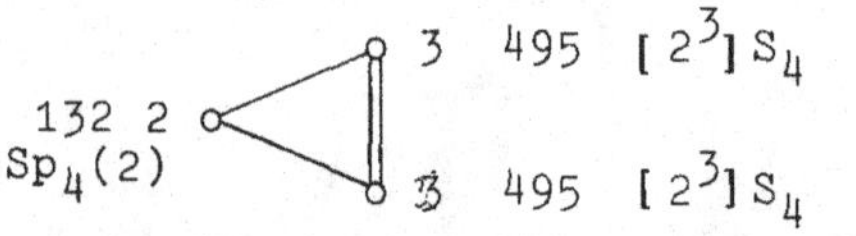

$B=2^3.2$

which satisfies (IP). The group $M_{12}.2$ does not act on the points and so it provides no dualities. Could there nevertheless be dualities? Assume a is one. Then it must centralize M_{12} and therefore it fixes each point in (14). Then it fixes also each 4-clique. Hence there are no dualities. Two of the truncations give (10) again. We did not analyze the third one.

We can slightly modify our construction. Instead of 132 points we take 66 points or cosets of some $M_{10}.2$, or pairs of points in the Steiner system. Here $M_8.S_4$ has an orbit of 4 points consisting of the 4 two-cycles of its central involution on Ω_{12}. By intersections, these sets of 4 points give also sets of 2 points and we can analyze this situation to get

(15) 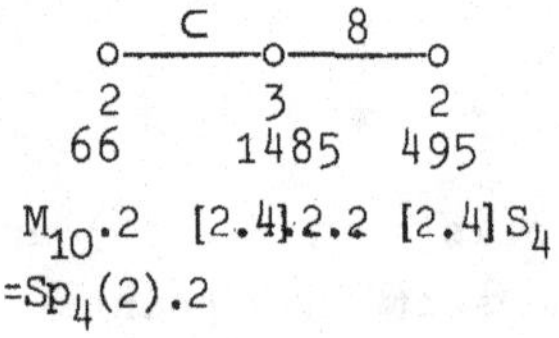

$B=2.4.2$

The relation between (14) and (15) is clear: identifying all hyperplanes in (14) we get the same diagram as in (15) and very likely, a covering of (15).

We did not analyze truncations here.

9. Another geometry of Ronan.

Ronan [personal communication] and G.Glauberman have observed the following geometry which is remarkably close to a generalized polygon

(16) 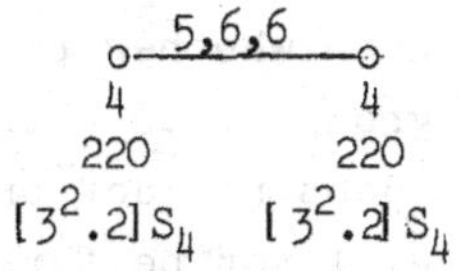

$B=3^{1+2}.2^2$

It has polarities with 20 absolute points. Here is the incidence graph, seen from a point (G.Glauberman and A.Katz, personal communication).

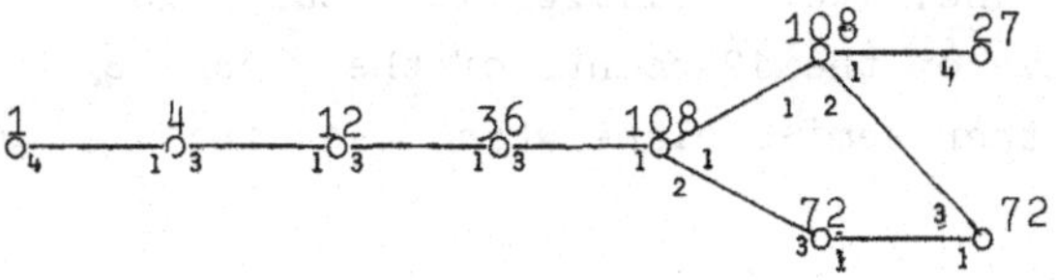

10. The lonesome $L_2(11)$.

It is striking that maximal subgroups $L_2(11)$ do not interfere very much with the other maximal subgroups of M_{12}. This is perhaps not completely hopeless for the smaller maximal subgroups, which we did not analyze much. Anyway $L_2(11)$ has an interesting life on itself.

Let us recall that there are 2 classes C_1, C_2 of subgroups $L_2(11)$ in M_{12}. The members of C_1 are maximal. Each element of C_2 is in a unique subgroup M_{11} of each of the two classes of such subgroups. In $M_{12}.2$, C_1 and C_2 are invariant and the normalizer of a subgroup $L_2(11)$ is now $PGL_2(11)$ which is maximal in $M_{12}.2$, in both cases.

Consider the action of M_{12} on the 144 cosets of some $L_2(11)$. Since the degree 144 is small and since an element of order 11 fixes a point, it is not hard to show that the subdegrees must be 1,11,11,55 and 66. The group $M_{12}.2$ acts on the 144 points and fuses the two suborbits of degree 11 since $PGL_2(11)$ cannot act on 11 points.

Let a plane be any subset of degree 11 which is a suborbit of a point stabilizer. Together with its transforms it generates a semi-biplane in the sense of D.Hughes [7].

(17)
(18) 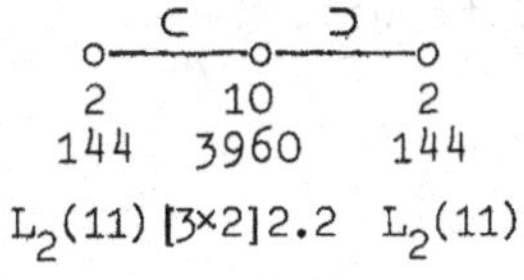

$B=S_3$

These two geometries are non isomorphic. We do not know whether there are polarities. Truncations were not analyzed. The group $M_{12}.2$ acts on the points with $PGL_2(11)$ as point-stabilizer. It does not act on the planes but rather on 288 planes whose geometry will not be further studied here.

Let now M_{11} act on the 144 points. This is a transitive action and it must necessarily lead to 12 blocks of imprimitivity of size 12. Consider such a block as a hyperplane. Its stabilizer is a subgroup $L_2(11)$ of M_{11} which is 2-transitive on the 12 points of the block. By intersections, this generates another semi-biplane which was discovered by D.Leonard [8].

(19) o —C— o —Ɔ— o B=5
2 11 2
144 4752 144
$L_2(11)$ [5] 2.2 $L_2(11)$

Here the stabilizer of a point is maximal and the stabilizer of a plane is not. Hence there are no dualities. However the action of M_{11} provides a parallelism on the planes which is invariant by M_{12} and since there are 2 classes of subgroups M_{11}, there are 2 invariant parallelisms. These two parallelisms are interchanged by $M_{12}.2$.

We did not analyze truncations.

11. Some other interesting situations.

If we dualize (9) on 495 points and if we let $M_{12}.2$ act on it we get 440 (instead of 220) lines of 9 points and a geometry having (IP):

(20) $M_{12}.2$ o —($3,d_0,d_1$)— o $B=2.4.S_3$
9 8
495 440
[2.4] $S_4.2$ $M_9.S_3$

We did not determine the parameters d_o,d_1.

The following are also given in such an incomplete way.

(21) M_{12} o —($4,d_o,d_1$)— o $B=2\times A_4$
3 10
396 1320
[2] S_5 $[A_4]\times S_3$

(22) $\overset{4,d_o,d_1}{o\!-\!-\!-\!-\!-\!o}$ $B=A_4\times 2$

3 8

495 1320

$2.4.S_4$ $[A_4]\times S_3$

12.<u>References</u>.

[1] Buekenhout,F. Diagrams for geometries and groups. J.Comb.Th.(A) 27 (1979) 121-151.

[2] Buekenhout,F. The basic diagram of a geometry. Lecture Notes Springer. 893-1981.

[3] Buekenhout,F. (g,d^*,d)-gons. Proceedings of a conference in honor of T.G.Ostrom. M.Dekker 1982.

[4] Conway,J.H. Three lectures on exceptional groups in Powell-Higman Finite simple groups. Academic Press, 1971, 215-247.

[5] Fischer,J. and McKay,J. The nonabelian simple groups $G, |G| < 10^6$- Maximal subgroups. Math.of Comput. 32, 1978 (1293-1302).

[6] Goldschmidt,D. Automorphisms of trivalent graphs. Ann.of Math. 111(1980) 377-406.

[7] Hughes,D. Biplanes and semibiplanes. Lecture Notes in Math.686 Springer-Berlin (1978), 55-58.

[8] Leonard,D. Semi-biplanes and semi-symmetric designs. Thesis. Ohio State University - 1980.

[9] Lüneburg,H. Transitive Erweiterungen endlicher Permutationsgruppen. Springer Lect.Notes, 84, 1969.

[10] McKay,J. The simple groups $G, |G| < 10^6$-Characters tables. Comm.Alg. 7 (1979) 1407-1445.

[11] Ronan,M. and Smith,S. 2-local geometries for some sporadic groups. Proc.Symp.Pure Math. Vol.37. AM.S.1981.

[12] Ronan,M. and Stroth,G. Sylow-2 geometries. (In preparation)

[13] Tits,J. Buildings of spherical type and finite BN-pairs. Lectures Notes in Math, 386, Springer-Verlag, Berlin, 1974.

[14] Tits,J. Buildings and Buekenhout Geometries. "Finite Simple Groups II" ed.M.Collins, Academic Press, New York, 1981, 309-320.

Orbits and Enumeration

Peter J. Cameron
Merton College, Oxford OXI 4JD, U.K.

1. Introduction

Sequences of natural numbers feature prominently in almost all parts of mathematics, as well as many areas outside the traditional boundaries of the subject. Neil Sloane's "Handbook of Integer Sequences" [11] bears testimony to this fact. It is a list of 2372 sequences, in lexicographic order, drawn from a wide range of topics. The only criteria for inclusion of a sequence of natural numbers in the Handbook are that enough terms should be known to distinguish it from its neighbours, and that somebody must have found it interesting enough to commit it to print in the scientific literature.

The sequences discussed in this article arise in the following way. G is a group of permutations of an infinite set X, having the property that G has only finitely many orbits on the set of all k-element subsets of X, for each natural number k; we let $n_k(G)$ denote the number of these orbits (with the convention that $n_0(G) = 1$).

For example, if G = S, the symmetric group on X, then $n_k(G) = 1$ for all k, and G realizes the sequence (1,1,1...) (which, for technical reasons, does not appear in the Handbook). If we take instead for G the direct product of two copies of S, acting on the disjoint union of the corresponding sets, then $n_k(G) = k+1$, since a k-set contains ℓ points from the first G-orbit (for some $\ell \in \{0,1,2,\ldots,k\}$), and the orbit containing the k-set is completely specified by the value of ℓ. Thus G realizes the sequence (1,2,3,4,5,...) of natural numbers (#173 in the Handbook).

This can be described concisely using generating functions. For any group G of the type we are considering, let $f_G(t)$ denote the formal power series $\sum_{k=0}^{\infty} n_k(G)t^k$. An easy argument shows that $f_{G\times H}(t) = f_G(t)f_H(t)$, where the direct product acts on the disjoint union of the two sets. Since $f_S(t) = (1-t)^{-1}$, we have $f_{S\times S}(t)=(1-t)^{-2}$, from which our above observation follows by the binomial theorem. More generally, we have $f_{S^m}(t) = (1-t)^{-m}$, so that $n_k(S^m) = \binom{m+k-1}{m-1}$; thus the groups $S^3, S^4, \ldots$ realize the sequences of triangular, tetrahedral, and higher figurate numbers (##1002, 1363, 1578, 1719, 1847, 1911, 1976, 2013, 2046 and 2073). (From now on, #N refers to sequence number N in the Handbook.)

Looked at another way, $n_k(S^m)$ is the number of partitions of k into m parts (of which some may be zero), where the order of the parts is significant; that is, the number of ways of placing k identical objects into m distinguishable boxes. We see that there is an enumeration problem intimately connected with the group. In this article, I shall examine which enumeration problems are connected with groups in this way, and what it tells us about an enumeration problem to know that it determines $(n_k(G))$ for some group G. Sections 2-4 treat the first topic, giving constructions of groups (and sequences) using wreath products, ultrahomogeneous models, and ad hoc methods. Section 5 discusses model-theoretic generalities about the sequences $(n_k(G))$, while section 6 examines the question of rate of growth of such sequences. The final section introduces a further sequence associated with certain groups G, in terms of a graded algebra A^G; this sequence enumerates "connected structures" of the type concerned.

Of course, more than one group can realize a given sequence. As well as S, the group $A = \mathrm{Aut}(\mathbb{Q},<)$ of all order-preserving permutations of the rational numbers satisfies $n_k(A) = 1$ for all k. However, whereas there is an element of S mapping a given k-tuple to another given k-tuple in any order, in A we may map only in one prescribed order. We say that S is highly transitive, and A is highly homogeneous but not highly transitive. We will exploit the difference in the next section.

2. Wreath products

The wreath product provides a means of constructing new groups (and hence new sequences) from old ones, that is considerably more flexible and subtle than the direct product construction mentioned in the preceding section. Among the sequences we will meet here are the Fibonacci sequence and the partition function.

Let H and K be permutation groups on sets Y and Z. We set $X = Y \times Z$, regarded as a family of copies of Y indexed by Z. The wreath product $G = H \,\mathrm{Wr}\, K$ is generated by (i) the cartesian product of $|Z|$ copies of H, one for each element of Z, where the copy of H indexed by $z \in Z$ acts on the copy of Y indexed by z and fixes all the others pointwise, and (ii) elements of K, permuting the copies of Y among themselves according to their given action on Z.

Examples. (i) Let S_m be the finite symmetric group of degree m. Then an orbit of $G = S \,\mathrm{Wr}\, S_m$ on k-sets is determined by a partition of k into m parts, some possibly empty, where the order of the parts is irrelevant; so $n_k(G)$ is the number of ways of placing k identical objects in m identical boxes, with generating function

$f_G(t) = (1-t)^{-1}(1-t^2)^{-1} \dots (1-t^m)^{-1}$.

For $m = 3,4,5,6$, we obtain sequences ##186, 229, 237, 243.

(ii) Similarly, $n_k(S_m \mathrm{Wr}\, S)$ is the number of partitions of k into parts of size at most m; by the familiar duality of partitions, this is the same as above.

(iii) $n_k(S\,\mathrm{Wr}\, S)$ is the number $p(k)$ of partitions of k (#244), with generating function $\prod_{i=0}^{\infty} (1-t^i)^{-1}$.

Example. Recall the group $A = \mathrm{Aut}(\mathbb{Q},<)$. An orbit of $G = S_2 \mathrm{Wr}\, A$ on k-sets is determined by a partition of k into parts of size 1 or 2, where the order of the parts is important; that is, an expression for k as a sum of ones and twos, in order. It is well-known that this implies that $n_k(G) = F_k$, the kth Fibonacci number (#256).

This example shows that $(n_k(G\,\mathrm{Wr}\, H))$ is not determined by $(n_k(G))$ and $(n_k(H))$; we require more information about H, which can be summarized in a formal power series in countably many indeterminates (see [2 III] for details).

The cycle index of a finite permutation group G is the polynomial in the indeterminates $s_1, s_2, \ldots,$ given by

$$Z(G; s_1, s_2, \ldots) = \frac{1}{|G|} \sum_{g \in G} s_1^{n_1(g)} s_2^{n_2(g)} \cdots,$$

where $n_i(g)$ is the number of cycles of lenght i in the cycle decomposition of g. If G is any permutation group with $n_k(G) < \infty$ for all k, select representatives $X_1, X_2, \ldots$ of the G-orbits on finite sets, and let G_i be the (finite) permutation group induced on X_i by its setwise stabilizer. The modified cycle index of G is defined by

$$\tilde{Z}(G; s_1, s_2, \ldots) = \sum_i Z(G_i; s_1, s_2, \ldots).$$

(Our convention is that the cycle index of the "group of degree zero" is 1.)

If G is finite, then $\tilde{Z}(G; s_1, s_2, \ldots) = Z(G; s_1 + 1, s_2 + 1, \ldots)$.

Let $N_k(G)$ be the number of orbits of G on ordered k-tuples of distinct elements, and $F_G(t) = \sum_{k=0}^{\infty} N_k(G) t^k/k!$. The next result summarizes properties of the modified cycle index and its connection with our problem.

Theorem 2.1. (i) $f_G(t) = \tilde{Z}(G; t, t^2, t^3, \ldots)$.

(ii) $F_G(t) = \tilde{Z}(G; t, 0, 0, \ldots)$

(iii) $\tilde{Z}(G \times H) = \tilde{Z}(G)\tilde{Z}(H)$.

(iv) $\tilde{Z}(G \operatorname{Wr} H)$ is obtained from $\tilde{Z}(H)$ by substituting $\tilde{Z}(G;s_k,s_{2k},s_{3k},\ldots)-1$ for each occurrence of s_k, for $k = 1,2\ldots$.

(v) If G is transitive, then $\tilde{Z}(G_x) = \frac{\partial}{\partial s_1}\tilde{Z}(G)$, where G_x is the stabilizer of x, acting on the points different from x.

Corollary 2.2. $f_{G \operatorname{Wr} H}(t) = \tilde{Z}(H\,;f_G(t)-1,\ f_G(t^2)-1,\ f_G(t^3)-1,\ldots)$.

Example. The group A has a single orbit on k-sets for each k, and the group induced on a k-set is trivial. So $\tilde{Z}(A;\ s_1,s_2,\ldots) = 1 + s_1 + s_1^2 + \ldots = 1/(1-s_1)$. We have $f_{S_2}(t) = 1 + t + t^2$, so $f_{S_2 \operatorname{Wr} A}(t) = 1/(1 - t - t^2)$, the well-known generating function for the Fibonacci sequence.

In the same way, we see that the sequences realized by $S_m \operatorname{Wr} A$ for $m = 3,4,5,6$ are the tribonacci, tetranacci, pentanacci and hexanacci numbers (##406, 423, 429, 431); while if G is the iterated wreath product of m copies of A, then G realizes the sequence of powers of m (for m = 2,3,...,9,11, these are ##432, 1129, 1428, 1620, 1765, 1874, 1937, 1992 and 2054).

The modified cycle index of the infinite symmetric group S is a bit more complicated, but familiar manipulations (see [8] p.52) put it in the form $\exp \sum_{j=1}^{\infty} (s_j/j)$. Further manipulation shows that, if $n_k(G) = m_k$, then

$$f_{G \operatorname{Wr} S}(t) = \prod_{j=1}^{\infty} (1-t^j)^{-m_j}.$$

(This is easily proved directly.)

Taking $m_j = 1$ for all j, we obtain the generating function for the partition function p(k). Taking $m_j = p(j)$, we see that S Wr S Wr S realizes #1019 in the Handbook, studied under the name "functional determinants" by Cayley. Sequences produced by further iteration do not appear in the Handbook.

Any non-trivial wreath product is imprimitive, in the sense that there is a proper equivalence relation left invariant by the group. Conversely, any imprimitive group can be embedded in a wreath product. The techniques of this section are useful in studying (and finding lower bounds for) sequences realized by imprimitive groups. In the next two sections, I give some constructions which yield primitive groups.

Incidentally, I do not have any example of a sequence realized by both a primitive and an imprimitive group.

3. Ultrahomogeneous models

The countable random graph Γ (see [6]) has the property that any isomorphism between finite induced subgraphs of Γ can be extended to an automorphism of Γ. Moreover, every finite graph occurs as an induced subgraph of Γ. Thus, if G is the automorphism group of Γ, then $n_k(G)$ is the number of isomorphism types of graphs with k vertices. So G realizes the sequence (1,1,2,4,11,34,...) (#479).

This is a special case of a very general phenomenon. Call a structure Γ ultrahomogeneous relative to an isomorphism-closed class C of finite structures if (i) every finite subset of Γ carries an "induced substructure" in C, and every member of C is isomorphic to an induced substructure of Γ; and (ii) any isomorphism between finite induced substructures of Γ can be extended to an automorphism of Γ. Thus, if G is the automorphism group of Γ, then $n_k(G)$ is the number of k-element structures in C, up to isomorphism. Furthermore, the modified cycle index of G (if it exists) is the sum of the cycle indices of the automorphism groups of all C-structures.

In what follows, it will be convenient to talk of classes of structures containing infinite as well as finite members (e.g.graphs). Necessary and sufficient conditions for the existence of a unique countable ultrahomogeneous structure in C (corresponding to all the finite members of C) are as follows:

(a) C has only countably many finite members, up to isomorphism;

(b) any finite subset of a C-structure carries an induced C-structure, and inclusion maps behave well (so that restricting to Y and then to $Z \subseteq Y$ yields the same result as restricting to Z);

(c) a structure of type C is determined by its finite substructures (in the sense that if every finite subset of X carries a C-structure, and if the inclusion maps behave well, then there is a unique compatible C-structure on X);

(d) (the amalgamation property) if M_0, M_1 and M_2 are C-structures, and $f_i : M_0 \to M_i$ are embeddings $(i = 1,2)$, then there is a C-structure M_3 and embeddings $g_i : M_i \to M_3$ $(i = 1,2)$ such that $f_1 g_1 = f_2 g_2$.

The reader is urged to check that conditions (a)-(d) hold for the class of graphs. A proof of the equivalence is given in [12], and further discussion in [4].

The amalgamation property says, loosely speaking, that if we are given two C-structures with a common substructure, then an amalgam exists in which the intersection of the two structures is at least the given substructure.

In many situations, the first three conditions are obvious, and only the fourth requires proof. (Thus, (a) - (c) hold if C is the class of all relational structures over a given first-order language with no function or constant symbols, or all those satisfying a given collection of universal sentences. For example, the structure of a graph is completely determined by the knowledge of all its 2-vertex subgraphs.)

Thinking of a complementary pair of graphs as a colouring of the edges of the complete graph with two interchangeable colours, we see that there is a group G for which $n_k(G)$ is the number of k-vertex graphs up to complementation (the average of ##479 and 780). (Alternatively, the random graph Γ is self-complementary; let G be the group of its automorphisms and anti-automorphisms.) Similarly, there is a group G for which $n_k(G)$ is the number of switching classes (Seidel equivalence classes) of graphs on k vertices (#321). We may allow the graphs to have loops (#646), or to be directed (##715, 784, 1229). Other examples are tournaments (#484), and various generalizations to relations of higher arity (##606, 872, 875).

Note that, if the group G and the class C are related in the above way, then $n_k(G)$ is the number of unlabelled k-element structures in C, while $N_k(G)$ (the number of orbits of G on ordered k-tuples) is the number of labelled k-element structures in C. However, in some cases, it is possible to construct a group G' for which $n_k(G')$ is the number of labelled k-element structures. The requirement is that the amalgamation in (d) can be performed without making any additional identifications. For example, let C be the class of graphs. Take C' to be the class of structures each of which consists of a graph together with a total ordering of the vertices Then C' still satisfies (a) - (d); and a finite C'-structure is essentially a labelled graph, so the corresponding group G' realizes the sequence (1,1,2,8,64,1024,...) enumerating labelled graphs.

Three further examples of this phenomenon:

(i) If C is the class of total orders, then $n_k(G')$ is the number of orderings of {1,2,...,k}, viz. k! (#659). This example can be modified by regarding the two total orders on a C'-structure as interchangeable, giving a group realizing the sequence (1,1,2,5,17,73,...), the average of ##469 and 659 (the number of inverse pairs in S_k).

(ii) If C corresponds to the group $G = S \operatorname{Wr} S$ (so that $n_k(G)$ is the number of partitions of k), then $n_k(G')$ is the number of partitions of the set {1,2,...,k}, the kth Bell number (#585).

(iii) If C is the class of k-uniform hypergraphs, then G' is a subgraph of Aut($\mathbb{Q}$,<) which is (k-1)-homogeneous but not k-homogeneous, answering a question of Glass ([6], p.248).

4. Other examples

There is a group G realizing the sequence #545 (1,1,2,5,11,26,...) enumerating graphs with k edges and no isolated vertices. For let X be the set of all 2-element subsets of an auxiliary infinite set Y, and G the symmetric group on Y, in its induced action on X. A k-subset of X consists of k distinct 2-subsets of Y, i.e. k edges of a graph on Y; two k-sets are in the same orbit if and only if the corresponding graphs (after deletion of isolated vertices) are isomorphic. In a similar way, we can realize the sequence enumerating t-uniform hypergraphs with k edges, for given t. Furthermore, the direct product of two symmetric groups, acting on the direct product of the underlying sets, realizes the sequence enumerating graphs with k edges having a named bipartite block, while its extension by a group of order 2 (interchanging the factors) realizes the sequence enumerating graphs with k edges having a named bipartition.

Further examples include the collineation groups of projective or affine spaces of infinite dimension over finite fields. In these cases, $n_k(G)$ is the number of different configurations of k points in projective or affine space, up to collineations of the space.

In the above example, the connection between the groups and the structures enumerated is fairly obvious. This is less so in the next three cases. They are #121 (sequences generated by a binary shift register), #298 (commutative bracketings - the Wedderburn-Etherington problem), and #122 (boron trees).

The binary sequences in the first case are defined by the condition $x_{i+k} = 1 + x_i$ for all i; two sequences are equivalent if they differ by a cyclic shift. The group G is constructed in [2 II]: it is a transitive extension of the subgroup of A fixing a dense subset of $\mathbb{Q}$ with dense complement. (The fact that all such subsets are equivalent under order-automorphisms of $\mathbb{Q}$ was proved by Skolem.) In another formulation, the structures being enumerated are local orders, or tournaments containing no 4-vertex subtournament consisting of a vertex dominating or dominated by a 3-cycle. Any local order gives rise to a circular ordering of the point set; the local orders giving a fixed circular order form a switching class. The equivalence of local orders and shift register sequences was shown by A.Brouwer, after he had noted that the sequence obtained in [2 II] for $(n_k(G))$ agreed with #121 in the Handbook - a good example of how the Handbook was intended to be used!

The Wedderburn-Etherington numbers count commutative bracketings of a sequence of k symbols, or words of length k in the free commutative non-associative structure on 1 generator. Equivalently, they count binary trees with k end vertices, where the distinction between left and right is not significant. The corresponding group was constructed in [2 IV] as a group of permutations of the set of finite sequences of rational numbers. (If the left-right distinction is significant, we obtain the sequence of Catalan numbers; I do not know any group realizing this sequence.)

A boron tree is a tree in which all vertices have valency 1 or 3. The sequence enumerating boron trees is realized by a group which is a transitive extension of the group described in the preceding paragraph. (This is to be expected, because of the correspondence between boron trees rooted at an end vertex and binary trees.) Inspection of the boron trees with up to 5 end vertices shows that this group is 5-homogeneous $(n_5(G) = 1)$ and 3-transitive but not 4-transitive. Thus it is a counterexample to an earlier conjecture of the author (see [4]).

It is notable that the last three groups were constructed in the context of problems of a purely permutation-group-theoretic nature: see [2, II , IV]. Another common feature of the three will be seen in Section 6.

5. Model-theoretic observations.

In this section and the next, we turn to the question: which sequences (n_k) of natural numbers are realized by a group G, in the sense that $n_k = n_k(G)$ for all k?

First, we observe that it is enough to consider groups of countable degree.

Proposition 5.1. If G is an infinite permutation group with $n_k(G) < \infty$ for all k, then there is a countable permutation group G_1 of countable degree satisfying $n_k(G) = n_k(G_1)$ for all k.

Proof. There are first-order formulae in a suitable language expressing the facts that G is a group, that G acts on X, and that G has n_k orbits on the set of k-element subsets of X for all k. Now the existence of a countable model follows from the downward Löwenheim-Skolem theorem ([1], p.10).

This argument can be modified to find a group G_1 of countable degree with the same modified cycle index as G. However, there is another way to proceed. Given the group G, construct a first-order language with a name (a k-ary relation symbol) for each orbit of G on ordered k-tuples, for each k.

The set X carries an ultrahomogeneous relational structure over this language, with G as an automorphism group acting "ultrahomogeneously". It follows easily that the class of finite substructures of X satisfies conditions (a)-(d) of Section 3. We deduce the existence of a countable ultrahomogeneous model on a set X_1, with automorphism group G_1. (If a countable group is desired, enumerate the pairs of k-tuples lying in the same orbit, select an element of G_1 mapping the first member of each pair to the second, and take the subgroup generated by these elements.) This argument shows that, in a certain sense, the construction of Section 3 is the most general one possible. However, it often tells us more about a structure to have an explicit construction than merely to have an existence theorem, as the last three examples of Section 4 show.

Next, we show that there is no upper bound for the rate of growth of the sequence $(n_k(G))$, with the consequence that there are uncountably many such sequences.

Proposition 5.2. Given a natural number t and a sequence (m_k) of natural numbers, there is a group G such that $n_k(G) = 1$ if $k < t$, while $n_k(G) \geq m_k$ if $k \geq t$.

Proof. A structure in the relevant class consists of a set, together with a colouring of its k-element subsets with m_k distinguished colours for each $k \geq t$. Clearly this class has properties (a)-(d) of Section 3. Now let G be the automorphism group of the countable ultrahomogeneous structure.

Corollary 5.3. There are uncountably many sequences $(n_k(G))$.

Proof. This is a simple diagonalization. If there were only countably many, say $(n_k(G_1))$, $(n_k(G_2))$,..., apply (5.2) with $t = 1$, $m_k = n_k(G_k)+1$, to obtain a contradiction.

For those who don't believe the Continuum Hypothesis, I remark that, in fact, there are $2^{\aleph_0}$ such sequences. (Given a subset V of $\mathbb{N}$, perform the construction of Proposition 5.2 with $m_k = 1$ if $k \notin V$, $m_k = p^{2^k}+1$ if $k \in V$, where $p = \sum_{j=0}^{k-1} n_j$: note that n_j depends only on $m_1, m_2, \ldots, m_j$. We have $m_k \leq p^{2^k}$ if $k \notin V$; so we can recover the set V from the sequence.)

By contrast to Proposition 5.2, we have the following:

Proposition 5.4. Let G be the automorphism group of an ultrahomogeneous structure with only finitely many relation symbols. Then $n_k(G)$ is bounded by the exponential of a polynomial in k.

Proof. If the relations have arities $m_1, m_2, \ldots, m_n$, then the number of isomorphism types of k-element structures is at most $2^{p(k)}$, where $p(k) = k^{m_1} + k^{m_2} + \ldots + k^{m_n}$

We can formulate the general existence question as follows. Suppose that we are given a sequence (n_k). As in the proof of (5.1), there is a first-order theory of groups realizing this sequence. Is this theory consistent? According to the Compactness Theorem ([1], p.10), the theory is consistent if, for every natural number ℓ, there is a group G_ℓ with $n_k(G_\ell) = n_k$ for all $k \leq \ell$. This suggests looking for "local" necessary conditions. Some conditions of this sort are given in the next section, as well as others of a more "global" character.

There is another way in which our problem is related to model theory. Let L be a countable first-order language with no function or constant symbols, T a complete consistent theory over L, and M an infinite model of T. A theorem of Engeler, Ryll-Nardzewski, and Svenonius ([1], p.81) asserts that T is $\aleph_0$-categorical if and only if Aut(M) has only finitely many orbits on k-subsets of M, for all k. (Another equivalent condition is that, for each k, there are only finitely many k-tuples of elements of M, up to elementary equivalence in M.)

6. Necessary conditions.

In this section we list a few known properties of the sequences $(n_k(G))$ arising from groups G. Obviously, a sequence must satisfy these conditions if we are to be able to find a group realizing it!

The most important property is that such sequences are non-decreasing.

Theorem 6.1. $n_{k+1}(G) \geq n_k(G)$.

This raises two questions: how fast must the sequence grow, and in what circumstances can consecutive terms be equal?

Examples in Sections 1 and 2 show that intransitive groups, or transitive but imprimitive groups, may exhibit polynomial growth rate. By contrast, Macpherson [9] showed the following:

Theorem 6.2. If G is primitive, then either $n_k(G) = 1$ for all k (G is highly homogeneous), or for any $\varepsilon > 0$, $n_k(G) > \exp(k^{1/2-\varepsilon})$ holds for all sufficiently large k.

It is conjectured that in fact, for primitive but not highly homogeneous groups, the sequence $(n_k(G))$ grows at least exponentially.

It is known [2 IV] that, in order to prove this, it is enough to show that a primitive group realizing slower than exponential growth rate is 3-homogeneous. Macpherson has refined the techniques used for Theorem 6.2 to an extent where a proof of the conjecture looks possible.

In fact, very few examples are known of primitive groups where the growth rate is exponential, that is, where $\log n_k(G) \sim c\,k$: only the last three examples in Section 4 and some related ones. (We have $c = \log 2$ for #121, and $c = \log 2.48...$ for ##122, 298. No example with $c < \log 2$ is known.) Perhaps there is a gap between exponential growth and the type exhibited by ##545, 659, that is, $\log n_k(G) \sim c\,k \log k$. If this were true, the sequences with exponential growth would be especially interesting.

Questions about growth rate can be asked even when no group is present. Here are some examples.

(i) In the Engeler-Ryll-Nardzewski-Svenonius theorem (see Section 5), is it true that if T is finitely axiomatisable, then $n_k(\mathrm{Aut}(M))$ is bounded by the exponential of a polynomial (cf.(5.4))?

(ii) A theorem of Pouzet [8] asserts that, if R is a relation on an infinite set, and m_k the number of restrictions of R to k-element subsets, then either (m_k) grows polynomially (i.e. $a\,k^n \leq m_k \leq b\,k^n$ for some natural number n and positive constants a and b), or (m_k) grows faster than any polynomial. Can this be extended to structures with arbitrarily many relations? What can be said about structures with polynomial growth? (For example, which infinite graphs have this property?) And if the growth rate is faster than polynomial, must it be at least fractional exponential?

Turning to the other question, groups with $n_k = n_{k+1}$, we observe that any $(k+1)$ homogeneous group (that is, a group with $n_{k+1} = 1$) has this property, by Theorem 6.1. A considerable amount is known about such groups (see, for example, [2 IV]), but for the present problem we regard them as "trivial". In [2 I] it is shown that, if $n_k(G) = n_{k+1}(G)$ and G is intransitive, then G fixes a set of size at most k and acts $(k+1)$-homogeneously on its complement. Thus, we need only consider transitive groups. A similar reduction in [2 II] allows us to consider only primitive groups. It is also shown there that no primitive group G has $n_2(G) = n_3(G) > 1$; while if G is transitive and $n_3(G) = n_4(G) > 1$, then G acts on a dense local order (so that G is a subgroup of the group associated with #121 if its degree is countable). Other known examples of primitive groups with $n_k(G) = n_{k+1}(G)$ are the infinite-dimensional affine group (k=4), the group associated with switching classes of tournaments (k=4), the group associated with boron trees (#122) (k=6), and a related group associated with "boron-carbon trees" (k=4). In all known cases, $n_k(G) = 2$ and G is (k-1)-homogeneous.

The investigation of groups with this property has raised a number of interesting combinatorial questions, related to Ramsey's theorem. For a discussion of some of these, see [3] and [4].

It is shown in [2 I] that, if G is transitive and $n_k(G) = n_{k+2}(G)$, then G is (k+2)-homogeneous. Evidence suggests that, for primitive groups, the sequence $(n_k(G))$ is nearly log-concave; that is, violations of the inequality $n_k(G)n_{k+2}(G) \geq n_{k+1}(G)^2$ are comparatively rare. (This reinforces the conjecture that $(n_k(G))$ grows at least exponentially.) The only result in this direction is

$$n_k(G)n_{k+2}(G) \geq (n_{k+1}(G)-1)(n_{k+1}(G) + n_k(G)-1)/(k+1)(k+2)$$

by Cameron and Saxl [5].

An interesting test case is k=1, assuming that $n_1(G) = 1$ (that is, G is transitive). Putting $n_2(G) = r$, $n_3(G) = s$, the above formula gives $s \geq \frac{1}{6}r(r-1)$. This bound is attained for G a finite elementary abelian 2-group acting regularly (and only for such a group). Even for infinite primitive groups, it is best possible, apart from a factor. (For example, let C-structures consist of colourings of the edges of complete graphs with r colours in such a way that any triangle has at least one edge of the last colour. The automorphism group G of the countable ultrahomogeneous C-structure is primitive and has $n_2(G) = r$, $n_3(G) = \frac{1}{2}r(r+1)$. For r=2, the subgraph of the first colour is that described by Woodrow [12].)

7. Algebras

We have already seen (in Section 2) the relation

$$\sum_{k=0}^{\infty} n_k t^k = \prod_{j=1}^{\infty} (1-t^j)^{-m_j} \qquad (*)$$

There, it connected the sequence $(m_k = n_k(H))$ with the sequence $(n_k = n_k(H \text{ Wr } S))$, where S is an infinite symmetric group. There are two other familar situations where (*) occurs:

(i) If m_k is the number of connected graphs with k vertices, then n_k is the total number of graphs with k vertices. Similarly for connected graphs and graphs with k edges, trees and forests with k vertices, etc.

(ii) If A is a graded algebra which is a polynomial ring generated by m_k homogeneous elements of degree k (for each k), then n_k is the dimension of the kth homogeneous component.

These three occurrences of the same relation are, of course, not unconnected.

Let X be a set (in our application, an infinite set). Let V_k be the rational vector space of functions from k-subsets of X to $\mathbb{Q}$, and $A = \bigoplus_{k=0}^{\infty} V_k$; define multiplication on A by setting, for $f \in V_k$, $g \in V_\ell$, $M \subseteq X$, $|M| = k+\ell$,

$$(fg)(M) = \sum_{\substack{K \subseteq M \\ |K|=k}} f(K)g(M-K),$$

and extending linearly. This makes A a (commutative and associative) graded algebra. If G is a permutation group on X, V_k^G the vector space of G-invariant functions in V_k, and $A^G = \bigoplus_{k=0}^{\infty} V_k^G$, then A^G is a subalgebra of A; and if $n_k(G)$ is finite, then $\dim V_k^G = n_k(G)$.

Remarks: (i) Theorem 6.1 follows from the fact that, if e is the constant function in V_1 with value 1, then e is not a zero-divisor, so that multiplication by e is a monomorphism from V_k^G to V_{k+1}^G.

(ii) Macpherson's result (Theorem 6.2) implies that, if G is primitive, then A^G cannot be finitely generated unless G is highly homogeneous (in which case A^G is the polynomial ring generated by e).

Under certain conditions, A^G is a polynomial ring. If G is the automorphism group of the countable ultrahomogeneous structure in a class C, then sufficient conditions can be formulated in terms of C. The requirements are that C possesses concepts of "connected structure" and "disjoint union" (so that any structure is uniquely the disjoint union of connected ones) and "involvement" (so that, if the point set of a C-structure M is partitioned, then M involves the disjoint union of the induced structures on the parts). We see immediately that for the groups G corresponding to the sequences #479 (graphs with k vertices) and #545 (graphs with k edges), the algebra A^G is a polynomial ring; and the sequences enumerating its polynomial generators by degree are #649 (connected graphs with k vertices) and #985 (connected graphs with k edges) respectively.

If G = H Wr S, then a member of the class enumerated by $(n_k(G))$ consists of a partition of k together with a structure from the class enumerated by $n_k(H)$ on each part. The connected structures are precisely those in which the partition has a single part. Thus A^G is a polynomial ring, and the sequence enumerating its generators is $(n_k(H))$.

It would be interesting to have further examples. A first step might involve computing the sequence (m_k) from a given sequence (n_k) using (*), and checking whether it occurs in the Handbook.

A test case is #321 (switching classes of graphs). Since there are equally many switching classes and even graphs, the sequence (m_k) enumerates connected even graphs (i.e. Eulerian graphs). However, I do not know whether A^G is a polynomial ring or not

Of course, A^G is not always a polynomial ring. For example, if H is a finite permutation group, then $A^{S \mathrm{Wr} H}$ is isomorphic to the ring of invariants of H, where H acts as a linear group via permutation matrices (see [2 II]), so that $f_{S \mathrm{Wr} H}(t)$ is the Molien series of H. (It is an interesting exercise to check that Corollary 2.2 agrees with Molien's theorem in this case.)

References

1. J.Barwise,(ed.) Handbook of Mathematical Logic, North-Holland, Amsterdam-New York-Oxford, (1977).

2. P.J.Cameron, Orbits of permutation groups on unordered sets, I, J.London Math.Soc.(2) 17 (1978), 410-414; II, ibid.(2) 23 (1981) 249-265; III, IV, ibid., to appear.

3. P.J.Cameron, Colour schemes, Annals Discr.Math., in press.

4. P.J.Cameron, Orbits, enumeration and colouring, Proc. 9th Australian Conf.Combinatorial Mathematics, (ed.A.P.Street), Lecture Notes in Math., Springer-Verlag, in press.

5. P.J.Cameron and J.Saxl, Permuting unordered subsets, Quart. J. Math. Oxford, to appear.

6. P.Erdös and J.Spencer, Probabilistic Methods in Combinatorics, Academic Press, New York, (1974).

7. A.M.W.Glass, Ordered Permutation Groups, London Math.Soc.Lecture Notes 55, Cambridge Univ.Press, Cambridge, (1981).

8. F.Harary and E.M.Palmer, Graphical Enumeration, Academic Press, New York-London (1973).

9. H.D.Macpherson, The action of an infinite permutation group on the unordered subsets of a set, J.London Math.Soc., to appear.

10. M.Pouzet, Caractérisation topologique et combinatoire des âges les plus simples, preprint.

11. N.J.A.Sloane, Handbook of Integer Sequences, Academic Press, New York-San Francisco-London (1973).

12. R.E.Woodrow, There are four countable ultrahomogeneous graphs without triangles, J.Combinatorial Theory (B), 27 (1979), 168-179.

PICTURES AND SKEW (REVERSE) PLANE PARTITIONS

Michael Clausen and Friedrich Stötzer
Lehrstuhl II für Mathematik, Universität Bayreuth
D-8580 Bayreuth, West-Germany

Pictures first appeared in papers by James/Peel [2] and Zelevinsky [5,6] in connection with the representation theory of symmetric groups. Roughly speaking a picture is a bijection $T:A \to B$ $(A,B \subseteq \mathbb{N}\times\mathbb{N})$ such that T and T^{-1} both satisfy the same standard property. We distinguish two different standard properties, which are defined by means of several orderings of $\mathbb{N}\times\mathbb{N}$:

$$(a,b) \leqslant_P (c,d) \quad :\Longleftrightarrow \quad (a \leqslant c \text{ and } b \leqslant d)$$
$$(a,b) \leqslant_C (c,d) \quad :\Longleftrightarrow \quad (a \leqslant c \text{ and } b \geqslant d)$$
$$(a,b) \leqslant_L (c,d) \quad :\Longleftrightarrow \quad (\text{either } a < c \text{ or } a = c \text{ and } b \leqslant d)$$
$$(a,b) \leqslant_J (c,d) \quad :\Longleftrightarrow \quad (\text{either } a < c \text{ or } a = c \text{ and } b \geqslant d)$$

Note that $\leqslant_L$ (resp. $\leqslant_J$) is a linearization of $\leqslant_P$ (resp. $\leqslant_C$). A map $f:X \to Y$ $(X,Y \subseteq \mathbb{N}\times\mathbb{N})$ is called PJ-standard (resp. PC-standard), iff f is an order morphism $(X,\leqslant_P) \to (Y,\leqslant_J)$ (resp. $(X,\leqslant_P) \to (Y,\leqslant_C)$). A bijection $T:A \to B$ $(A,B \subseteq \mathbb{N}\times\mathbb{N})$ is said to be a PJ-picture (resp. PC-picture) of shape $A =: |T|$ and content B, if T and T^{-1} are PJ-standard (resp. PC-standard). Let $PJ(A,B)$ (resp. $PC(A,B)$) denote the set of all PJ-pictures (resp. PC-pictures) of shape A and content B.

Since $\leqslant_J$ is a linearization of $\leqslant_C$ we get immediately $PC(A,B) \subseteq PJ(A,B)$.

Example.

$T =$ 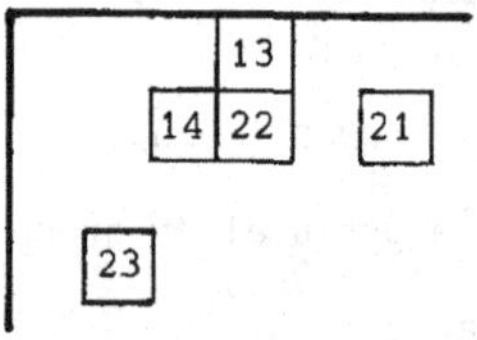

illustrates the bijection T of shape $A = \{(1,4),(2,3),(2,4),(2,6),(4,2)\}$ and content $B = \{(1,3),(1,4),(2,1),(2,2),(2,3)\}$ with

$T(1,4)=(1,3)$, $T(2,3)=(1,4)$, $T(2,4)=(2,2)$, $T(2,6)=(2,1)$, and $T(4,2)=(2,3)$.
T is PC-standard as well as

$T^{-1} =$ 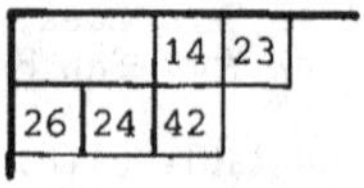

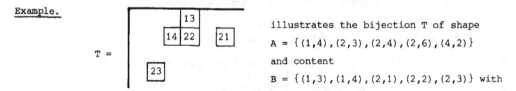

Hence T is a PC-picture.

U = 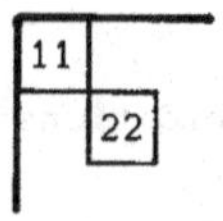

is a PJ-picture but not a PC-picture. □

A problem which arises in representation theory is to compute explicitly all pictures between two skew diagrams A and B. Before we establish an algorithm which constructs all those pictures by suitable hook deformations we give some characterizations of pictures.

The reader is referred to [1] for a more detailed exposition.

1. Pictures

We have learnt the following useful "geographical" notation from A. Zelevinsky [5,p. 157].

Every point $(c,d) \in \mathbb{N}\times\mathbb{N}$ decomposes $\mathbb{N}\times\mathbb{N}$ into disjoint subsets:

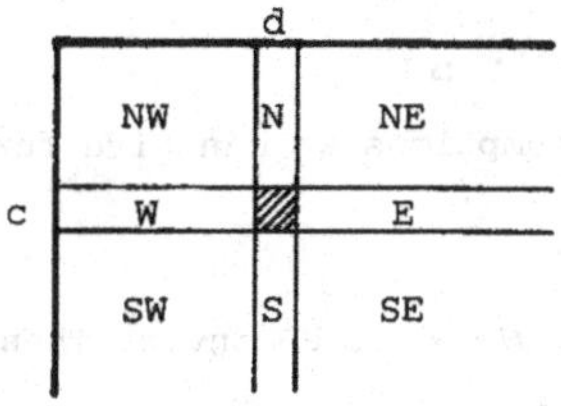

We write (a,b) $(X,Y,\ldots,Z)$ (c,d) iff $(a,b) \neq (c,d)$ and (a,b) lies in one of the regions $X,Y,\ldots,Z$ with respect to (c,d).

<u>Example.</u> $(a,b) \underset{P}{\lneq} (c,d) \Leftrightarrow (a,b)\ (N,NW,W)\ (c,d)$. □

We frequently have to deal with bijections $T:A\to B$ satisfying for all $x,y \in A$ some of the following "geographical" conditions.

name	geographical condition
(E)	$x\ (E)\ x \Rightarrow T(x)\ (W,SW)\ T(y)$
(S,SE)	$x\ (S,SE)\ y \Rightarrow T(x)\ (SW,S,SE)\ T(y)$
(S)	$x\ (S)\ y \Rightarrow T(x)\ (SW,S)\ T(y)$
(SE)	$x\ (SE)\ y \Rightarrow T(x)\ (SW)\ T(y)$
(SW)	$x\ (SW)\ y \Rightarrow T(x)\ (SW,S,SE,E,NE)\ T(y)$

By definition, a subset A of $\mathbb{N}\times\mathbb{N}$ is <u>P-convex</u> iff $x \underset{P}{\lneq} y \underset{P}{\lneq} z$ and $x,z \in A$ implies $y \in A$. Finite P-convex sets will be called <u>skew diagrams.</u> A skew diagram containing the point $(1,1)$ will be called a <u>diagram.</u> $X \subseteq A$ is <u>A-regular</u> iff there is a P-convex set D with $(1,1) \in D$ and $X = A \setminus D$. π_1 and π_2 denote the natural projections $\mathbb{N}\times\mathbb{N}\to\mathbb{N}$ given by $\pi_i : (a_1,a_2)\mapsto a_i$. For the chain $\{x_1 \underset{J}{\lneq} \ldots \underset{J}{\lneq} x_n\}$ we often use the short-hand notation $\{x_1,\ldots,x_n\}_{\underset{J}{\lneq}}$. $A \mathbin{\dot\cup} B$ denotes the disjoint union of A and B.

Now we can state

<u>Theorem 1.</u> For a bijection $T:A\to B$ $(A,B \subseteq \mathbb{N}\times\mathbb{N})$ the following conditions are equivalent

(1) $T:A\to B$ is a PJ-picture.

(2) For all $x \in A$: $T[\{y \in A \mid x \underset{J}{\leqslant} y\}]$ is B-regular and for all $z \in B$: $T^{-1}[\{y \in B \mid z \underset{J}{\leqslant} y\}]$ is A-regular.

(3) For all $x,y \in A$ the bijection T satisfies the geographical conditions (E),(S,SE) and (SW). □

A similar characterization holds for PC-pictures:

<u>Theorem 2.</u> For a bijection $T:A\to B$ $(A,B \subseteq \mathbb{N}\times\mathbb{N})$ the following conditions are equivalent

(1) $T:A\to B$ is a PC-picture.

(2) For all $x \in A$: $T[\{y \in A \mid x \underset{C}{\leqslant} y\}]$ is B-regular and for all $z \in B$: $T^{-1}[\{y \in B \mid z \underset{C}{\leqslant} y\}]$ is A-regular.

(3) For all $x,y \in A$ the bijection T satisfies the geographical conditions (E),(S),(SE),(SW). □

Under additional assumptions we can give further characterizations.

(Compare with [5,p.157].)

<u>Theorem 3.</u> Let $A,B \subseteq \mathbb{N}\times\mathbb{N}$ be P-convex. Then for a bijection $T:A\to B$ the following conditions are equivalent.

(1) T is a PJ-picture.

(2) T is a PC-picture.

(3) For all $x,y \in A$ the bijection T satisfies the geographical conditions (E),(S) and (SW).

(4) $\pi_1\circ T$ and $\pi_1\circ T^{-1}$ are column strict skew reverse plane partitions and $\pi_2\circ T$ as well as $\pi_2\circ T^{-1}$ are row strict skew plane partitions.
[So the entries in $\pi_1\circ T$ are non-decreasing from left to right in each row and strictly increasing down the columns. The entries in $\pi_2\circ T$ are strictly decreasing in the rows and non-increasing down the columns.]. □

Next we answer the question which subsets of $\mathbb{N}\times\mathbb{N}$ can be the shape or content of a picture. Define $\underline{k} := \{1,2,\ldots,k\}$.

<u>Lemma.</u> Let $\emptyset \neq A \subseteq \mathbb{N}\times\mathbb{N}$.

(1) If the bijection $T:A\to B$ satisfies (E) then A is <u>row-finite</u>, i.e. for all $i \in \mathbb{N}$: $|A \cap \{(i,j) \mid j \in \mathbb{N}\}| < \infty$.

(2) If the bijection $T:A\to B$ satisfies (E) and (SE) then $A \subseteq \mathbb{N}\times\underline{k}$ for some $k \in \mathbb{N}$.

(3) $PJ(A,B) \neq \emptyset$ for suitable $B \subseteq \mathbb{N}\times\mathbb{N}$ iff A is row-finite.

(4) $PC(A,B) \neq \emptyset$ for suitable $B \subseteq \mathbb{N}\times\mathbb{N}$ iff $A \subseteq \mathbb{N}\times\underline{k}$ for some $k \in \mathbb{N}$.

Proof.

"(1)" Note that if x (E) x' and $T(x) =: (a,b)$, $T(x') =: (a',b')$ then by (E): $b < b'$.

"(2)" If there is no such k then by (1) there exists an infinite sequence $(x_1, x_2, \dots)$ of elements in A such that x_i (NW) x_{i+1}. Then $T(x_1)$ (NE) $T(x_2)$ (NE) $T(x_3)$ ..., which is impossible.

"(3)" "⇒" is clear by (1) and Theorem 1.

" ⇐ " If A is row-finite then there is a unique order isomorphism $N_A : (A, \leq_L) \to (A, \leq_J)$.

By Theorem 1,(3), $N_A \in PJ(A,A)$.

[N_A will be called the natural PJ-picture with respect to A.]

"(4)" "⇒" is clear by (2) and Theorem 2.

" ⇐ " Let $A \subseteq K := \mathbb{N} \times \{1,\dots,k\}$. The natural PJ-picture N_K is even a PC-picture. The "restriction" of N_K to A is a PC-picture of shape A.

□

Let $\underline{\underline{R}}$ be the set of all row-finite subsets of $\mathbb{N} \times \mathbb{N}$ and let $\underline{\underline{R}}^{o} := \{A \mid \exists k : A \subseteq \mathbb{N} \times \underline{k}\}$.

Now we introduce equivalence relations on $\underline{\underline{R}}$ and $\underline{\underline{R}}^{o}$.

$A, A' \in \underline{\underline{R}}$ are said to be PJ-equivalent iff there exists a bijection $f : A' \to A$ such that for all $B \in \underline{\underline{R}}$: $PJ(A,B) \circ f := \{T \circ f \mid T \in PJ(A,B)\} = PJ(A',B)$.

Similarly, $A, A' \in \underline{\underline{R}}^{o}$ are PC-equivalent iff there exists a bijection $f : A' \to A$ such that for all $B \in \underline{\underline{R}}^{o}$: $PC(A,B) \circ f = PC(A',B)$.

Of course, if $PJ(A,-) \circ f = PJ(A',-)$ and $PJ(B,-) \circ g = PJ(B',-)$ then $PJ(A',B') = g^{-1} \circ PJ(A,B) \circ f$.

We give more handy characterizations of these equivalence relations.

Theorem 4. For $A, A' \in \underline{\underline{R}}$ the following conditions are equivalent.

(1) A and A' are PJ-equivalent.

(2) There exists a bijection $f : A' \to A$ such that for all $x, y \in A'$:

x (E) y ⟺ f(x) (E) f(y)

x (S,SE) y ⟺ f(x) (S,SE) f(y)

x (SW) y ⟺ f(x) (SW) f(y).

(3) There exists a bijection $f : A' \to A$ such that f is an order isomorphism $f : (A', \leq_P) \to (A, \leq_P)$ as well as an order isomorphism $f : (A', \leq_J) \to (A, \leq_J)$.

□

Since $\leq_J$ is a total order there exists at most one bijection $f : A' \to A$ satisfying the conditions above.

Theorem 5. For $A, A' \in \underline{\underline{R}}^{o}$ the following conditions are equivalent.

(1) A and A' are PC-equivalent.

(2) There exists a bijection $f:A'\to A$ such that for all $x,y \in A'$ and all directions $R \in \{E,S,SE,SW\}$ the following holds:

$$x\ (R)\ y \iff f(x)\ (R)\ f(y).$$

(3) There exists a bijection $f:A'\to A$ such that f is an order isomorphism $f:(A',\leq_P)\to(A,\leq_P)$ as well as an order isomorphism $f:(A',\leq_C)\to(A,\leq_C)$. □

If we cancel in $A \in \underline{\underline{R}}$ all "empty rows" and "empty columns" then we get a set cpr(A), the <u>compression</u> of A.

<u>Example.</u>

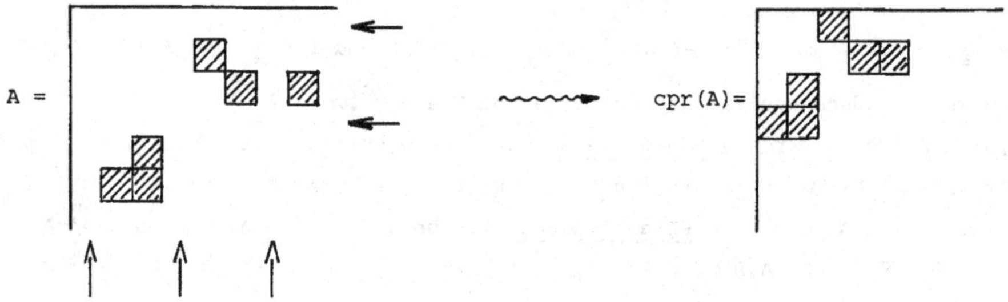

<u>Corollary.</u> $\underline{\underline{T}}^{o} := \{A \in \underline{\underline{R}}^{o} \mid A = cpr(A)\}$ is a transversal with respect to PC-equivalence. □

The specification of a PJ-transversal is more complicate. For $(i,j) \in \mathbb{N}\times\mathbb{N}$, $j \geq 2$, we define

$$N_{ij} := \{(k,j) \mid k < i\} \cup \{(h,j-1) \mid h \geq i\}.$$

<u>Theorem 6.</u> $\underline{\underline{T}} := \{A \in \underline{\underline{R}} \mid A = cpr(A)$ and for all $(i,j) \in A,\ j \geq 2,\ A \cap N_{ij} \neq \emptyset\}$ is a transversal with respect to PJ-equivalence. □

Now we are going to define orderings on $\underline{\underline{T}}^{o}$ and $\underline{\underline{T}}$ allowing estimates for PC(A,B) and PJ(A,B).

For $A,C \in \underline{\underline{T}}$ (resp. $A,C \in \underline{\underline{T}}^{o}$) we write $A \leq_{PJ} C$ (resp. $A \leq_{PC} C$) iff there exists a bijection $f:C\to A$ such that for all $B \in \underline{\underline{R}}$ (resp. $B \in \underline{\underline{R}}^{o}$): $PJ(A,B)\circ f \subseteq PJ(C,B)$ (resp. $PC(A,B)\circ f \subseteq PC(C,B)$).

<u>Theorem 7.</u> $(\underline{\underline{T}},\leq_{PJ})$ and $(\underline{\underline{T}}^{o},\leq_{PC})$ are partially ordered sets (= posets). □

If $A \leqslant_{PJ} B$ (resp. $A \leqslant_{PC} B$) then necessarily $|A| = |B|$. This suggests the following partitions.

$\underline{\underline{T}} = \underline{\underline{T}}_\infty \uplus \biguplus_n \underline{\underline{T}}_n$ and $\underline{\underline{T}}^O = \underline{\underline{T}}^O_\infty \uplus \biguplus_n \underline{\underline{T}}^O_n$,

where

$\underline{\underline{T}}_\infty := \{A \in \underline{\underline{T}} \mid |A| = \infty\}$; $\underline{\underline{T}}^O_\infty := \{A \in \underline{\underline{T}}^O \mid |A| = \infty\}$

$\underline{\underline{T}}_n := \{A \in \underline{\underline{T}} \mid |A| = n\}$; $\underline{\underline{T}}^O_n := \{A \in \underline{\underline{T}}^O \mid |A| = n\}$.

The maximal and minimal elements in the posets above are characterized in the following.

Theorem 8.

(1) $\{\{(i,n+1-i) \mid i \in \underline{n}\} \mid n \in \mathbb{N}\}$ is the set of all maximal elements in $(\underline{\underline{T}}, \leqslant_{PJ})$ as well as in $(\underline{\underline{T}}^O, \leqslant_{PC})$.

(2) $A \in \underline{\underline{T}}$ is minimal in $(\underline{\underline{T}}, \leqslant_{PJ})$ iff $(A, \lesseqqgtr_P)$ is linear.

(3) $\{\{(i,i) \mid i \in \underline{n}\} \mid n \in \mathbb{N}\}$ is the set of all minimal elements in $(\underline{\underline{T}}^O, \leqslant_{PC})$.

□

According to the last theorem $(\underline{\underline{T}}^O_n, \leqslant_{PC})$ is a poset with 0- and 1-element.

Example. $(\underline{\underline{T}}^O_3, \leqslant_{PC})$

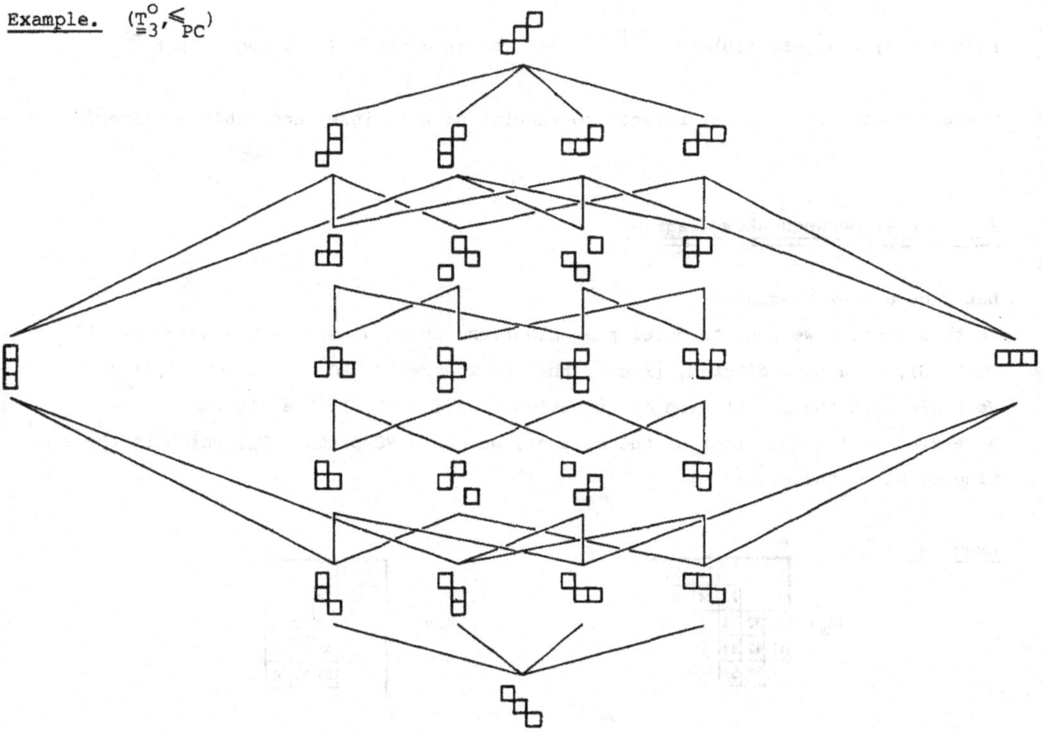

Note that $(\underline{\underline{T}}_3^{\circ},\leqslant_{PC})$ is neither a lattice nor ranked. The same is true for all $n\geqslant 3$. This example also suggests several "geographical" dualisms and automorphisms of $(\underline{\underline{T}}_n^{\circ},\leqslant_{PC})$, which we are going to describe now.

The dihedral group $D_8 = \langle \sigma,\tau \mid \sigma^4=\tau^2=1,\ \tau\sigma\tau^{-1}=\sigma^3 \rangle = \{1,\sigma,\sigma^2,\sigma^3,\tau,\sigma\tau,\sigma^2\tau,\sigma^3\tau\}$ acts on every $\underline{\underline{T}}_n^{\circ}$ in the following obvious way.
For $A \in \underline{\underline{T}}_n^{\circ}$, $\sigma A \in \underline{\underline{T}}_n^{\circ}$ is the 90° rotation of A and $\tau A \in \underline{\underline{T}}_n^{\circ}$ is the transpose of A.

Example.

A = 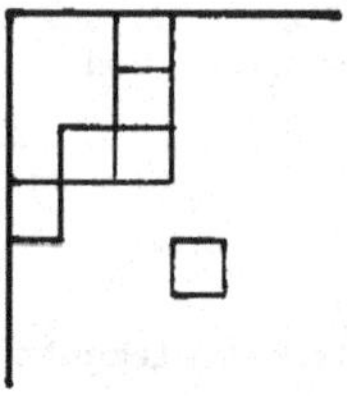 σA = 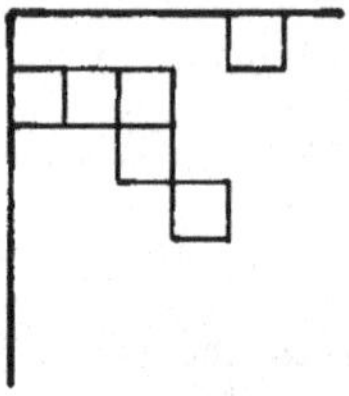 τA = 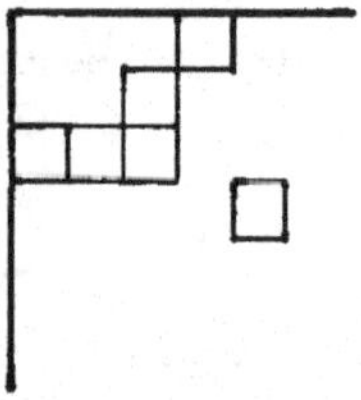

Obviously, D_8 acts faithfully on $\underline{\underline{T}}_n^{\circ}$ iff $n\geqslant 3$. For $n\geqslant 3$, the Kleinian 4-group $\{1,\sigma^2,\tau,\sigma^2\tau\}$ is a group of automorphisms of $(\underline{\underline{T}}_n^{\circ},\leqslant_{PC})$. $\sigma,\sigma\tau,\sigma^3,\sigma^3\tau$ act as anti-automorphisms on $(\underline{\underline{T}}_n^{\circ},\leqslant_{PC})$. Note that σ^2 is a central symmetry and $\sigma^2\tau$ is a reflection fixing a line parallel to the axis $\{(i,-i) \mid i\in\mathbb{Z}\}$.
For $\delta \in D_8$ and $A \in \underline{\underline{T}}_n^{\circ}$ let δ_A denote the obvious bijection $\delta_A: A\to\delta A$.
Then a straightforward computation shows that for all $A,B\in\underline{\underline{T}}_n^{\circ}$ the following holds:

$$PC(\sigma^2A,\sigma^2B) = \sigma^2{}_B\circ PC(A,B)\circ(\sigma^2{}_A)^{-1} \quad \text{and} \quad PC(\tau A,\sigma^2\tau B) = (\sigma^2\tau)_B\circ PC(A,B)\circ\tau_A^{-1}.$$

These formulae are closely related to results of Zelevinsky and Schützenberger/Knuth.

2. Pictures between skew diagrams

Let S be a skew diagram.
In this section we want to develop an algorithm which generates the union of all PC(S',S), S' a skew diagram. [Recall that by Theorem 3 PC(S',S) = PJ(S',S).]
We begin with the computation of the union of all PC(D,S), D a diagram.
Reversing in id_S the order of the columns, we get a PC-picture T_S, which is the starting point of the algorithm.

Example.

id_S = 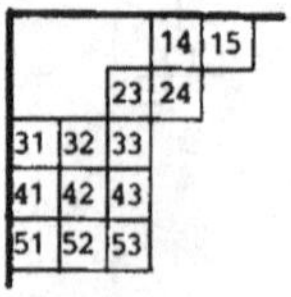

T_S = 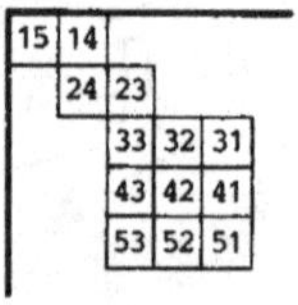

Now we can formulate the algorithm in its first version.

Let $S \in \underline{\underline{T}}$ be a skew diagram.

<u>Algorithm I</u>

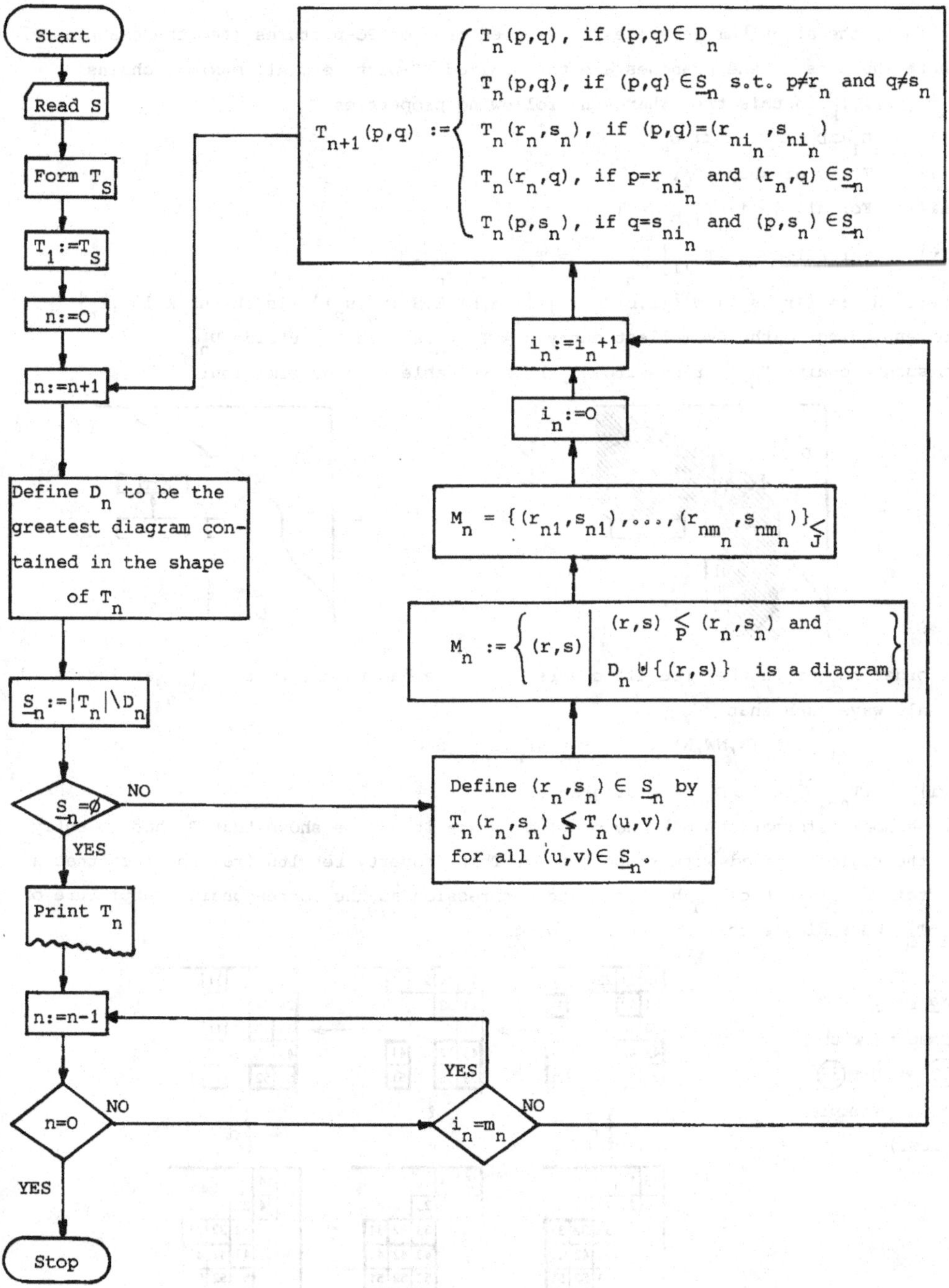

Theorem 9. Given a skew diagram S, the algorithm above will produce

$$\bigcup_{D\,:\,D \text{ a diagram}} PC(D,S).$$

In addition, there are no repetitions in the list. □

In fact, the algorithm generates a directed tree of PC-pictures (see the example below) where the tops of the branches are the desired PC-pictures. All maximal chains $T_S = T_1, \ldots, T_k$ in this tree share the following properties:

(i) $D_1 \subset D_2 \subset \ldots \subset D_k$.

(ii) T_k is of shape D_k.

(iii) For all $j > i$: $T_j\big|_{D_i} = T_i\big|_{D_i}$.

(iv) For all $n < k$: $T_{n+1}\Big|_{|T_n| \setminus H_n} = T_n\Big|_{|T_n| \setminus H_n}$,

where $H_n := \{(r_n, s_n)\} \uplus \{(a,b) \in |T_n| \mid (a,b)(E,S)(r_n,s_n)\}$ is the <u>hook</u> in $|T_n|$ corresponding to the J-smallest entry $x = T_n(r_n, s_n)$ in T_n outside D_n.
In such a chain, T_{n+1} arises from T_n by a suitable hook deformation:

$T_n =$ 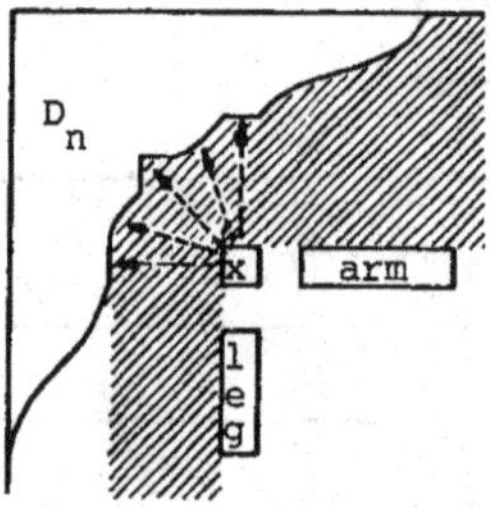

$T_{n+1} =$ 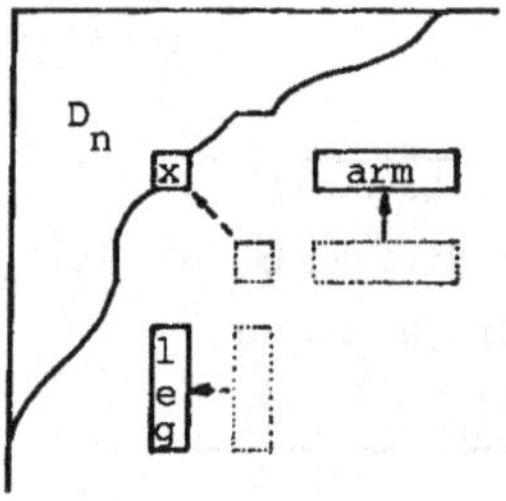

In order to get in the tree all covers of T_n one has to deformate $T_n\big|_{H_n}$, as indicated, in all ways such that

(v) $T_{n+1}^{-1}(x)\ (N,NW,W)\ T_n^{-1}(x) = (r_n, s_n)$ and

(vi) $\{T_{n+1}^{-1}(x)\} \uplus D_n$ is a diagram.

These hook deformations make no problems since it can be shown that T_n has no entry in the region shaded with respect to x. This property results from the fact that a certain subpicture of T_n has the same compression as the corresponding subpicture of $T_1 = T_S$. We indicate this in the following.

<u>Example.</u>

(Compare with (30)→(31)→(32) in the example below.)

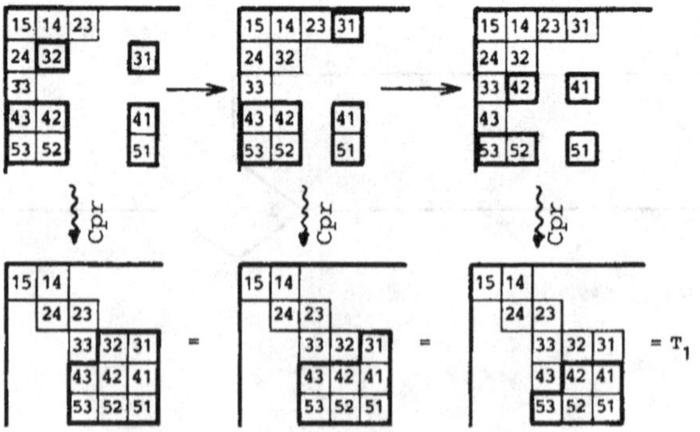

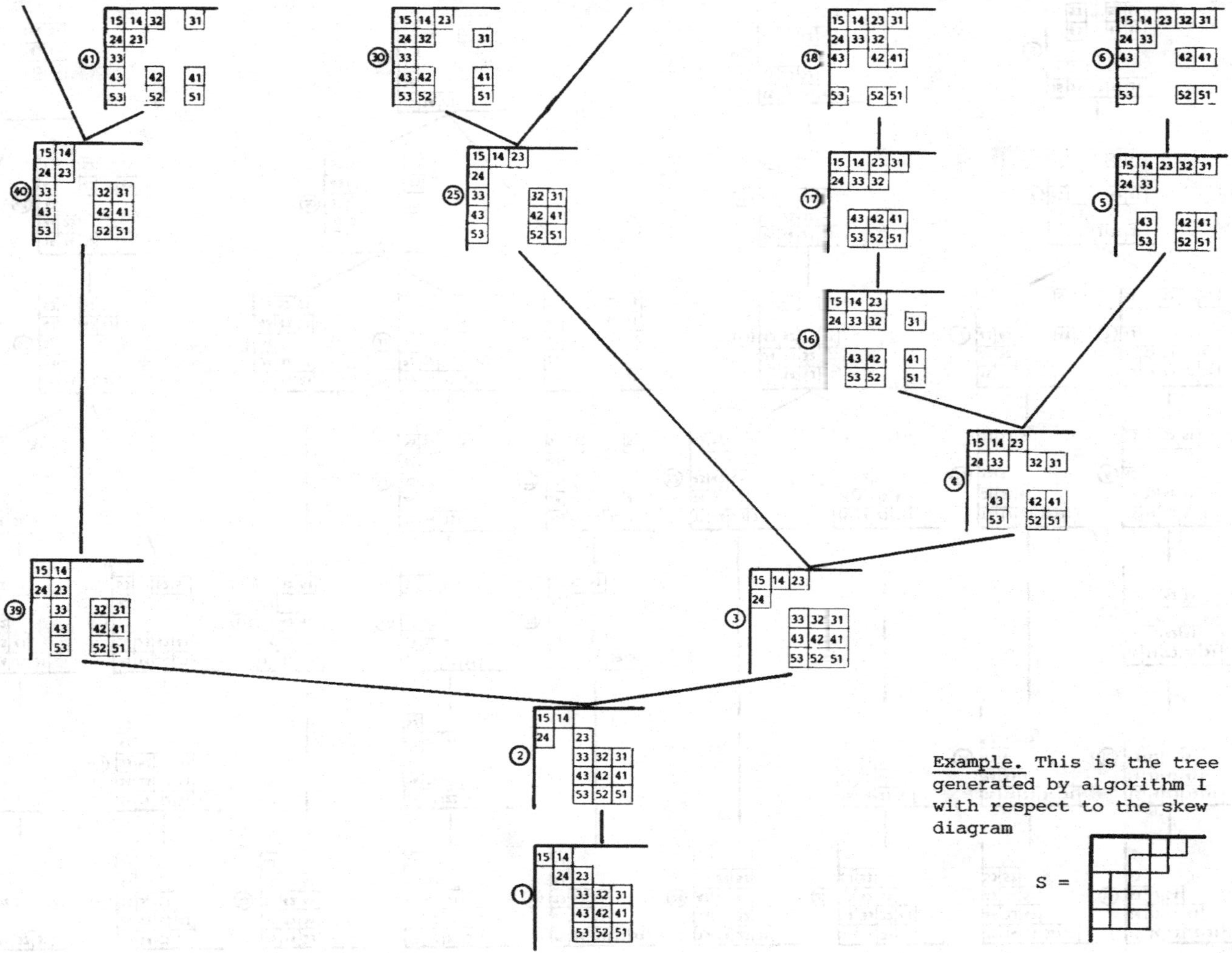

Example. This is the tree generated by algorithm I with respect to the skew diagram

S =

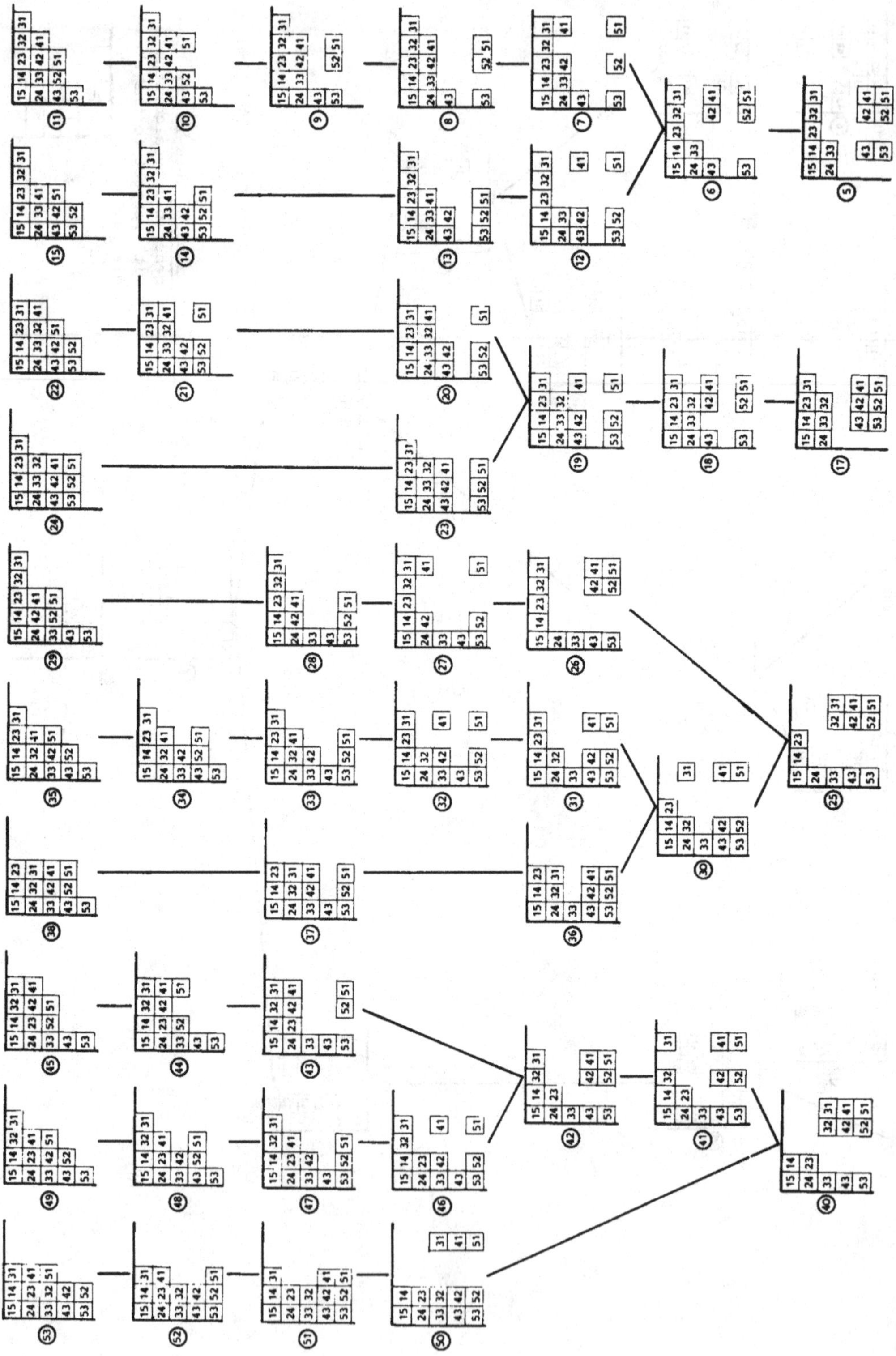

We will now deal with the second version of the algorithm, which for a given skew diagram S will produce the union of all PC(S',S), S' a skew diagram. Again, beginning with T_S, a directed tree of PC-pictures will recurrently be generated by hook deformations. But in order to get skew diagrams as shapes empty rows and/or empty columns eventually have to be filled in at suitable positions.

To skip formal details in the formulation of the second algorithm we are going to describe some of them now.

If A is a finite subset of $\mathbb{N}\times\mathbb{N}$ then the smallest diagram containing A will be called the <u>diagram-closure</u> of A.

Let $U \in \underline{\underline{T}}^{o}_{n+1}$, $n \in \mathbb{N}$, be a skew diagram and $x = (a,b) \in U$ a U-regular point (i.e. $\{x\}$ is U-regular). [Note that then $(a,b) \neq (1,1)$.] The <u>type vector</u>

$$Z(U,x) := \begin{cases} (1,1,0,0), & \text{if } a = 1 \\ (1,0,1,0), & \text{if } b = 1 \\ (1,0,0,0), & \text{if } (a-1,b-1) \in U \\ (1,1,1,1), & \text{if } (a-1,b-1) \notin U \text{ and } a>1,\ b>1 \end{cases}$$

will tell us whether, or in what order of succession empty rows and/or empty columns have to be filled in.

If T_n has been constructed by the algorithm below then, according to the algorithm, one has to associate to T_n and $x := (r_{ni_n}, s_{ni_n})$ the type vector $Z(n,i_n) := Z(U,x)$, where $U := \{x\} \uplus T_n^{-1}[\{(a_i,b_i) \mid i \in \underline{n}\}]$.

In the course of the algorithm the ones in $Z(n,i_n)$ will step by step be replaced by zeroes. Such a (possibly modified) $Z(n,i_n)$ describes how to get the PC-picture $T := Z(n,i_n) * T_n$ out of T_n. The "*-product" is defined as follows. [Let $a,b,c \in \{0,1\}$.]

$Z(n,i_n)$	$T = Z(n,i_n) * T_n$
(1,a,b,c)	$T := T_n$
(0,1,a,b)	T arises from T_n by inserting in T_n an empty row between the rows of index $r_{ni_n}-1$ and r_{ni_n}.
(0,0,1,a)	T arises from T_n by inserting in T_n an empty column between the columns of index $s_{ni_n}-1$ and s_{ni_n}.
(0,0,0,1)	$T := (0,1,0,0) * ((0,0,1,0) * T_n)$.

<u>Example.</u> Assume T_4 (see the table) has been constructed by the algorithm below. Then $M_4 = \{(1,4),(2,3),(3,2),(4,1)\}$. If $i_4 = 2$ then $Z(4,2) = (1,1,1,1)$ and we get:

$T_4 = (1,1,1,1) * T_4$	$(0,1,1,1) * T_4$	$(0,0,1,1) * T_4$	$(0,0,0,1) * T_4$

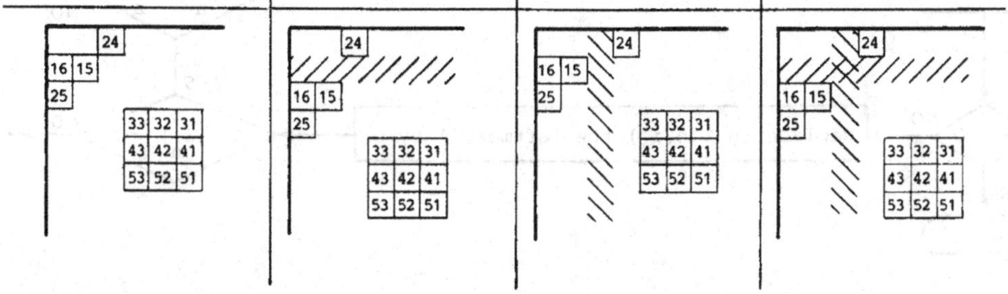

Let $S = \{(a_1,b_1),\ldots,(a_k,b_k)\}_{\leq_J} \in \underline{\underline{T}}^{O}_{k}$ be a skew diagram.

Algorithm II

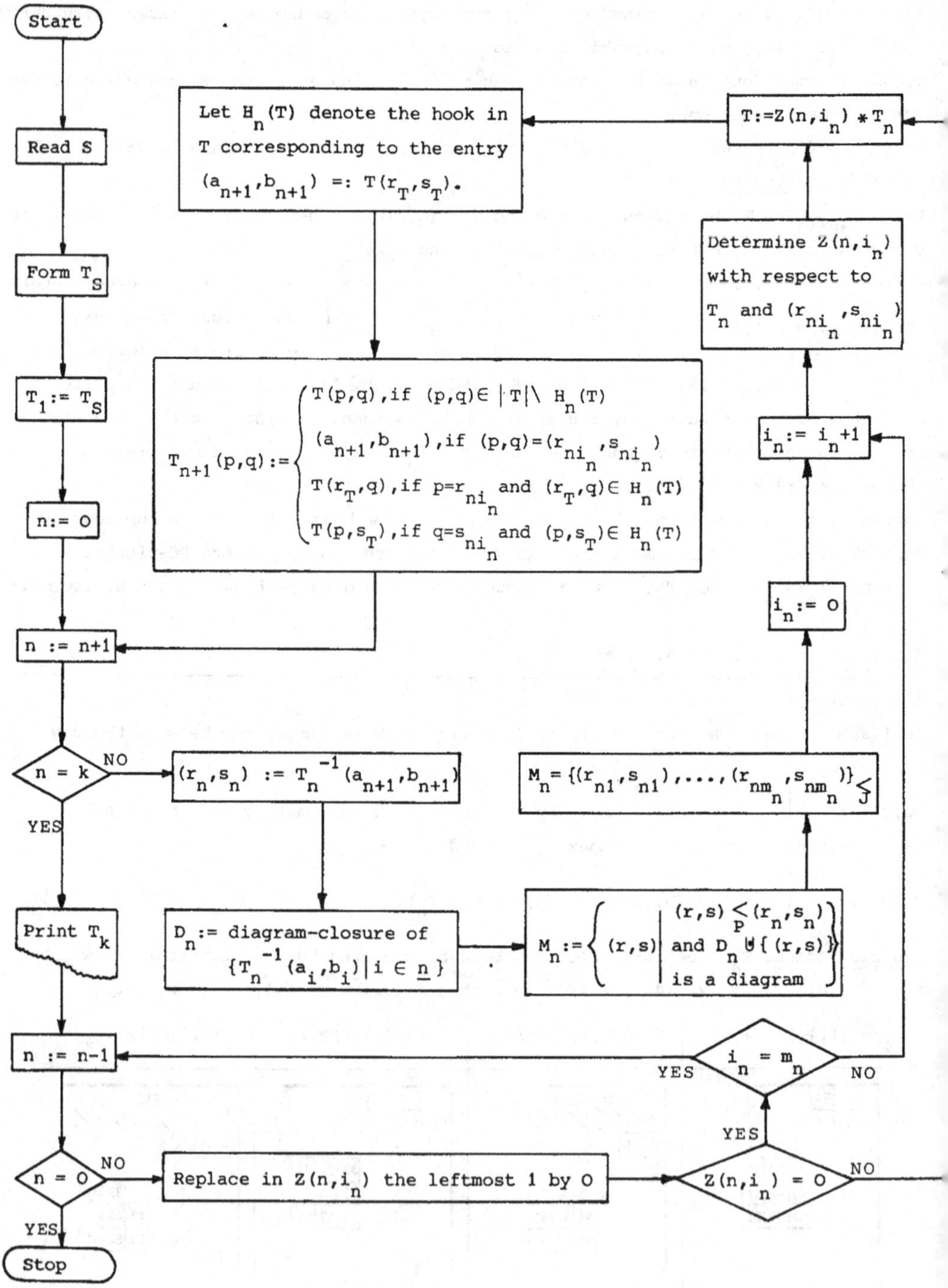

Theorem 10. Given a skew diagram $S \in \underline{\underline{T}}_k^o$, algorithm II will produce $\cup\, PC(S',S)$, union over all skew diagrams $S' \in \underline{\underline{T}}_k^o$.

□

Since trees generated by algorithm II are rather extensive even for small skew diagrams S, we restrict ourselves to the following.

Example. Let S be the content of T_4 (see the last example). Then T_4 actually occurs in the tree generated by algorithm II with respect to S. We describe all covers of T_4 in this tree.

According to $|M_4| = 4$ these covers naturally decompose into 4 classes.

i_4	(r_{4i_4}, s_{4i_4})	$z(4,i_4)$	(1,a,b,c)	(0,1,a,b)	(0,0,1,a)	(0,0,0,1)
1	(1,4)	(1,1,0,0)	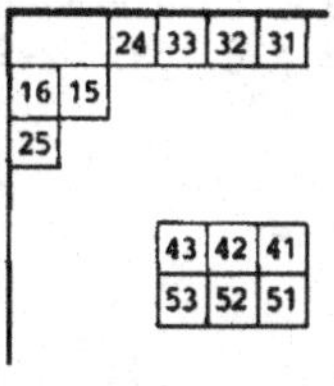	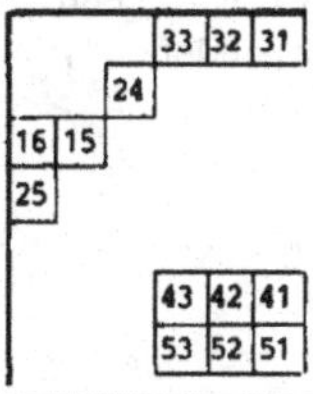		
2	(2,3)	(1,1,1,1)	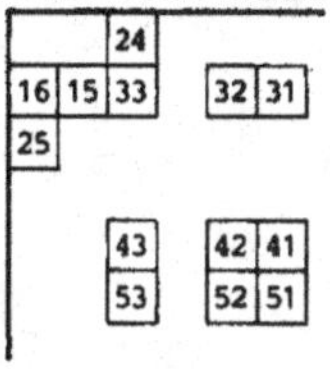	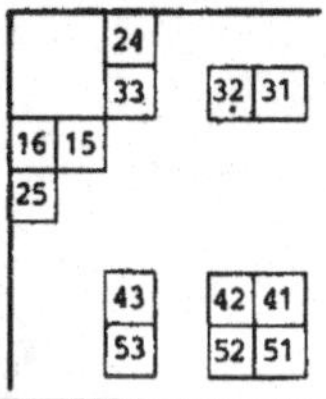	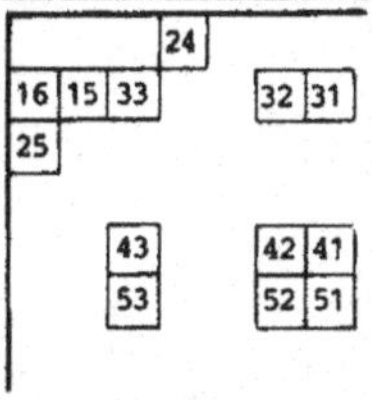	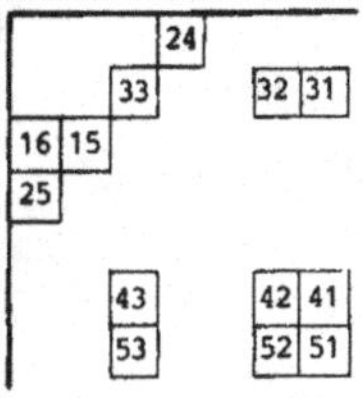
3	(3,2)	(1,0,0,0)	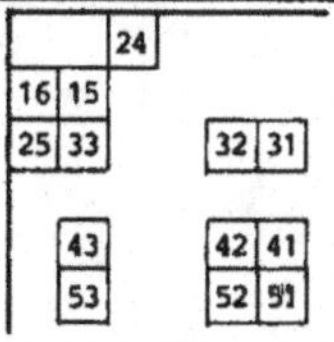			
4	(4,1)	(1,0,1,0)	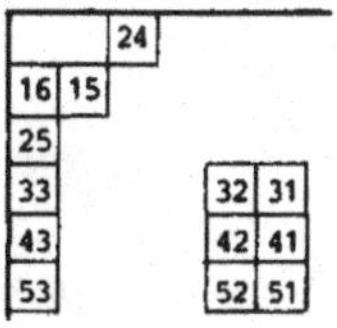		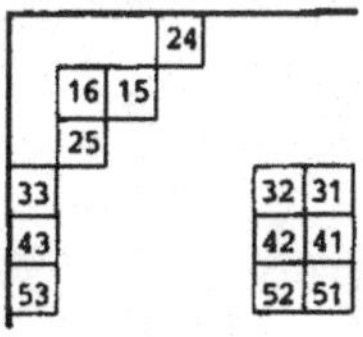	

Starting with a skew diagram $S \in \underline{\underline{T}}_k^o$, a suitable modification of algorithm II will generate a tree consisting essentially of all pictures of content S, i.e. algorithm III constructs $\cup\, PC(A,S)$, union over all $A \in \underline{\underline{T}}_k^o$.

These algorithms can be applied to various problems in representation theory and combinatorics (see [1,2,5,6]).

References

1. M. Clausen, F. Stötzer, "A constructive approach to pictures, standard tableaux and skew modules" (in preparation).

2. G.D. James, M.H. Peel, "Specht series for skew representations of symmetric groups", J. of Alg. 56, 343-364 (1979).

3. M. Schützenberger, "La correspondence de Robinson", in "Combinatoire et Représentation du Groupe Symmetrique", Strasbourg, 1976, ed. by D. Foata, Lecture Notes in Math. No. 579, Springer-Verlag, Berlin.

4. R.P. Stanley, "Theory and application of plane partitions", Stud. Appl. Math. 50, Part 1: 167-188, Part 2: 259-279 (1971).

5. A.V. Zelevinsky, "Representations of finite classical groups - a Hopf algebra approach", Lecture Notes in Math. No. 869, Springer-Verlag, Berlin.

6. A.V. Zelevinsky, "A generalization of the Littlewood-Richardson rule and the Robinson-Schensted-Knuth correspondence", J. of Alg. 69, 82-94 (1981).

A CANONICAL PARTITION THEOREM FOR CHAINS IN REGULAR TREES

W. Deuber, H.J. Prömel, B. Voigt

Fakultät für Mathematik
Universität Bielefeld
4800 Bielefeld 1
West-Germany

Abstract

In this paper we prove a generalization of the Erdös-Rado canonization theorem to regular trees.

§ 1 Introduction

In 1950 Erdös and Rado proved the following theorem:

Theorem 1.1 [2] (*Erdös-Rado canonization theorem*).
Let $\Delta : [\mathbb{N}]^k \to \mathbb{N}$ *be a coloring of the* k*-element subsets of* $\mathbb{N}$ *(the nonnegative integers) with arbitrarily many colors. Then there exists an infinite subset* $X \in [\mathbb{N}]^\omega$ *and there exists a* 0-1 *sequence* $I = (i_0,\dots,i_{k-1}) \in 2^k$ *such that every two* k*-element subsets* $A = \{a_0,\dots,a_{k-1}\}_<$ *and* $B = \{b_0,\dots,b_{k-1}\}_<$ *of* X *are colored the same iff*

$$a_\nu = b_\nu \quad \text{for every} \quad \nu < k \quad \text{with} \quad i_\nu = 1 .$$

This result generalizes the wellknown theorem of Ramsey [4]: if $\Delta : [\mathbb{N}]^k \to \delta$ is a coloring using only finitely many colors, then necessarily $I = (0,\dots,0)$, viz. all k-element subsets of X are colored the same.

Recall that the formulation of the Erdös-Rado canonization theorem involves an

ordering on the ground-set, here the nonnegative integers:
subsets A and B of X are colored the same iff they agree on the subsets given by the sequence I .

In this paper we consider a generalization of the Erdös-Rado canonization theorem to certain partially ordered sets, where the coloring acts on k-chains, i.e. totally ordered k-element subsets.
We prove also some apparently new partition results for chains in d-regular trees.

The paper is organized as follows:
The main results are presented in section 2. In section 3 the partition results for chains in d-regular trees are proved. Section 4 contains some technical tools that are used in section 5 in order to prove the canonical partition theorem for chains in d-regular trees.

§ 2 Results

A tree is a partially ordered set $(P,\leq)$ possessing a minimum such that every interval $[x,y] = \{z \mid x \leq z \leq y\}$ is a totally ordered set.

A d-regular tree , where d is a positive integer, is a tree $(P,\leq)$ such that every non-maximal element $x \in P$ possesses precisely d immediate successors.

Notation: By "T(d)" we denote the d-regular tree of height ω without any maximal nodes.

For our purposes the following explicit representation of T(d) is useful:

- elements of T(d) are finite $\{0,\dots,d-1\}$ - sequences, including $\emptyset$, the empty sequence.
- $(a_0,\dots,a_{m-1}) \leq (b_0,\dots,b_{n-1})$ iff $(a_0,\dots,a_{m-1})$ is an initial sequence of $(b_0,\dots,b_{n-1})$, i.e. $m \leq n$ and $a_\nu = b_\nu$ for every $\nu < m$. Particularly $\emptyset$ is the minimum of T(d) .

In diagram 1 the first 4 levels of T(2) are depicted:

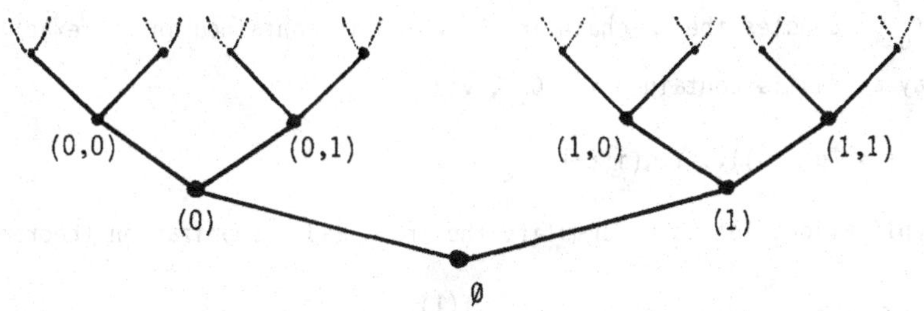

diagram 1

Notation: By "C_k" we denote the chain of length k .

Here it is convenient to represent C_k by nonnegative integers less than k , viz. $\{0,\dots,k-1\}$ with the natural order of the integers.

Definition: Let $(P,\leq)$ be a tree. A subset $\hat{P} \subseteq P$ is a subtree (with the order $\leq$ coming from P) iff the infima with respect to P and $\hat{P}$ agree, more precisely

$$\inf_P(x,y) = \inf_{\hat{P}}(x,y) \quad \text{for all } x,y \in \hat{P} .$$

Compare also diagram 2 .

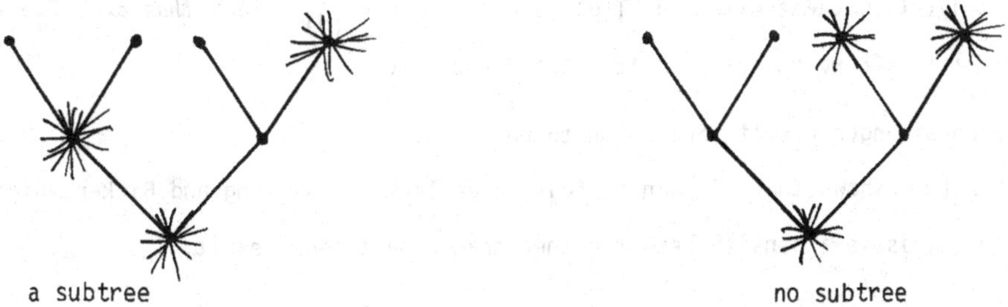

a subtree | no subtree

diagram 2

For trees R and T the binomial coefficient $\binom{T}{R}$ denotes the set of subtrees of T which are isomorphic to R .

Finally we introduce the following convention:

If $g \in \binom{T}{C_k}$ is a k-chain in T , say

$$g = \{g(0),\dots,g(k-1)\}_<$$,

and if $i \in \binom{C_k}{C_\ell}$ is an ℓ-chain in C_k , say

$$i = \{i(0),\dots,i(\ell-1)\}_<$$,

then $g \cdot i \in \binom{T}{C_\ell}$ denotes the ℓ-chain in T which is contained in g exactly in the same way as i is contained in C_k, viz.

$$g \cdot i = \{g(i(0)),\ldots,g(i(\ell-1))\}_< \quad .$$

As a first "application" let us reformulate the Erdös-Rado canonization theorem:

Theorem 1.1* [2] *For every coloring* $\Delta : \binom{T(1)}{C_k} \to \mathbb{N}$, *where* k *is a positive integer, there exists a* $T(1)$*-subtree* $\hat{T} \in \binom{T(1)}{T(1)}$, *i.e.* $\hat{T}$ *is given by an infinite subset of* $T(1)$, *and there exists* $\ell \le k$ *and a subchain* $i \in \binom{C_k}{C_\ell}$ *of* C_k *such that each two* k*-element subchains* $g,h \in \binom{\hat{T}}{C_k}$ *of* $\hat{T}$ *are colored the same iff* $g \cdot i = h \cdot i$.

Before we state the main result of this paper, namely a canonization result for colorings $\Delta : \binom{T(d)}{C_k} \to \mathbb{N}$, let us study partition properties of the trees $T(d)$ with respect to colorings of chains.

First a positive result:

Theorem 2.1 [3] *Let* d *and* δ *be positive integers and let* $\Delta : \binom{T(d)}{C_1} \to \delta$ *be a coloring. Then there exists a* $T(d)$*-subtree* $\hat{T} \in \binom{T(d)}{T(d)}$ *such that all* C_1*-subchains, i.e. all points of* $\hat{T}$ *are colored the same.*

Here even stronger results are known to be valid:

Milliken [3] shows that $\hat{T}$ can be found even level-preserving and Bicker, Voigt [1] show that this is a density result rather than a partition result.

For k-chains with $k > 1$ one obtains negative results:

Theorem 2.2 *Let* d *and* k *be positive integers larger than* 1. *Then there exists a coloring* $\Delta : \binom{T(d)}{C_k} \to 2$, *such that every* $T(d)$*-subtree* $\hat{T} \in \binom{T(d)}{T(d)}$ *contains* k*-chains* $g,h \in \binom{\hat{T}}{C_k}$ *that are colored differently.*

The reason for the negative result 2.2 is that to each C_k-chain in $T(d)$ there may be associated a type in such a way, that types are hereditary under subtrees. Let us visualize this for the particular case $d = 2$ and $k = 2$:

Proof of 2.2 for $d = 2, k = 2$: A 2-chain $g \in \binom{T(2)}{C_2}$ is given by two 0-1 se-

quences $a = (a_0,\dots,a_{m-1})$ and $b = (b_0,\dots,b_{n-1})$, where $m < n$ and $a_\nu = b_\nu$ for all $\nu < m$. Let us call the chain $\{a,b\}$ a "chain of type 0" iff $b_m = 0$ and let us call $\{a,b\}$ "chain of type 1" otherwise, i.e. iff $b_m = 1$.
Compare the following diagram 3.

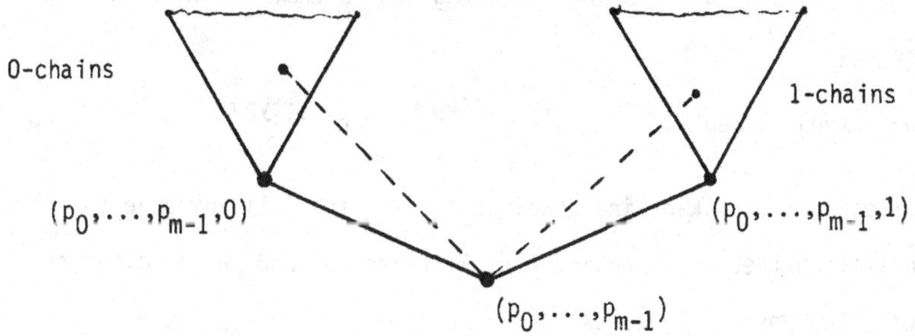

diagram 3

Finally let $\Delta(g) = 0$ iff g is a chain of type 0 and $\Delta(g) = 1$ iff g is a chain of type 1. Obviously Δ has the desired properties. □

Next we use the ideas of the preceding proof in order to associate to each chain in $T(d)$ a type, where a type of k-chains is a $\{0,\dots,d-1\}$ - sequence of length $k-1$, i.e. an element of d^{k-1} .

Definition: Let $g \in \binom{T(d)}{C_k}$ be a k-chain in $T(d)$, say

$$g = \{(a_0,\dots,a_{m_1-1}) , (a_0,\dots,a_{m_2-1}) ,\dots, (a_0,\dots,a_{m_k-1})\} ,$$

where $m_1 < m_2 <\dots< m_k$. Then define

$$\mathrm{typ}(g) = (a_{m_1},\dots,a_{m_{k-1}}) \in d^{k-1} ,$$

where particularly $\mathrm{typ}(g) = \emptyset$ for $g \in \binom{T(d)}{C_1}$.

The next result shows that playing around with the types of k-chains in $T(d)$ is the only possibility in order to get negative results like 2.2 , viz.

Theorem 2.3 *Let δ,k and d be positive integers. Then for every coloring $\Delta : \binom{T(d)}{C_k} \to \delta$ there exists a $T(d)$ - subtree $\hat{T} \in \binom{T(d)}{T(d)}$ such that each two k-chains $g,h \in \binom{\hat{T}}{C_k}$ of same type (i.e. $\mathrm{typ}\,g = \mathrm{typ}\,h$) are colored the same.*

Now we can state the main result of this paper:

Theorem 2.4 *Let* $\Delta : \binom{T(d)}{C_k} \to \mathbb{N}$ *be a coloring, where* d *and* k *are positive integers. Then there exists a* T(d)-*subtree* $\hat{T} \in \binom{T(d)}{T(d)}$, *for every type* $\hat{\xi} \in d^{k-1}$ *there exist an integer* $\ell(\hat{\xi})$ *and a subchain* $i^{(\hat{\xi})} \in \binom{C_k}{C_{\ell(\hat{\xi})}}$ *and there exists an equivalence relation* π *on* d^{k-1} *such that every two* k-*chains* $g,h \in \binom{\hat{T}}{C_k}$ *are colored the same iff*

$$\text{typ}\, g \approx \text{typ}\, h \pmod{\pi} \quad \textit{and} \quad g \cdot i^{(\text{typ}\, g)} = h \cdot i^{(\text{typ}\, h)} \quad .$$

This result implies that for k-chains of given type $\hat{\xi}$ the full analogue to the Erdös-Rado canonization theorem is valid. For k-chains g and h of different type the following two possibilities exist:

- either all chains of type typ g are colored differently from those chains of type typ h (i.e. $\text{typ}\, g \not\approx \text{typ}\, h \pmod{\pi}$)
- or eventually these chains are colored the same (i.e. $\text{typ}\, g \approx \text{typ}\, h \pmod{\pi}$) , viz. $\Delta(g) = \Delta(h)$ iff $g \cdot i^{(\text{typ}\, g)} \approx h \cdot i^{(\text{typ}\, h)}$.

§ 3 Partition results

Convention: For the remainder of this paper let d be a fixed positive integer. The d-regular tree T(d) is abbreviated by T .

Notation: For an element x of T the maximal subtree of T which is rooted in x , i.e. the subtree given by $\{z \in T \mid x \leq z\}$ is denoted by T_x .
For subtrees $\hat{T} \in \binom{T}{T}$ containing x the expression '$\text{ISuc}_{\hat{T}}(x)$' denotes the set of immediate successors of x with respect to the tree $\hat{T}$.

Analogously $\text{ISuc}_{\hat{T}}(S) = \cup \{\text{ISuc}_{\hat{T}}(x) \mid x \in S\} \setminus S$ for subsets $S \subseteq \hat{T}$.
We shall omit the subscript $\hat{T}$ when no confusion can arise.
For subtrees $R \in \binom{\hat{T}_x}{T}$ we denote by "$\hat{T}(\hat{T}_x \to R)$" the tree which is obtained from $\hat{T}$ by replacing $\hat{T}_x$ by R , i.e. $\hat{T}(\hat{T}_x \to R)$ is obtained from $\hat{T}$ by deleting all

elements $\hat{T}_x \setminus R$.

Analogously, we denote by $\hat{T}(\hat{T}_y \to R(y) \mid y \in \mathrm{ISuc}_{\hat{T}}(S))$ the tree which is obtained from $\hat{T}$ by replacing every $\hat{T}_y$ by $R(y)$.

Let $\hat{\xi} \in d^{k-1}$ be a type. Then $\binom{T}{C_k, \hat{\xi}} = \{g \in \binom{T}{C_k} \mid \mathrm{typ}\ g = \hat{\xi}\}$ denotes the set of k-chains of T which have type $\hat{\xi}$.

<u>Lemma 3.1</u> *Let* δ *and* k *be positive integers and let* $\hat{\xi} \in d^{k-1}$ *be a type. Then for every coloring* $\Delta : \binom{T}{C_k, \hat{\xi}} \to \delta$ *there exists a* T-*subtree* $\hat{T} \in \binom{T}{T}$ *such that* $\Delta(g) = \Delta(h)$ *for all* k-*chains* $g, h \in \binom{\hat{T}}{C_k, \hat{\xi}}$.

<u>Proof:</u> Proceed by induction on k. The case $k = 1$ has been established by Milliken [3], see also [1]. Assume that 3.1 is valid for some k. We prove it for $k+1$. Let $\hat{\xi} \in d^k$.

The crucial point is how to reduce the problem to the case $k = 1$, this invokes the inductive assumption.

(3.1.1) *For every coloring* $\Delta : \binom{T}{C_{k+1}, \hat{\xi}} \to \delta$ *there exists a* T-*subtree* $\hat{T} \in \binom{T}{T}$ *such that* $\Delta(g) = \Delta(h)$ *for all* (k+1)-*chains* $g, h \in \binom{T}{C_{k+1}, \hat{\xi}}$ *with* $\min g = \min h$.

This may be proved by a straightforward recursive construction which is based on the following observation:

(3.1.2) *Let* $S \subseteq T$ *be a finite downward closed set and let* $q \in \mathrm{ISuc}(S)$. *For every coloring* $\Delta : \binom{T}{C_{k+1}, \hat{\xi}} \to \delta$ *there exists a subtree* $\hat{T} \in \binom{T_q}{T}$ *such that* $T^* = T(T_q \to \hat{T})$ *satisfies:*

the element root $\hat{T}$, *which plays the role of* q *in* T^* *satisfies:*

$\Delta(g) = \Delta(h)$ *for all* $g, h \in \binom{T^*}{C_{k+1}, \hat{\xi}}$ *with* $\mathrm{typ}\ g = \mathrm{typ}\ h$ *and* $\min g = \min h =$ root $\hat{T}$.

Assertion (3.1.2) follows immediately from the inductive assumption on k: let $\hat{\xi}' = (\xi_1, \ldots, \xi_{k-1}) \in d^{k-1}$ where $\hat{\xi} = (\xi_0, \ldots, \xi_{k-1}) \in d^k$ and let r be the ξ_0-th immediate successor of q. More precisely, say that $q = (a_0, \ldots, a_{m-1})$ with respect to the representation of $T = T(d)$, then let $r = (a_0, \ldots, a_{m-1}, \xi_0)$. Consider the coloring $\Delta^* : \binom{T_r}{C_k, \hat{\xi}'} \to \delta$ which is defined as

$$\Delta^*(g) = \Delta(q \circledast g) \quad ,$$

where $q \circledast g$ is the $(k+1)$-chain with minimal element q and k-tail g. By the inductive assumption there exists a T-subtree $\hat{T}' \in \binom{T_r}{T}$ such that $\Delta^*(g) = \Delta^*(h)$ for all $g,h \in \binom{\hat{T}'}{C_k, \hat{\xi}'}$. Obviously then $T^* = T(T_r \to \hat{T}')$ has the desired properties.

Once we have established (3.1.1) we can restrict our consideration to colorings $\Delta : \binom{T}{C_{k+1}, \hat{\xi}} \to \delta$ such that $\Delta(g) = \Delta(h)$ for all $g,h \in \binom{T}{C_{k+1}, \hat{\xi}}$ with $\min g = \min h$. This induces the coloring $\Delta^* : \binom{T}{C_1} \to \delta$ with $\Delta^*(x) = \Delta(g)$ for any $g \in \binom{T}{C_{k+1}, \hat{\xi}}$ with $x = \min g$. The assertion then follows from the case $k = 1$. □

<u>Remark 3.2</u> The reader is asked to recall how assertion (3.1.1) has been proved. The basic idea was to obtain $\hat{T}$ by a recursive construction - which has not been carried out explicitly. However, the main tool for that recursive construction is provided by (3.1.2) which implies that any downward closed finite configuration $S \subseteq T$ may be extended to a strictly larger configuration such that the new points - here the point q, resp. the point $\operatorname{root} \hat{T}$ playing the role of q - satisfy some property P. Provided this property is hereditary under subtrees, viz. once $x \in T$ has this property with respect to T then also x has this property with respect to every subtree $\hat{T}$ of T containing x, one can easily construct a T-subtree T^* of T such that every point of T^* satisfies this property. All what is left is to prove assertion (3.1.2). This idea will be tacitly used many times throughout this paper.

The next lemma is an immediate but useful application of Lemma 3.1:

<u>Lemma 3.3</u> *Let* $\hat{\xi} \in d^{k-1}$ *be a type and let* $W \subseteq [\mathbb{N}]^{<\omega}$ *be a finite set of nonnegative integers. Then for every one-to-one coloring* $\Delta : \binom{T}{C_k, \hat{\xi}} \to \mathbb{N}$ *there exists a* T*-subtree* $\hat{T} \in \binom{T}{T}$ *such that* $\Delta(g) \notin W$ *for every* $g \in \binom{\hat{T}}{C_k, \hat{\xi}}$.

<u>Proof:</u> Consider the coloring $\Delta^* : \binom{T}{C_k, \hat{\xi}} \to W \cup \{*\}$ which is defined as

$$\begin{aligned} \Delta^*(g) &= \Delta(g) \quad \text{if} \quad \Delta(g) \in W \\ &= * \quad\quad \text{if} \quad \Delta(g) \notin W \quad . \end{aligned}$$

By Lemma 3.1 there exists a monochromatic $\hat{T} \in \binom{T}{T}$, but as Δ is one-to-one $\hat{T}$ is necessarily monochromatic in color '*' . □

Lemma 3.4 *Let* $k \leq \ell$ *be positive integers and let* $\hat{\xi} \in d^{k-1}$ *, resp.* $\hat{\zeta} \in d^{\ell-1}$ *be types. Let* T^0 *and* T^1 *be* $T(d)$ *- trees.*
Then for every pair $\Delta_0 : \binom{T^0}{C_k, \hat{\xi}} \to \mathbb{N}$ *and* $\Delta_1 : \binom{T^1}{C_\ell, \hat{\zeta}} \to \mathbb{N}$ *of one-to-one colorings there exist* $T(d)$ *- subtrees* $\hat{T}^0 \in \binom{T^0}{T}$ *, resp.* $\hat{T}^1 \in \binom{T^1}{T}$ *such that* $\Delta_0(g) \neq \Delta_1(h)$ *for every* $g \in \binom{\hat{T}^0}{C_k, \hat{\xi}}$ *and* $h \in \binom{\hat{T}^1}{C_\ell, \hat{\zeta}}$ *.*

Proof: The proof can be performed by a zig-zag recursive construction using Lemma 3.3. At even steps the tree T^0 is constructed and at odd steps the tree T^1 is constructed. The recursive construction follows the pattern as given in remark 3.2. We perform the main tool for the even steps:

Let $S^0 \subseteq T^0$ be a finite downward closed set and let $q \in \mathrm{ISuc}(S^0)$, additionally let $S^1 \subseteq T^1$ be a finite downward closed set. We show that there exists a subtree $\hat{T}^1 \in \binom{T^1}{T}$ with $S^1 \subseteq \hat{T}^1$ such that $\Delta_0(g) \neq \Delta_1(h)$ for every $g \in \binom{S^0 \cup \{q\}}{C_k, \hat{\xi}}$ with $\max g = q$ and every $h \in \binom{\hat{T}^1}{C_\ell, \hat{\zeta}}$ with $\max h \notin S^1$.

Let $W = \{\Delta_0(g) \mid g \in \binom{S^0 \cup \{q\}}{C_k, \hat{\xi}}$ and $\max g = q\}$. By Lemma 3.3 for every immediate successor $r \in \mathrm{ISuc}(S^1)$ there exists a T-subtree $T^{(r)} \in \binom{T^1_r}{T}$ such that $\Delta_1(h) \notin W$ for every $h \in \binom{T^1}{C_\ell, \hat{\zeta}}$ with $\max h \in T^{(r)}$.
Thus $\hat{T}^1 = T^1\{T^1_r \to T^{(r)} \mid r \in \mathrm{ISuc}(S^1)\}$ has the desired properties. □

§ 4 Diversification

The result of this section (viz. Theorem 4.1) is the combinatorial core of our proof. Loosely speaking it states that given a pair $\Delta_1 : \binom{T}{C_k, \hat{\xi}} \to \mathbb{N}$ and $\Delta_2 : \binom{T}{C_\ell, \hat{\zeta}} \to \mathbb{N}$ of one-to-one colorings there exists a T-subtree $\hat{T} \in \binom{T}{T}$ such that the restrictions $\Delta_1 \upharpoonright \binom{\hat{T}}{C_k, \hat{\xi}}$ and $\Delta_2 \upharpoonright \binom{\hat{T}}{C_\ell, \hat{\zeta}}$ have images that are as disjoint as possible, viz. $\Delta_1(g) = \Delta_2(h)$ implies that $g = h$. This will help us to construct the equivalence relation π as stated in Theorem 2.4 .

<u>Theorem 4.1</u> *Let* $k \leq \ell$ *be positive integers and let* $\hat{\xi} \in d^{k-1}$ *and* $\hat{\zeta} \in d^{\ell-1}$ *be types of* k *, resp. of* ℓ*-chains. Then for every pair* $\Delta_1 : \binom{T}{C_k, \hat{\xi}} \to \mathbb{N}$ *and* $\Delta_2 : \binom{T}{C_\ell, \hat{\zeta}} \to \mathbb{N}$ *of one-to-one mappings there exists a* T*-subtree* $\hat{T} \in \binom{T}{T}$ *such that* $\Delta_1(g) \neq \Delta_2(h)$ *for all* $g \in \binom{\hat{T}}{C_k, \hat{\xi}}$ *and* $h \in \binom{\hat{T}}{C_\ell, \hat{\zeta}}$ *with* $g \neq h$.

We need two preliminary lemmas:

<u>Notation:</u> Let $k \leq \ell$ be positive integers and let $h \in \binom{T}{C_\ell}$. By "$h^{<k>}$" we denote the k-tail of h , i.e. the subchain of h consisting of the last k elements of h , viz. $h^{<k>} = \{h(\ell-1-k) , h(\ell-k) ,\dots, h(\ell-1)\}$.

<u>Lemma 4.2</u> *Let* $k < \ell$ *be positive integers and let* $\hat{\zeta} \in d^{\ell-1}$ *be a type of* ℓ*-chains.*
Then for every pair $\Delta_1 : \binom{T}{C_k} \to \mathbb{N}$ *and* $\Delta_2 : \binom{T}{C_\ell, \hat{\zeta}} \to \mathbb{N}$ *of one-to-one mappings there exists a* T*-subtree* $\hat{T} \in \binom{T}{T}$ *such that* $\Delta_1(h^{<k>}) \neq \Delta_2(h)$ *for every* $h \in \binom{\hat{T}}{C_\ell, \hat{\zeta}}$.

<u>Proof:</u> Let $S \subseteq T$ be a downward closed finite set and let $q \in ISuc(S)$. We show that there exists a T-subtree $\hat{T} \in \binom{T_q}{T}$ such that $\Delta_1(h^{<k>}) \neq \Delta_2(h)$ for every $h \in \binom{\hat{T}}{C_\ell, \hat{\zeta}}$ with $\min h = \text{root } \hat{T}$.

Let r be the ζ_0 - th immediate successor of q , let $\hat{\zeta}' = (\zeta_1,\dots,\zeta_{\ell-2}) \in d^{\ell-2}$ be the type of $(\ell-1)$-chains consisting of the last $(\ell-2)$ elements of $\hat{\zeta}$. Consider the coloring $\Delta^* : \binom{T_r}{C_{\ell-1}, \hat{\zeta}'} \to \{0,1\}$ which is defined by

$$\begin{aligned} \Delta^*(h) &= 0 \quad \text{if} \quad \Delta_1(h^{<k>}) = \Delta_2(q \circledast h) \\ &= 1 \quad \text{if} \quad \Delta_1(h^{<k>}) \neq \Delta_2(q \circledast h) \quad , \end{aligned}$$

where '$\circledast$' refers to the concatenation of chains.

By Lemma 3.1 there exists a T-subtree $\hat{T} \in \binom{T_r}{T}$ with all its $C_{\ell-1}$ - subchains of tpye $\hat{\zeta}'$ monochromatic in color 0 or monochromatic in color 1 . We consider these two cases separately:

- there exists a T-subtree $\hat{T} \in \binom{T_r}{T}$ which is monochromatic in color 0 . Then $\Delta_1(h^{<k>}) \neq \Delta_2(h)$ for every $h \in \binom{\hat{T}}{C_\ell, \hat{\zeta}}$, because otherwise $\Delta_2(h) = \Delta_2(q \circledast h^{<\ell-1>})$, contradicting that Δ_2 is a one-to-one coloring.

- there exists a T-subtree $\hat{T}' \in \binom{T_r}{T}$ which is monochromatic in color 1 . Then obviously $\hat{T} = T_q(T_r \to \hat{T}')$ has the desired properties. □

Lemma 4.3 *Let* $2 \leq k \leq \ell$ *be positive integers. Let* $\hat{\xi} \in d^{k-2}$, $\hat{\zeta} \in d^{\ell-k}$ *and let* $\varphi, \Theta \in d$. *Then for every pair* $\Delta_1 : \binom{T}{C_k, \varphi \circledast \hat{\xi}} \to \mathbb{N}$ *and* $\Delta_2 : \binom{T}{C_\ell, \hat{\zeta} \circledast \Theta \circledast \hat{\xi}} \to \mathbb{N}$ *of one-to-one mappings*

- *where* "$\circledast$" *refers to concatenation of sequences, viz.* $\varphi \circledast \hat{\xi} \in d^{k-1}$ *and* $\hat{\zeta} \circledast \Theta \circledast \hat{\xi} \in d^{\ell-1}$ -

there exists a T-*subtree* $\hat{T} \in \binom{T}{T}$ *such that* $\Delta_1(g) \neq \Delta_2(f \circledast g^{<k-1>})$ *for every* $g \in \binom{\hat{T}}{C_k, \varphi \circledast \hat{\xi}}$ *and* $f \in \binom{\hat{T}}{C_{\ell-k+1}, \hat{\zeta}}$ *with* max f < min g *and* $f \circledast g^{<k-1>} \in \binom{\hat{T}}{C_\ell, \hat{\zeta} \circledast \Theta \circledast \hat{\xi}}$

Proof: Let $S \subseteq T$ be a downward closed finite subset and let $q \in ISuc(S)$. Let $C = \{g \in \binom{S}{C_{\ell-k+1}, \hat{\zeta}} \mid \max g < q\}$ be the set of $(\ell-k+1)$-chains of type $\hat{\zeta}$ occuring strictly below q . We show that there exists a T-subtree $\hat{T} \in \binom{T_q}{T}$ such that $\Delta_1(g) \neq \Delta_2(f \circledast g^{<k-1>})$ for every $g \in \binom{\hat{T}}{C_k, \varphi \circledast \hat{\xi}}$ with min g = root $\hat{T}$ and every $f \in C$ with $f \circledast g^{<k-1>} \in \binom{T}{C_\ell, \hat{\zeta} \circledast \Theta \circledast \hat{\xi}}$.

Let r be the φ.th immediate successor of q . Consider the coloring $\Delta : \binom{T_r}{C_{k-1}, \hat{\xi}} \to P(C)$ which is defined by

$$\Delta(g) = \{f \in C \mid f \circledast g \in \binom{T}{C_\ell, \hat{\zeta} \circledast \Theta \circledast \hat{\xi}} \text{ and } \Delta_1(q \circledast g) = \Delta_2(f \circledast g)\} \quad .$$

By Lemma 3.1 there exists a T-subtree $\hat{T}' \in \binom{T_r}{T}$ which is monochromatic in some color $C^* \subseteq C$.

We consider two cases separately:

- there exists a T-subtree $\hat{T}' \in \binom{T_r}{T}$ which is monochromatic in color $\emptyset$. Then obviously $\hat{T} = T_q(T_r \to \hat{T}')$ has the desired properties.

- there exists a T-subtree $\hat{T}' \in \binom{T_r}{T}$ which is monochromatic in color $C^* \subseteq C$, where $C^* \neq \emptyset$. Then $\Delta_1(g) \neq \Delta_2(f \circledast g^{<k-1>})$ for every $g \in \binom{\hat{T}'}{C_k, \varphi \circledast \hat{\xi}}$ and every $f \in C^*$, because otherwise $\Delta_1(g) = \Delta_1(q \circledast g^{<k-1>})$, contradicting that Δ_1 is a one-to-one coloring.

  Consider $T^* = T(T_q \to \hat{T}')$ and apply the same argument as before, but now with

$C \setminus C^*$ instead of C and with root $\hat{T}'$ instead of q. After finitely many steps, i.e. by induction, the process necessarily ends with some T-subtree which is monochromatic in color $\emptyset$. □

Proof of Theorem 4.1: We proceed by induction on k.

First we consider the case $k = 1$, i.e. $\hat{\xi} = \emptyset$. Let $S \subseteq T$ be a downward closed finite set and let $q \in ISuc(S)$. By Lemma 4.2 we can assume that

(4.1.1) $\Delta_1(h^{<1>}) \neq \Delta_2(h)$ for every $h \in \binom{T}{C_{\ell},\hat{\zeta}}$.

We show that there exist T-subtrees $T(x) \in \binom{T_x}{T}$, $x \in ISuc(S)$, such that $\hat{T} = T(T_x \rightarrow T(x) \mid x \in ISuc(S))$ satisfies that

$$\Delta_1(\min T(q)) \neq \Delta_2(h) \quad \text{for every} \quad h \in \binom{T}{C_{\ell},\hat{\zeta}} \quad .$$

Let $\hat{q} \in T_q$ be such that

(4.1.2) $\Delta_1(\hat{q}) \neq \Delta_2(h)$ for every $h \in \binom{S}{C_{\ell},\hat{\zeta}}$.

For every $x \in ISuc(S \cup \{\hat{q}\} \setminus \{q\})$ let $T(x) \in \binom{T_x}{T}$ be such that

(4.1.3) $\Delta_1(\hat{q}) \neq \Delta_2(f \circledast g)$ for every $j < \ell$, $f \in \binom{S \cup \{\hat{q}\} \setminus \{q\}}{C_j,(\zeta_0,\ldots,\zeta_{j-2})}$, and every $g \in \binom{T_x}{C_{\ell-j},(\zeta_j,\ldots,\zeta_{\ell-2})}$ with $f \circledast g \in \binom{T}{C_{\ell},\hat{\zeta}}$.

Such trees exist according to Lemma 3.3.

By (4.1.1), (4.1.2) and (4.1.3) then $\hat{T} = T(T_x \rightarrow T(x) \mid x \in ISuc(S))$, where $T(q) = T_{\hat{q}}(T_x \rightarrow T(x) \mid x \in ISuc(\hat{q}))$, has the desired properties.

Next comes the inductive step. Assume that the assertion of Theorem 4.1 is valid for every $j < k$, where k is a positive integer larger than 1. Let $S \subseteq T$ be a downward closed finite set and let $q \in ISuc(S)$. We show that there exist T-subtrees $T(x) \in \binom{T_x}{T}$, $x \in ISuc(S)$, such that $\hat{T} = T(T_x \rightarrow T(x) \mid x \in ISuc(S))$ satisfies that

$$\Delta_1(g) \neq \Delta_2(h) \quad \text{for every} \quad g \in \binom{\hat{T}}{C_k,\hat{\xi}} \quad \text{with} \quad \min g = \text{root}\, T(q)$$
$$\text{and for every} \quad h \in \binom{\hat{T}}{C_{\ell},\hat{\zeta}} \quad .$$

By Lemma 4.2 and Lemma 4.3 we can assume that

(4.1.4) $\Delta_1(g) \neq \Delta_2(f \circledast g^{<k-1>})$

for every $g \in \binom{T}{C_k,\hat{\xi}}$ with $q = \min g$ and every $f \in \binom{T}{C_{\ell-k+1}}$ with $\max f < q$ and $f \circledast g^{<k-1>} \in \binom{T}{C_\ell,\hat{\xi}}$.

Let r be the ξ_0-th immediate successor of q . Let $T^1 \in \binom{T_r}{T}$ be such that

(4.1.5) $\Delta_1(q \circledast g) \neq \Delta_2(h)$ for every $g \in \binom{T^1}{C_{k-1},(\xi_1,\ldots,\xi_{k-2})}$ and for every $h \in \binom{S \cup \{q\}}{C_\ell,\hat{\xi}}$.

Such a tree T^1 exists according to Lemma 3.3 . Next let $T^2 \in \binom{T^1}{T}$ be such that

(4.1.6) $\Delta_1(q \circledast g) \neq \Delta_2(f \circledast h)$ for every $g \in \binom{T^2}{C_{k-1},(\xi_1,\ldots,\xi_{k-2})}$ and for every $j < \ell$, $f \in \binom{S \cup \{q\}}{C_j,(\zeta_0,\ldots,\zeta_{j-2})}$ and $h \in \binom{T^2}{C_{\ell-j},(\zeta_j,\ldots,\zeta_{\ell-2})}$ with $f \circledast h \in \binom{T}{C_\ell,\hat{\zeta}}$.

Such a tree T^2 exists according to the inductive assumption.

Finally let $T^3 \in \binom{T^2}{T}$ and for each $x \in \mathrm{ISuc}(S) \setminus \{q\}$ let $T(x) \in \binom{T_x}{T}$ be such that

(4.1.7) $\Delta_1(q \circledast g) \neq \Delta_2(f \circledast h)$ for every $g \in \binom{T^3}{C_{k-1},(\xi_1,\ldots,\xi_{k-2})}$ and for every $j < \ell$, $f \in \binom{S \cup \{q\}}{C_j,(\zeta_0,\ldots,\zeta_{j-2})}$ and $h \in \binom{T(x)}{C_{\ell-j},(\zeta_j,\ldots,\zeta_{\ell-2})}$ with $f \circledast h \in \binom{T}{C_\ell,\hat{\zeta}}$.

Such a tree T^3 , resp. such trees $T(x)$ exist according to Lemma 3.4 .

By (4.1.4) up to (4.1.7) then $\hat{T} = T(T_x \to T(x) \mid x \in \mathrm{ISuc}(S))$, where $T(q) = T_q(T_r \to T^3)$, has the desired properties. □

§ 5 Proof of the main theorem

First we prove a special case of Theorem 2.3:

Theorem 5.1 *Let* $\hat{\xi} \in d^{k-1}$ *be a type of* k*-chains. Then for every coloring* $\Delta : \binom{T}{C_k,\hat{\xi}} \to \mathbb{N}$ *there exists a* T*-subtree* $\hat{T} \in \binom{T}{T}$ *and there exists* $\ell \leq k$ *and a subchain* $i \in \binom{C_k}{C_\ell}$ *such that every two* k*-chains* $g,h \in \binom{T}{C_k,\hat{\xi}}$ *are colored the same iff* $g \cdot i = h \cdot i$.

<u>Proof:</u> We proceed by induction on k. First we prove the case $k = 1$. Recall that the theorem asserts that $\hat{T}$ is colored monochromatically or that $\hat{T}$ is colored one-to-one.

Thus let $S \subseteq T$ be a downward closed finite subset and let $q \in ISuc(S)$. We show that there exists a T-subtree $\hat{T} \in \binom{T_q}{T}$ such that either $\hat{T}$ is colored monochromatically or such that $\Delta(x) \neq \Delta(y)$ for every $x \in \hat{T} = \binom{\hat{T}}{C_1}$ and every $y \in S = \binom{S}{C_1}$.

Consider the coloring $\Delta^* : \binom{T_q}{C_1} \to \{\Delta(y) | y \in S\} \cup \{*\}$ with $\Delta^*(x) = \Delta(x)$ if $\Delta(x) \in \{\Delta(y) | y \in S\}$ and $\Delta^*(x) = *$ if $\Delta(x) \notin \{\Delta(y) | y \in S\}$.

According to Lemma 3.1 there exists a monochromatic $\hat{T} \in \binom{T}{T}$ which obviously has the desired properties.

Next assume that 5.1 is valid for some k. We prove it for $k + 1$. Let $\hat{\xi} = (\xi_0, \ldots, \xi_{k-1}) \in d^k$ and let $\hat{\xi}' = (\xi_1, \ldots, \xi_{k-1}) \in d^{k-1}$ consist of the last $(k-1)$ entries in $\hat{\xi}$.

(5.1.1) *There exists a* T-*subtree* $\hat{T}^1 \in \binom{T}{T}$ *such that for every* $x \in \hat{T}^1$ *there exists an* $i(x) \in \cup \{\binom{C_k}{C_\ell} | \ell \leq k\}$ *satisfying that every two* $(k+1)$-*chains* $g, h \in \binom{\hat{T}^1_x}{C_{k+1}, \hat{\xi}}$ *with* $\min g = \min h = x$ *are colored the same iff* $g^{<k>} \cdot i(x) = h^{<k>} \cdot i(x)$.

Let $S \subseteq T$ be a downward closed finite set and let $q \in ISuc(S)$. Let r be the ξ_0-th immediate successor of q and consider the coloring $\Delta^* : \binom{T_r}{C_k, \hat{\xi}'} \to \mathbb{N}$ with $\Delta^*(g) = \Delta(q \circledast g)$. According to the inductive assumption on k there exists a subchain $i(q) \in \cup \{\binom{C_k}{C_\ell} | k \leq \ell\}$ and there exists a T-subtree $T^* \in \binom{T_r}{T}$ such that every two k-subchains $g, h \in \binom{T^*}{C_k, \hat{\xi}'}$ are colored the same under the coloring Δ^* iff $g \cdot i(q) = h \cdot i(q)$.

Since $\cup \{\binom{C_k}{C_\ell} | \ell \leq k\}$ is a finite set by Lemma 3.1 the following assertion is an immediate consequence of (5.1.1):

(5.1.2) *There exists a* T-*subtree* $\hat{T}^2 \in \binom{\hat{T}^1}{T}$ *and there exists a subchain* $i' \in \cup \{\binom{C_k}{C_\ell} \ell \leq k\}$ *such that every two* $(k+1)$-*chains* $g, h \in \binom{\hat{T}^2}{C_{k+1}, \hat{\xi}}$ *with* $\min g = \min h$ *are colored the same iff* $g^{<k>} \cdot i' = h^{<k>} \cdot i'$.

(5.1.3) *There exists a T-subtree* $\hat{T}^3 \in \binom{\hat{T}^2}{T}$ *such that* $g^{<k>} \cdot i' = h^{<k>} \cdot i'$ *holds whenever* $g,h \in \binom{\hat{T}^3}{C_{k+1},\hat{\xi}}$ *are* $(k+1)$*-chains in* $\hat{T}^3$ *which are colored the same.*

Let S be a downward closed finite set and let $q \in \mathrm{ISuc}(S)$. We show that for every $x \in \mathrm{ISuc}(S\cup\{q\})$ there exists a T-subtree $T(x) \in \binom{\hat{T}^2}{T}$ such that

$g^{<k>} \cdot i' = h^{<k>} \cdot i'$ holds whenever $g \in \binom{\hat{T}^2_q}{C_{k+1},\hat{\xi}}$ with $\min g = q$ and $h \in \binom{T^2(T_x \to t(x)\,|\,x \in \mathrm{ISuc}(S\cup\{q\}))}{C_{k+1},\hat{\xi}}$ are colored the same, i.e whenever $\Delta(g) = \Delta(h)$.

Let r be the ξ_0 - th immediate successor of q . Let $T'(r) \in \binom{\hat{T}^2}{T}$ be such that

$g \cdot i' = (f \circledast h)^{<k-1>} \cdot i'$ holds whenever $g \in \binom{T'(r)}{C_k,\hat{\xi}'}$, $f \in \binom{S}{C_j,(\xi_0,\dots,\xi_{j-2})}$, $h \in \binom{T'(r)}{C_{k+1-j},(\xi_j,\dots,\xi_{k-1})}$ with $f \circledast h \in \binom{T}{C_{k+1},\hat{\xi}}$, where $j \leq k+1$, are such that $\Delta(q \circledast g) = \Delta(f \circledast h)$.

Such a tree $T'(r)$ exists according to repeated applications of Lemma 3.3 and Theorem 4.1 .

Next let $T(x) \in \binom{\hat{T}^2}{T}$ for $x \in \mathrm{ISuc}(S\cup\{q\}) \setminus \{r\}$ and let $T(r) \in \binom{T'(r)}{T}$ be such that

$\Delta(q \circledast g) \neq \Delta(f \circledast h)$ for every $g \in \binom{T'(r)}{C_k,\hat{\xi}'}$, $f \in \binom{S}{C_j,(\xi_0,\dots,\xi_{j-2})}$, $h \in \binom{T(x)}{C_{k+1-j},(\xi_j,\dots,\xi_{k-1})}$ with $f \circledast h \in \binom{T}{C_{k+1},\hat{\xi}}$, where $j \leq k$.

Such trees exist to repeated applications of Lemma 3.4 . Then obviously the trees $T(x)$, $x \in \mathrm{ISuc}(S\cup\{q\})$ have the desired properties.

(5.1.4) *There exists a T-subtree* $\hat{T}^4 \in \binom{\hat{T}^3}{T}$ *such that for every 2-chain* $f \in \binom{\hat{T}^4}{C_2,(\xi_0)}$ *one of the following two alternatives is valid:*

- $\Delta(f(0) \circledast g) = \Delta(f(1) \circledast g)$ *for every* $g \in \binom{\hat{T}^4}{C_k,\hat{\xi}'}$ *with* $f(0) \circledast g \in \binom{\hat{T}^4}{C_{k+1},\hat{\xi}}$ *and* $f(1) \circledast g \in \binom{\hat{T}^4}{C_{k+1},\hat{\xi}}$.

Let $S \subseteq \hat{T}^3$ be a downward closed finite subset and let $q \in \mathrm{ISuc}(S)$. Let r be the ξ_0 - th immediate successor of q .

We show that there exists a T-subtree $T^* \in \binom{\hat{T}^3_r}{T}$ such that for every 2-chain

$f \in \binom{SU\{q\}}{C_2,(\xi_0)}$ with $\max f = q$ either

- $\Delta(f(0) \otimes g) = \Delta(f(1) \otimes g)$ for every $g \in \binom{T^*}{C_k,\hat{\xi}'}$

or

- $\Delta(f(0) \otimes g) \neq \Delta(f(1) \otimes g)$ for every $g \in \binom{T^*}{C_k,\hat{\xi}'}$.

Consider the coloring $\Delta^* : \binom{\hat{T}^3_r}{C_k,\hat{\xi}'} \to P(\{y \in S | y < q\})$ with

$$\Delta^*(g) = \{y \in S | y < q \text{ and } y \otimes g \in \binom{T}{C_{k+1},\hat{\xi}} \text{ and } \Delta(y \otimes g) = \Delta(q \otimes g)\} .$$

According to Lemma 3.1 there exists a monochromatic T-subtree $T^* \in \binom{\hat{T}^3_r}{T}$ which obviously has the desired properties.

Applying 3.1 once more yields the following strenghtening of (5.1.4):

(5.1.5) *There exists a* T-*subtree* $\hat{T} \in \binom{\hat{T}^4}{T}$ *such that one of the following two alternatives is valid:*

(5.1.6) $\Delta(f(0) \otimes g) = \Delta(f(1) \otimes g)$ *for every* $f \in \binom{\hat{T}}{C_2,(\xi_0)}$ *and every* $g \in \binom{\hat{T}}{C_k,\hat{\xi}'}$ *with* $f(1) \otimes g \in \binom{\hat{T}}{C_{k+1},\hat{\xi}}$

or

(5.1.7) $\Delta(f(0) \otimes g) \neq \Delta(f(1) \otimes g)$ *for every* $f \in \binom{\hat{T}}{C_2,(\xi_0)}$ *and every* $g \in \binom{\hat{T}}{C_k,\hat{\xi}'}$ *with* $f(1) \otimes g \in \binom{\hat{T}}{C_{k+1},\hat{\xi}}$.

We claim that $\hat{T}$ satisfies the requirements of assertion 5.1 . However, it remains to define the subchain i .

Say $i' = \{i'(0),\ldots,i'(\ell-1)\}$, where i' has been introduced in (5.1.2) .

If (5.1.6) is valid then let

$$i \in \binom{C_{k+1}}{C_\ell} \text{ be with } i = \{i'(0)+1,\ldots,i'(\ell-1)+1\} ,$$

if (5.1.7) is valid then let

$$i \in \binom{C_{k+1}}{C_{\ell+1}} \text{ be with } i = \{0,i'(0)+1,\ldots,i'(\ell-1)+1\} .$$

By (5.1.3) and (5.1.6), resp. by (5.1.3) and (5.1.7) then $\hat{T}$ and i have the desired properties.

□

<u>Proof of Theorem 2.3</u> By Theorem 5.1 we can assume that for every type $\hat{\xi} \in d^{k-1}$ there exists a nonnegative integer $\ell(\hat{\xi}) \leq k$ and a subchain $i^{(\hat{\xi})} \in \binom{C_k}{C_{\ell(\hat{\xi})}}$ such

that each two k-chains $g,h \in \binom{T}{C_k,\hat{\xi}}$ of type $\hat{\xi}$ are colored the same iff $g \cdot i^{(\hat{\xi})} = h \cdot i^{(\hat{\xi})}$.

By Theorem 4.1 there exists a T-subtree $T' \in \binom{T}{T}$ such that $\Delta(g) = \Delta(h)$ for k-chains $g,h \in \binom{T'}{C_k}$ always implies that $g \cdot i^{(\text{typ}\,g)} = h \cdot i^{(\text{typ}\,h)}$.

It remains to define the equivalence relation π on d^{k-1}.

For every type $\hat{\xi} \in d^{k-1}$ consider the coloring

$$\Delta_{\hat{\xi}} : \binom{T}{C_k,\hat{\xi}} \to P(d^{k-1}) \text{ which is defined as}$$

$$\Delta_{\hat{\xi}}(g) = \{\hat{\zeta} \in d^{k-1} \mid \text{there exists an } h \in \binom{T}{C_k,\hat{\zeta}} \text{ with } \Delta(g) = \Delta(h)\} .$$

By Lemma 3.1 we can assume that $\Delta_{\hat{\xi}}(f) = \Delta_{\hat{\xi}}(g)$ for every $f,g \in \binom{T}{C_k,\hat{\xi}}$. By abuse of language let us denote this common color by $\Delta_{\hat{\xi}}$.

The equivalence relation π is defined via these colorings, viz. put

$$\hat{\xi} \approx \hat{\zeta} \pmod{\pi} \quad \text{iff} \quad \hat{\zeta} \in \Delta_{\hat{\xi}} .$$

Obviously then π together with the family $(i^{(\hat{\xi})} \mid \hat{\xi} \in d^{k-1})$ has the desired properties. □

§ 6 Concluding remark

Using a compactness-argument (e.g. Königs - lemma) Theorem 2.3 implies the following 'finite' version:

Theorem 6.1 *Let* R *be a finite* d*-regular tree and let* k *be a positive integer Then there exists a finite* d*-regular tree* S *such that for every coloring* $\Delta : \binom{S}{C_k} \to \mathbb{N}$ *there exists an* R*-subtree* $\hat{R} \in \binom{S}{R}$ *and for every type* $\hat{\xi} \in d^{k-1}$ *there exists a subchain* $i^{(\hat{\xi})} \in \cup \{\binom{C_k}{C_\ell} \mid \ell \leq k\}$ *and there exists an equivalence relation* π *on* d^{k-1} *such that every two* k*-chains* $g,h \in \binom{\hat{R}}{C_k}$ *are colored the same iff* typ g ≈ typ h (mod π) *and*

$$g \cdot i^{(\text{typ}\,g)} = h \cdot i^{(\text{typ}\,h)} .$$

For $d = 1$ this yields the finite version of Erdös-Rado canonization theorem.

References:

[1] R. Bicker, B. Voigt — A density theorem for finitistic trees, Bielefeld 1982.

[2] P. Erdös, R. Rado — A combinatorial theorem, Journal London Math Soc. 25(1950), 249 - 255.

[3] K. Milliken — A Ramsey theorem for trees, JCT(A) 26(1979), 215 - 237.

[4] F.P. Ramsey — On a problem of formal logic, Proc. London Math. Soc. 30(1930), 264 - 286.

Incidence Algebras, Exponential Formulas and Unipotent Groups

Arne Dür and Ulrich Oberst
Institut für Mathematik der Universität Innsbruck
Innrain 52, A-6020 Innsbruck, Österreich

The main objectives of these notes are the following: (a) to give a new, directly applicable setting for and a new version of the exponential formula (Theorem 3.52) for Krull-Schmidt categories, (b) to extend the combinatorial treatment of partitions of finite sets by means of the "Faa di Bruno" bialgebra (cf. [8], pp.100, and [24], pp.36) to arbitrary sheaf-like categories (Theorem 3.74), and (c) to give an application of homological results to the determination of incidence algebras (§3, section B). The basic references for us, due to G.-C. Rota and his school, have been [8] for incidence algebras and [24] concerning Hopf algebras in combinatorics Other papers in this direction are [6], [18], [29]. Similar and very interesting relations between combinatorics and Hopf algebras (in a different language: affine and formal groups) are contained in [15] and [19]. Detailed proofs will appear elsewhere.

Essentially, a Krull-Schmidt category $\underline{K}$ is one in which the Krull-Schmidt theorem holds: Each object X of $\underline{K}$ has a KS-(Krull-Schmidt) decomposition, i.e. admits a finite direct sum (dually: direct product) decomposition $X = \amalg X_i$ into indecomposables X_i, which is moreover unique up to isomorphism. This type of category is abundant in combinatorics since the necessary finiteness assumptions for the proof of a KS-theorem are trivially satisfied in finite enumeration problems. The simplest examples are the categories of finite sets, vector spaces, topological spaces, ordered sets, graphs and groups. Whole classes of examples are the sheaf-like categories of this paper (see below) generalizing the category of finite sets, and the "finite" abelian categories generalizing the categories of finite vector spaces. In sheaf-like categories the KS-decomposition is unique up to the order of the summands and not only unique up to isomorphism. In combinatorics, this unique KS-decomposition is usually called "the partition of a structure into its connected components". KS-categories give rise to generalized exponential structures, introduced by Stanley [27] as the right frame work for exponential formulas. The corresponding formula of these notes (Theorem 3.52) is an identity of power series with rational coefficients and in variables X_P, where P runs over a (possibly infinite) system $\underline{P}$ of representatives of the isomorphism classes of "connected", i.e. indecomposable objects of a KS-category, and involves the orders of automorphism groups, replacing the customary factorials, as a distinguishing feature. The examples of the literature are obtained for categories with only one indecomposable object, up to isomorphism, such as finite sets and vector spaces respectively (compare [27] and [28]). The simplest standard example is the identity $\exp(\exp(X)-1) = \sum_{n=0}^{\infty} B_n \frac{X^n}{n!}$, where B_n are the Bell-numbers of all partitions of a set with n elements. As an application we obtain the typical relation between numbers of all structures=objects and those of "connected" structures=indecomposable objects over a given base. In §4, section A, we apply the exponential formula to subobjects and equivalence relations respectively in sheaflike categories.

The distinguishing sheaf-axiom of a sheaf-like category $\underline{K}$ - in addition to some standard assumptions on direct sums etc. and a combinatorially obvious finiteness condition-is the universality of finite coproducts: If $f:X \longrightarrow Y$ is a morphism and $Y = \amalg Y_i$ is a finite direct sum(coproduct) decomposition then $X = \amalg f^{-1}(X_i)$ is the direct sum of the inverse images. It is easily seen that this axiom is connected with the distributivity of lattices of subobjects. The basic example is the category Set f of finite sets. Combinatorial standard examples are obtained as the presheaf categories of all functors from a category I with finitely many objects into Set f. If, for instance, I is the monoid with one generator g satisfying a relation $g^m=g^n$, $1 \le m \le n < \infty$, e.g. $g^2=g$ or $g^n=1$, then one obtains the category of all pairs (M,s) of a finite set M with an endomorphism s satisfying the same relation. Of course, all categories of sheaves or toposes, e.g. that of finite sets with a group operation, but also combinatorially interesting categories like ordered sets, graphs etc. are sheaf-like. However, the latter are no toposes. The guiding idea, well established in algebra and sheaf theory and useful in combinatorics is that "combinatorial problems concerning finite sets can also be formulated and solved for sheaf-like categories". We demonstrate this philosophy with the counting of (effective) equivalence relations, corresponding to partitions for $\underline{K}$=Set f. The sheaf-axiom ensures that the relevant incidence algebras are bialgebras and, suitably modified, the contravariant bialgebras of affine unipotent group schemes (Theorem 3.74). The exponential map which in the case of rational coefficients defines a group isomorphism between the Lie Algebra of the unipotent group, supplied with the Campbell-Hausdorff composition, and the group itself, gives a new type of exponential formula. A simple modification of our construction shows that Rota's hereditary bialgebras for matroids give rise to unipotent groups in the same fashion (cf. [24], pp.89). We think that the affine algebras of unipotent groups are those "bialgebras with a simple axiomatic definition" which Rota calls for in [loc.cit.], p.95. The structure theory for unipotent algebraic groups over the rationals (compare, for instance, [7], pp.485) and over the integers [30] and not only an algebraic language can thus be applied to combinatorial problems. The new examples and detailed calculations of §4 indicate the combinatorial usefulness of sheaf-like categories and their derived unipotent groups.

The structure of these notes is the following. The first two paragraphs contain the necessary purely algebraic preliminaries on abstract incidence algebras and unipotent groups. The third paragraph develops the combinatorial applications of these notions and is the heart of these notes. The last section consists of the longer new examples.

In §1 we develop the notion of abstract incidence algebra (AIA) over a groundring k mainly over $\mathbb{Z}$, the integers, or $\mathbb{Q}$, the rationals, in combinatorics. Essentially, this is a complete topological associative algebra H with a separated filtration $H=H(0)\subseteq H(1)\subseteq H(2)\subseteq\ldots$ of two sided, closed ideals and which is topologically free as a k-module. The prototype is a power series algebra $k[[T]]=k[[T_1,\ldots,T_r]]$ in finitely many variables T_i with the monomials $T_1^{n(1)}\ldots T_r^{n(r)}$ as topological basis and the $\underline{m}$-adic filtration, $\underline{m}$ the maximal ideal, which also defines the topology. In combinatorics, the filtration is given by a dimension (rank) function. The notion of AIA is general enough to include all examples from [6], [8], [18], [24]. On the other hand,it is sufficiently special to admit useful algebraic and combinatorial consequences. The main new feature is the filtration which implies a tight connection between incidence algebras and unipotent groups (Theorems 2.31, 2.34 and 2.35). This relation is similar to that between complete discrete valuation rings and commutative unipotent groups in local class field theory (cf. [7], appendix). In the combinatorial applications (see §3) the multiplication constants of an AIA with respect to a topological basis (called section coefficients by Rota in [24], pp.10) are given as natural numbers from an enumeration problem. This method to transform combinatorial data into algebraic ones is due Ph.Hall (compare [19], p.88, and [8], p.110-111). A similar approach, but without the filtration and the topology, is taken by Joyal in [18]. The substitution of elements of AIA's into power series is possible, and can be used in the same fashion as the "generating function calculus" in ordinary combinatorics. In particular, one can define and calculate Möbius, characteristic functions and, in characteristic zero, exponential formulas (Theorem 1.29).
In §2 we define and prove some results on unipotent affine groups over a principal ideal domain, mainly $\mathbb{Z}$ and $\mathbb{Q}$ in combinatorics. It is interesting to note that here affine, non-algebraic unipotent groups over a ring, namely $\mathbb{Z}$, which is not a field, appear naturally. The reason for this is that counting problems deal with natural numbers, and that the customary use of fields like the rational, real or complex numbers has mainly technical reasons. The restriction to objects of bounded dimension gives rise to <u>algebraic</u> groups for which there is a detailed structure theory (see above).
The third paragraph is divided into five sections A to E. In the first we construct <u>combinatorial incidence algebras</u>. These are algebras k[[T]] (suggestive notation) of all k-valued functions on sets of types T (Rota's terminology in [8], p.100). The types are equivalence classes of suitable epimorphisms (dually: monomorphisms) with respect to an equivalence relation. The multiplication in k[[T]] is defined by multiplication constants which count certain sets of types. The prototype of this construction is the Hall algebra (cf. [19], pp.88, and [8], p.110-111) Our main observation is a set of simple axioms for the types which are satisfied in many cases and then easily verified and which ensure that the algebras k[[T]] are AIA's in the

sense of §1, i.e. have a rich and applicable algebraic structure. Combinatorial standard calculations known for power series algebras and incidence algebras of ordered sets can thus be extended to more difficult combinatorial situations. The algebras k[[T]] render the distinction between standard, reduced and large incidence algebras in [8] unnecessary, and can also be considered as a suitable reduced version of the incidence algebra of a category (cf. [6] and [24]). The combinatorial usefulness of generalized incidence algebras has already been shown in the basic paper [8].

In special situations (section B) the multiplication constants can be interpreted as 2-cocycles of a small category. The Koszul complex calculations for $H^2(\mathbb{N}_0{}^{(I)},A)$, where $\mathbb{N}_0{}^{(I)}$ operates trivially on an Abelian group A (cf. [3], pp.192), and the result $H^2(L,A)=0$, where L is a countable directed set [22], admit the determination of all incidence algebras of full Dirichlet type (cf. [8], pp.116) and of $L\times L$ - triangular type ([loc.cit.], pp.127) respectively.

The sections C and D of §3 are devoted to sheaf-like and Krull-Schmidt categories (see above).

In part E we consider the incidence for isomorphism classes (called types again) of effective epimorphisms (dually: strict monomorphisms) of a sheaf-like category. Due to the sheaf-axiom, this is a topological bialgebra, i.e. the multiplication constants are bisection coefficients in Rota's sense (cf. [24], p.11). The group of those multiplicative functions (cf. [8], p.40) which have the value one on the types of isomorphisms, is the already mentioned unipotent group, and contains all the essential combinatorial informations. In many cases this group is isomorphic to a group of parameter transformations, i.e. of automorphisms of power series algebras (3.81, 4.18, 4.22). For the "Faa di Bruno algebra" this has been shown in [8], p.102. In (3.81) we explain how the Butcher group used in the numerical treatment of differential equations (cf. [31]) can be interpreted as the unipotent group derived from monomorphisms of rooted forests. Example (4.18) contains the theory for the category of finite sets with an operation of a finite Abelian, e.g. a cyclic, group. In (4.22) we mention an application to representations of ordered sets.

Notations and abbreviations: $\mathbb{N}_0$ respectively $\mathbb{N}$ = the natural numbers including respectively excluding zero; $\mathbb{Z}$ = the ring of integers; $\mathbb{Q}$ = the field of rational numbers; $\mathbb{N}_0{}^{(I)}$ = the free additive monoid of all families $n=(n(i);i\in I)\in\mathbb{N}_0{}^I$ such that almost all $n(i)$ are 0; $a^n = \Pi\{a(i)^{n(i)}; i\in I\}$ if $a(i)$ is contained in a commutative ring; $\#(X) = |X|$ = number of elements of a set X; U(A) = the group of invertible elements of an associative ring A; lim = the inverse (projective) limit in a category or the limit of a convergent sequence; $\amalg$ = direct sum, coproduct; Π = direct product; ker = kernel; cok = cokernel; im = image; $\Rightarrow$ = "implies"; $\Leftrightarrow$ = "logically equivalent"; $\square$ = end of an argument.

The main results of these notes are called "theorems", the others are indicated as "propositions", "lemmas" or "corollarries".

§1. Abstract incidence algebras

Let k be a commutative (ground-) ring. For combinatorial purposes the ring $\mathbb{Z}$ of integers and the field $\mathbb{Q}$ of rational numbers are most suitable since enumeration problems deal with integers. We consider k as topological ring with the discrete topology.

Let X be a topological k-module; X is called linear topological if X has a basis of neighborhoods of zero (called a 0-basis) consisting of submodules. If $X=Y \oplus Z$ is a topological direct sum decomposition we write $X=Y \hat{\oplus} Z$ and $Y \hat{\uparrow} X$. A family $(x(i);\ i\in I)$ of elements of X is called a topological basis if the map

$$k^I \longrightarrow X\ ,\ (r(i);\ i\in I) \longrightarrow \Sigma\ r(i)x(i)$$

is defined and a topological isomorphism. Here k^I has the product topology. Then X is called topologically free. In general, if X is Hausdorff and complete and if $x=(x(i);\ i\in I)$ is a 0-family, i.e. if for each neighborhood U of 0 in X almost al $x(i)$ lie in U, then $\Sigma\{x(i);\ i\in I\}$ exists, i.e. x is summable.

(1.1) Definition: An abstract incidence algebra (AIA) over k is an associative topological algebra H with a filtration

(1.2) $$H=H(0)\supseteq H(1)\supseteq H(2)\supseteq\ \ldots$$

such that the following conditions are satisfied:

(1.3)(Topology) H is Hausdorff and complete and has a 0-basis of two-sided ideals.

(1.4)(Filtration) (i) The $H(d)$ are closed two-sided ideals.

(ii) $\lim_d H(d)=0$.

(iii) For $d_1,d_2\ge 0$: $H(d_1)H(d_2)\subseteq H(d_1+d_2)$.

(iv) For $d\ge 0$: $H(d+1) \hat{\uparrow} H(d)$, and $H(d)/H(d+1)$ is topologically free.

Condition (ii) means that for any neighborhood of 0 there is a d with $H(d)\subseteq U$. In the sequel H denotes an AIA over k. The structure introduced above has many consequences which admit combinatorial interpretations.

(1.5) Dimension, rank: Since H is Hausdorff and $\lim_d H(d)=0$, also $\cap_d H(d)=0$. Thus for $x\in H$, $x\neq 0$, the number

(1.6) $\dim(x):=\mathrm{Max}\ \{n\in\mathbb{N}_0;\ x\in H(n)\}$,

called the dimension of x, is well defined.

The filtration condition (1.3)(iii) implies

(1.7) $\dim(s(1)s(2))\ge\dim(s(1))+\dim(s(2))$, $s(i)\in H$.

(1.8) Topological nilpotence and power series: The conditions (1.3),(ii) and (iii), imply $\lim_d H(1)^d=0$, i.e. H(1) is a topologically nilpotent ideal. In particular, also $\lim_d x^d=0$ for all $x\in H(1)$. Thus, if

$$g=\Sigma\{g(d)X^d;\ d\in\mathbb{N}_0\}\in k[[X]]$$

is a power series in one variable X, then

(1.9) $g(x):=g(0)1_H+\Sigma\{g(d)x^d;\ d\in\mathbb{N}\}\in g(0)1_H+H(1)$

exists. In particular,

$$(1+x)^{-1}=\Sigma\{(-1)^d x^d;\ d\in\mathbb{N}_0\}\in 1_H+H(1).$$

This implies that H(1) is contained in the <u>Jacobson radical</u> of H, thus

(1.10) $x\in U(H) \Leftrightarrow \bar{x}\in U(H/H(1))$.

Special cases of this general fact appear at different places in the combinatorial literature (cf. [8], p.89 ; [18], p.67, Th.7).

(1.11) <u>Structure constants=section coefficients</u>: Since $H(d+1)\hat{\subset} H(d)$ for $d\geq 0$ there is a non-unique topological direct sum

(1.12) $H(d)=H^{(d)}\hat{\oplus}H(d+1),\ d\geq 0$, hence $H^{(d)}\cong H(d)/H(d+1)$.

By assumption (1.3) (iv), $H(d)/H(d+1)$, hence $H^{(d)}$, is topologically free. Let $(e(t);\ t\in T^{(d)})$ be a topological k-basis of $H^{(d)}$, i.e.

(1.13) $H^{(d)}=\Pi\{ke(t);\ t\in T^{(d)}\},\ d\geq 0.$

Obviously, $\dim e(t)=d$ for $t\in T^{(d)}$. From (1.3) (ii), follows that

(1.14) $\Pi\{H^{(d)};\ d\in\mathbb{N}_0\}\longrightarrow H,\ (x^{(d)};\ d\geq 0)\longrightarrow\Sigma x^{(d)}$

is a well-defined map and a topological isomorphism, i.e. $H=\Pi H^{(d)}$ by identification.

(1.15) With $T:=\dot{\cup}\{T^{(d)},\ d\geq 0\}$, $T(n):=\dot{\cup}\{T(d),\ d\geq n\}$

there result topological k-bases $(e(t);\ t\in T(n))$ of $H(n)$ for $n\geq 0$. In particular, $t\in T(n)\Leftrightarrow \dim e(t)\geq n$. With respect to the fixed basis $e(t),\ t\in T$, of H one obtains <u>multiplication constants</u>

$G(t;\ t(1)\ \dots\ t(r))\in k,\ t,t(i)\in T,\ r\geq 0$, by the formula

(1.16) $e(t(1))\ \dots\ e(t(r))=\Sigma\{G(t;\ t(1)\ \dots\ t(r))e(t),\ t\in T\}.$

Rota in [24], p.10, calls these numbers section coefficients. For $r=2$ the notations

$$G\binom{t}{t(1)\ t(2)}=G_{t(1)t(2)}(t)=\left\{{t \atop t(1)t(2)}\right\}=\binom{t}{t(1)t(2)}=(t;\ t(1)t(2))$$

are used instead of $G(t;\ t(1)t(2))$ in the combinatorial literature. Our notation is the one of Macdonald ([19], p.88) for the Hall algebra which is one of the prototypes of incidence algebras.

(1.17) <u>Remark</u>: Just to give section coefficients=structure constants with the obvious properties means to consider a topological k-algebra with a topological basis or, dually, an (abstract) coalgebra with a k-basis. Even for a <u>field</u> k there are no useful combinatorial consequences. The axioms of Joyal ([18], pp.62) in this context furnish H(1), but not the whole filtration $H(d),\ d\geq 0$. It is however this filtration which connects combinatorics with unipotent groups (§2).

The numbers $G(t;\ t(1)t(2))$ depend on the basis $e(t),\ t\in T$. For power series algebras in finitely many parameters (=indeterminates)

$X(1),\ \dots,\ X(r)$ the change of the basis $X^n=X(1)^{n(1)}\dots X(r)^{n(r)},\ n\in\mathbb{N}_0{}^r$, to another such basis is interesting and treated by means of the Lagrange inversion formula (see [17] and [2], due to Abhyankar). Rota et al. [23] call this the transfer formula and demonstrate its combinatorial usefulness (for the case of one variable). This is the reason why we require only the existence of a basis $e(t)$, $t \in T$, but do not incorporate a distinguished one into the structure as in [24], p.3 below. □

(1.18) Graded abstract incidence algebras: An AIA H over k is called graded if a decomposition $H = \Pi\{H^{(d)};\ d \geq 0\}$ is given and satisfies

(1.19) $H^{(d(1))}H^{(d(2))} \subseteq H^{(d(1)+d(2))}$.

In combinatorics this corresponds to the Jordan-Dedekind chain condition. If H has no zero-divisors then (1.19) is equivalent to

$$\dim(xy) = \dim(x)+\dim(y), x,y \in H.$$

The two standard examples are the power series algebra

$$k[[X]] = k[[X(i);\ i \in I]] = \Pi\{kX^n;\ n \in \mathbb{N}_0^{(I)}\}$$

in indeterminates $X(i)$, $i \in I$ an index set, with the standard grading

$$k[[X]]\ (d) = \Pi\{kX^n;\ |n| := \Sigma\{n(i);\ i \in I\} = d\},$$

and the corresponding non-commutative topological word algebra

(1.20) $\hat{Ass}(I) := \Pi\{\hat{Ass}^{(d)}(I);\ d \geq 0\}$ ([25], LA 4.13) where

$$\hat{Ass}^{(d)}(I) := \Pi\{kw;\ w = (i(1)\ \dots\ i(d)) \in I^d\}$$

and the multiplication is the composition of words. The topology is the product topology which for infinite I is coarser than the $\hat{Ass}(I)(1)$-adic topology. If H is an AIA and if $x = (x(i);\ i \in I)$ is a family of elements in $H(1)$ with $\lim_I x(i) = 0$, then there is a unique continuous k-algebra homomorphism

(1.21) $x^{\#} : \hat{Ass}(I) \longrightarrow H,\ i \longmapsto x(i)$, namely

$$f(x) := x^{\#}(f) = x^{\#}(\Sigma\{f(w)w;\ w = (i(1)\ \dots\ i(d))\}) = \Sigma f(w)x(i(1))..x(i(d)).$$

(1.22) Base ring extension: If X and Y are linear topological k-modules then the completed tensor product $X \hat{\otimes} Y = X \hat{\otimes}_k Y$ of X and Y is the Hausdorff completion of $X \otimes_k Y$ with respect to the topology given by the 0-basis $X' \otimes Y + X \otimes Y'$, where X' and Y' run over a 0-basis of X and Y respectively. If $X = k^I$, I an index set, then $k^I \hat{\otimes} Y = Y^I$ with an obvious identification. If finally $k \longrightarrow l$ is a ring homomorphism, then $l \hat{\otimes}_k Y$, l with the discrete topology, is a linear topological l-module. If $H = \Pi\{ke(t);\ t \in T\}$ is an AIA over k then $l \hat{\otimes}_k H = \Pi\{le(t);\ t \in T\}$ is an AIA over l.

(1.23) Power series calculations: The explicit calculation of substitutuions is customary and useful in combinatorics. Assume that $(e(t);\ t \in T(1))$ is a fixed basis of $H(1)$ obtained in the non-unique way of (1.11).

We identify

(1.24) $k^{T(1)} = H(1)$, i.e. $f = \Sigma\{f(t)e(t);\ t \in T(1)\}$.

In combinatorics this is the usual identification of "sequences" f with their "generating functions" $\Sigma f(t)e(t)$. For $r \geq 1$ define

(1.25) <u>type</u>: $T(1)^r \longrightarrow \mathbb{N}_0^{(T(1))}$, $\underline{t} = (t(1)\ldots t(r)) \longrightarrow m$,

where $m = (m(t);\ t \in T(1))$ and $m(t) = {}^{\#}\{i; t(i) = t\}$.

For a given $m \in \mathbb{N}_0^{(T(1))} - \{0\}$ and $t \in T(1)$ we define

(1.26) $GS(t;m) := \Sigma\,\{G(t;\underline{t});\ \text{type}(\underline{t}) = m\}$.

The combinatorial interpretation of this number is given in (3.19). If H is commutative then $G(t;\underline{t})$ does not depend on the order of the $t(i)$ in $\underline{t}$ and we define

(1.27) $G(t;m) := G(t;\underline{t})$, type $(\underline{t}) = m$. Then

(1.28) $GS(t;m) = ((m))G(t;m)$, where

$$((m)) := |m|\,!\ \Pi\{(m(t)!)^{-1};\ t \in T(1)\},\quad |m| := \Sigma\, m(t)$$

is the multinomial coefficient. Reordering terms in the expression g(f) of (1.9) one obtains the easy, but useful

(1.29) <u>Theorem</u>: Let

$$f = (f(t);\ t \in T(1)) = \Sigma\ f(t)e(t) \in k^{T(1)} = \Pi\ ke(t) = H(1), \text{ and}$$

$g = \Sigma\,\{g(d)X^d;\ d \geq 0\} \in k[[X]]$ be a power series in a variable X. Then

(1.30) $g(f) = g(0)1 + \Sigma_{t \in T(1)}\,\Sigma\,\{g(|m|)GS(t;m)f^m;\ m \in \mathbb{N}_0^{(T(1))} - \{0\}\}e(t)$.

There are two important special cases.

(1.31) $(1 + \Sigma_{t \in T(1)} f(t)e(t))^{-1} = 1 + \Sigma_{t \in T(1)}\ \Sigma\,\{(-1)^{|m|}GS(t;m)f^m;\ m \text{ as above}\}\ e(t)$.

(1.32) (<u>Exponential formula</u>) If $\mathbb{Q} \subseteq k$ and H is commutative then

$$\exp(\Sigma f(t)e(t)) = 1 + \Sigma_t(\ \Sigma_m G(t;m)(m!)^{-1}f^m)e(t).$$

If moreover $a \in U(k)^{T(1)}$ then

(1.33) $\exp(\Sigma_t f(t)\ \frac{e(t)}{a(t)}) = 1 + \Sigma_t(\ \Sigma_m \frac{a(t)G(t;m)}{a^m\ m!}\ f^m\ )\ \frac{e(t)}{a(t)}$. □

The combinatorial interpretation of the number $a(t)a^{-m}(m!)^{-1}$ as a number of partitions is given in (3.53). Calculations as those of (1.29) are of course known from combinatorics (compare, for instance, [5], pp.36 and [27], Th. 3.2). The generic case of formula (1.31) with indeterminates f(t) is treated in (3.77). The simplest, but interesting case is that of the <u>Möbius function</u> μ, depending on the basis e(t), $t \in T(1)$, where one defines functions

$$\eta := \text{const}_1 = \Sigma\,\{e(t);\ t \in T(1)\}\ ,\ \xi := 1 + \eta\ ,\ \text{and}$$

(1.34) $\mu := \xi^{-1} = 1 + \Sigma_{t \in T(1)}\ [\Sigma\,\{(-1)^{|m|}GS(t;m);\ m \in \mathbb{N}_0^{(T(1))} - \{0\}\}]\ e(t)$.

<u>§2. Connections with unipotent group schemes</u>.

This section is inspired by the theory of Witt vectors and λ-rings, (see, for instance, [20] pp.179; [7], §5, pp.119; [19], Ch.1).

Let k be a groundring and H an AIA over k. Notations as in §1. The properties of H are inherited by the group 1+H(1), and one obtains

(2.1) Theorem: Situation as above. Then
(i) With the topology induced from H the group 1+H(1) is a Hausdorff, complete topological group with a basis of neighborhoods of 1 consisting of normal subgroups.
(ii) The filtration $1+H(1) \supseteq 1+H(2) \supseteq \ldots$ of 1+H(1) consists of closed normal subgroups and satisfies $\lim_d(1+H(d)) = 1$. This implies

(2.2) $$1+H(1) = \lim(1+H(1)/1+H(d); d \geq 1)$$

where the first lim means convergence as in (1.4) (ii), and the second the inverse limit for topological groups.
(iii) The map $x \longrightarrow 1+x$ induces for $1 \leq m \leq n$ homeomorphisms

$$\begin{array}{ccccc} \prod \{ke(t); t \in T(m,n)\} & \cong & H(m)/H(n) & \longrightarrow & 1+H(m)/1+H(n) \\ \Sigma f(t)e(t) & \longrightarrow & \overline{\Sigma f(t)e(t)} & \longrightarrow & \overline{1+\Sigma f(t)e(t)} \end{array}$$

where $T(m,n) := \dot{\bigcup} \{T^{(k)}; k = m,\ldots,n-1\}$. For $n = m+1$ these maps are topological group isomorphisms. Hence for $d \geq 1$ the sequences

(2.3) $$0 \to \prod \{ke(t);\ t \in T^{(d)}\} \longrightarrow 1+H(1)/1+H(d+1) \xrightarrow{\text{can}} 1+H(1)/1+H(d) \longrightarrow 1$$

of groups are exact, and the groups 1+H(1)/1+H(d) have normal series with factors of type k^I, I a set.
(iv) If $\mathbb{Q} \subseteq k$ the functions exp and log induce inverse homeomorphisms

(2.4) $$H(1) \underset{\log}{\overset{\exp}{\rightleftarrows}} 1+H(1), \qquad \begin{array}{l} x \longrightarrow \exp(x) := \Sigma \{(n!)^{-1}x^n; n \geq 0\} \\ \log(1+x) := \Sigma \{(-1)^{n-1}n^{-1}x^n; n \geq 1\} \longleftarrow 1+x \end{array}$$ □

The ideal H(1) is a topological nilpotent Lie algebra with bracket $[x,y] := xy-yx$ in the sense of the following definition.

(2.5) Definition: Consider a topological Lie algebra $\underline{g}$ over a ring k and assume that $\underline{g}$ has a 0-basis of ideals. Then $\underline{g}$ is called topologically nilpotent if the descending central series $C^n(\underline{g})$, $n \geq 0$, converges to 0.
(See [25], LA5.3 and (1.4) (ii), for the definitions).

The standard example for this notion is the topologically free Lie algebra

$$\hat{L}(X) = \hat{L}(X(1), \ldots, X(r)) \subset \widehat{Ass}(X(1), \ldots, X(r)) = \widehat{Ass}(X)$$

on r generators ([25], LA 4.13). If $\underline{g}$ is any such Lie algebra and $x = (x(1), \ldots, x(r)) \in \underline{g}^r$ there is a unique continuous Lie algebra homomorphism

(2.6) $x^\# : \hat{L}(X(1), \ldots, X(r)) \longrightarrow \underline{g}$, $X(i) \longrightarrow x(i)$.

For $\underline{g} = H(1)$ this is the restriction of the corresponding map for the associative algebras (1.21). If, in particular, $\mathbb{Q} \subseteq k$ and $h(X,Y) \in \hat{L}(X,Y)$ is the Lie Power series from the Campbell-Hausdorff formula with

(2.7) $\exp(X)\exp(Y) = \exp(h(X,Y))$ in $\widehat{Ass}(X,Y)$([loc.cit.], LA 4.14), and if $x,y \in \underline{g}$ then $h(x,y) := (x,y)^\#(h)$ is well defined. The part (iv) of Theorem (2.1) can thus be improved to

(2.8) Corollary: In the situation of Theorem (2.1)(iv) the map $\exp: H(1) \longrightarrow 1+H(1)$

is a topological group isomorphism, where the composition on the left is given by $(x,y) \longrightarrow h(x,y)$. □

If R is any commutative k-algebra the results (2.1) and (2.8) can be applied to the AIA $R \hat{\otimes}_k H$ over R instead of H itself.
In the sequel we use the standard terminology for affine groups from [7], Ch.II.
Let $\underline{Al}_k := \underline{Al}$ denote the category of commutative k-algebras and $\underline{Gr}$ that of groups.
A representable functor $G:\underline{AL} \longrightarrow \underline{Gr}$ is called an affine group (-scheme, -functor) over k. If

(2.9) $\quad F_R: Al_k(A,R) \cong G(R) \quad , R \in \underline{Al} \quad ,$

is a functorial isomorphism then $A := A(G)$ is the affine algebra of G, unique up to isomorphism, and $x := F_A(id_A)$ is the corresponding universal element.
Then F is given by

(2.10) $\quad F_R(f)=(Gf)(x) \quad , \quad f:A \longrightarrow R.$

In the same fashion one defines affine monoids, rings etc. For example, the AIA H induces the affine ring

(2.11) $\quad \underline{H}: \ \underline{AL} \longrightarrow$ {associative rings} $, R \longrightarrow R \hat{\otimes} H$.

More important for the purposes of this paper is the subgroup $\wedge_H$ of $\underline{H}$, defined in

(2.12) Theorem: (i) The group functor

(2.13) $\quad \wedge_H:\underline{Al} \longrightarrow \underline{Gr} \ , \quad R \longrightarrow 1+R \hat{\otimes} H(1) \subset R \hat{\otimes} H$

is an affine k-group which is represented by the polynomial algebra $k[X(t);\ t \in T(1)]$, $X(t)$ indeterminates, through

(2.14) $\quad Al(k[X(t);\ t \in T(1)],R) \cong \wedge(R),\ f \longrightarrow 1+ \Sigma\{f(X(t))e(t);\ t \in T(1)\}$

and has the universal element

(2.15) $\quad 1+ \Sigma\{X(t)e(t);\ t \in T(1)\} \in \wedge(k[X(t);\ t \in T(1)])$.

(ii) The ideals $H(d)$ of H, $d \geq 1$, induce a filtration

(2.16) $\quad \wedge := \wedge(1) \supseteq \wedge(2) \supseteq \ldots \ , \ \wedge(d)(R) = 1+R \hat{\otimes} H(d)$,

by closed , affine , normal subgroups of $\wedge$ such that

(2.17) $\quad \wedge \cong \lim_d \wedge / \wedge (d)$.

The quotient $\wedge / \wedge (d)$ is taken in the category of all group functors, is affine and $can: \wedge \longrightarrow \wedge / \wedge (d)$ is faithfully flat [7].
(iii) Let G_a be the additive groups, represented by polynomial algebras,

(2.18) $$0 \longrightarrow G_a^{T(d)} \longrightarrow \wedge / \wedge (d+1) \xrightarrow{can} \wedge / \wedge (d) \longrightarrow 1,\ d \geq 1,$$

with faithfully flat can. In particular, $\wedge$ is an inverse limit of groups with normal series whose factors are groups G_a^I, I a set. □

Over an algebraically closed field k the latter property characterizes unipotent affine groups ([7], p.355,487). We define unipotent for k a principal ideal domain, e.g. $\mathbb{Z}$, such that the same result is true. We show moreover that unipotent affine groups give rise to AIA's in a canonical fashion.

(2.19) Assumption: In the remainder of this section we assume that k is a principal ideal domain, e.g. $\mathbb{Z}$ or $\mathbb{Q}$, unless explicitly stated otherwise. For combinatorics this is not a serious restriction. For group schemes over rings this assumption is often necessary (more generally: k Dedekind)(see, for instance,[1] and [30]). Some results hold for arbitrary noetherian k.

Assume now that, without loss of generality, $G \cong Al(A,-)$, $G = Al(A,-)$ is an affine flat k-monoid (i.e. A is flat as k-module) with comultiplication

(2.20) $\Delta : A \longrightarrow A \otimes A, \quad a \longrightarrow \Sigma\, a(1) \otimes a(2)$ (see [7], p.145 and [29], p.7)

and counit $\varepsilon : A \longrightarrow k$, $A^{+} := \ker(\varepsilon)$.

Flatness is a suitable technical condition ([26], [1], [30]), for instance, when going from $\mathbb{Q}$-groups to $\mathbb{Z}$-groups. For any k-algebra R with structure map $\eta : k \longrightarrow R$ the k-module $\mathrm{Hom}_k(A,R)$ of k-linear maps is an associative R-algebra with the convolution multiplication

(2.21) $fg = \mu(f \otimes g)\,\Delta$, i.e. $(fg)(a) = \Sigma\, f(a(1))g(a(2))$; $f,g : A \longrightarrow R$,

where μ denotes the multiplication on R (compare,for instance, [29], pp.14).

The unit of $\mathrm{Hom}_k(A,R)$ is $\eta\varepsilon =: \varepsilon$. Then

(2.22) $G(R) = Al(A,R) \subset \mathrm{Hom}(A,R)$

is a multiplicative submonoid. Moreover, Hom(A,R) is a linear topological R-module with the finite topology, a 0-basis of which is given by the Hom(A/A',R), where A' runs over the finitely generated k-submodules of A. We write

(2.23) $H := \mathrm{Hom}(A,k) = A^{*}$, the dual module, $H^{+} := \mathrm{Hom}(A/k,k)$.

If A is k-free with basis $x(t)$, $t \in T$, and if $e(t) \in H = \mathrm{Hom}(A,k)$, $t \in T$, is the dual basis then

(2.24) $\mathrm{Hom}(A,R) \cong R \mathbin{\hat{\otimes}} H, \quad f \longrightarrow \Sigma\, f(x(t))e(t)$

is a topological R-isomorphism. The methods of [26] give

(2.25) Theorem: Situation as above. The R-algebras Hom(A,R) are topological, Hausdorff, complete and have a 0-basis of two-sided ideals. If A is k-free then (2.24) is an algebra isomorphism. □

Define

(2.26) $\mathrm{Id}-\eta\varepsilon : A \longrightarrow A, \quad a \longrightarrow a - \varepsilon(a)1_A =: a^{+}$ and

(2.27) $\Delta : A \longrightarrow A^{\otimes n}, \quad a \longrightarrow \Sigma\, a(1) \otimes \ldots \otimes a(n)$

corresponding to the multiplication $G^n \longrightarrow G$. The canonical filtration of A is then defined by

(2.28) $A(n) := \ker(A \xrightarrow{\Delta} A^{\otimes n} \xrightarrow{(\mathrm{Id}-\eta\varepsilon)^{\otimes n}} A^{\otimes n}) =$

$$= \{a \in A;\ \Sigma\, a(1)^{+} \otimes \ldots \otimes a(n)^{+} = 0\}.$$

(2.29) Proposition: (i) The A(n) from (2.28) satisfy the recursive relations $A(1) = k = k1_A$, $A(n+1) = \{a;\ \Delta(a) - a \otimes 1 \in A(n) \otimes A\}$.

(ii) The A(n) form an increasing sequence of subcoalgebras of A, and satisfy $A(m)A(n) \subseteq A(m+n)$.

(iii) $A_u := \cup\{A(n);\ n \geq 1\}$ is a subbialgebra of A, and the injection $A_u \subset A$ is faithfully flat, i.e.

(2.30) $G = \mathrm{Al}(A,-) \longrightarrow G_u := \mathrm{Al}(A_u,-)$

is a faithfully flat epimorphism of flat affine monoids. □

(2.31) Theorem: Let $G = \mathrm{Al}(A,-)$ be a flat affine monoid. Notations as above. Then the following assertions are equivalent:

(i) The filtration A(n) is exhaustive, i.e. $A = \cup\ A(n)$.

(ii) For any $R \in \underline{\mathrm{Al}}$ the ideal Hom(A/k,R) of Hom(A,R) is topologically nilpotent.

If in addition A is k-free, then

(ii') The ideal $H^+ = \mathrm{Hom}(A/k,k)$ of H is topologically nilpotent.

(iii) If V is a finitely generated non-zero G-module, then also the fix module ${}^{G}V$ is non-zero. □

If (i) and (ii) are satisfied, the monoid G is a group and called unipotent. See [7], pp.169, for the definition of ${}^{G}V$. If k is a field the preceding definition of unipotence coincides with the usual one ([7], p.487). A similar notion is that of linear unipotence in [30], p.765. That G is a group in (2.31) follows in the same fashion as that an infinitesimal formal monoid is a group ([11], p.528). This is not surprising since commutative infinitesimal respectively unipotent groups a field are in (Cartier-) duality.

(2.32) Standard example: The additive group $G_a, G_a(R) = (R,+)$, is unipotent. □

(2.33) Corollary: The faithfully flat epimorphism (2.30) is the universal homomorphism from a monoid G to a unipotent group. □

For unipotent groups as defined above there is a stability theorem as in [7], p.485. This result, Example (2.32) and Theorem (2.14) imply

(2.34) Theorem: If H is an AIA over k then Λ_H, $\Lambda_H(R) = 1+R \otimes H(1)$, is k-free and unipotent. □

On the other hand one has

(2.35) Theorem: If $G = \mathrm{Al}(A,-)$ is k-free and unipotent then $H := A^* = \mathrm{Hom}_k(A,k)$ with $H(n) := \mathrm{Hom}(A/A(n),k)$, $n \geq 0$, $A(0) = 0$ is an AIA. Moreover, the unit η and the multiplication μ of A induce, by duality, maps $\varepsilon := \eta^* : A^* = H \longrightarrow k$ and $\Delta := \mu^* : H \longrightarrow H \hat{\otimes} H = (A \otimes A)^*$ respectively, such that H becomes a cocommutative topologically k-free topological Hopf algebra, called the covariant Hopf algebra of G. □

Commutative such algebras are the affine algebras of formal groups, they are treated in [11] and [15], pp.492.

We finally consider the Lie algebra of the k-free affine monoid $G = \mathrm{Al}(A,-)$. The algebra H is also a topological Lie algebra with bracket $[x,y] = xy-yx$.

(2.39) Proposition: Situation as above. Then

$$\mathrm{Lie}(G) := \mathrm{Der}\,(A,k) = \mathrm{Hom}(A/k+(A^+)^2, k) = \mathrm{Hom}(A^+/(A^+)^2, k)$$

is a closed sub Lie algebra of H, where Der contains the derivations D with $D(ab) = \varepsilon(a)D(b) + \varepsilon(b)D(a)$. The topology induced from H is the finite topology of $\mathrm{Hom}(A^+/(A^+)^2,k)$. With this topology Lie(G) is the (topological) Lie algebra of G. □

If $\omega(a/k) := A^+/(A^+)^2$ ([7], p.215) is k-free then

$$(2.40)\quad \begin{aligned} R \hat{\otimes} \mathrm{Lie}(G) &= R \hat{\otimes} \mathrm{Hom}(A^+/(A^+)^2,k) \cong \mathrm{Hom}(A^+/(A^+)^2,R) \cong \\ &\cong \mathrm{Hom}_R((R\otimes A^+)/(R\otimes A^+)^2, R) = \mathrm{Lie}(R\otimes G) \subset R \hat{\otimes} H. \end{aligned}$$

We identify all objects in this sequence and obtain the following

(2.41) Theorem: Notations as above. Assume $\mathbb{Q} \subseteq k$.
(i) If $G = \mathrm{Al}(A,-)$ is a k-free, affine, unipotent group and if $A^+/(A^+)^2$ is k-free, then Lie(G) is a topologically nilpotent (2.5) Lie algebra. The maps exp and log induce inverse group isomorphisms

$$(2.42)\quad \begin{array}{ccc} R \hat{\otimes} H^+ & \rightleftarrows & \wedge_H(R) = 1 + R \hat{\otimes} H^+ \\ \cup & & \cup \\ R \hat{\otimes} \mathrm{Lie}(G) & \underset{\log}{\overset{\exp}{\rightleftarrows}} & G(R) \end{array},$$

where on the left the group structure is the Campbell-Hausdorff composition (2.7)
(ii) If $A^+/(A^+)^2$ has tha basis $c(i)$, $i \in I$, then there are algebraically independent generators $x(i)$ of A, $i \in I$, i.e. $A = k[x(i);\ i\in I]$ is a polynomial algebra with $x(i)+(A^+)^2 = c(i)$.
(iii) The functor $G \longrightarrow \mathrm{Lie}(G)$ is an equivalence between the categories of groups as in (i) and that of topologically nilpotent, k-free Lie algebras. □

The preceding theorem and the finer structure of G is known if G is algebraic over a field k ([7], IV. §2); the generalization here is that to arbitrary affine groups instead of algebraic groups and to topologically nilpotent Lie algebras. If k is Dedekind (e.g. $k = \mathbb{Z}$ as needed in combinatorics) such a structure theory is contained in [30]. We plan to study its combinatorial implications. Theorem (3.74) below shows that there is a big class of combinatorially interesting unipotent affine, not algebraic, groups over $\mathbb{Z}$ (not only $\mathbb{Q}$) whose affine algebras are polynomial.

§3. Combinatorial incidence algebras and unipotent groups

A. Abstract incidence algebras in combinatorics

The incidence algebras constructed in this section are a variant of the Hall algebra (see, for instance, [19], pp.88), and the incidence algebra of a category (compare [6] and [24], p.43), adapted to the theory of §1. They include many new classes of examples and almost all examples of the literature. The main observation is the connection between sheaf theory and unipotent groups, developed in section E of this paragraph.

Let $\underline{K}$ be a category, M a class of epimorphisms of $\underline{K}$ and $\sim$ an equivalence relation on M (compare [24], p.43). The elements of $M/\sim =: T$ are called types (by Rota) and denoted by $\bar{s}$, $s \in M$. The main example for M is the class of epimorphisms in categories of sheaves. Dually one obtains a theory for monomorphisms. We assume that $\underline{K}$ is skeletal-small, i.e.

$$Ob(\underline{K})/\cong \;,\quad Ob(\underline{K}) := \text{class of objects of } \underline{K},\quad \cong \text{ isomorphism},$$

is a set. These data are supposed to satisfy the following condition:

(3.1)(Isomorphism): All isomorphisms are in M. □

(3.2)(Multiplication): If s(1),s(2) are composable morphisms, then $s(1),s(2) \in M \Rightarrow s(1)s(2) \in M \Rightarrow s(1) \in M$. □

Let $S_2(M)$ be the set of singular 2-simplices of M, i.e.

(3.3) $\quad S_2(M) := \{(s(1),s(2));\ X(0) \leftarrow s(1) — X(1) \leftarrow s(2) — X(2),\ s(i) \in M\}$.

Two 2-simplices (s(1),s(2)) and (r(1),r(2)) are equivalent $\approx$ if there is an isomorphism h with $s(1)h = r(1)$ and $s(2) = hr(2)$. This h is then unique.

Define

(3.4) $\quad S_2[M] = S_2(M)/_{\approx} \ni [s(1),s(2)]$. Then

(3.5)(Local finiteness): For all $s \in M$ the set $\{[s(1),s(2)] \in S_2[M];\ s(1)s(2) = s\}$ is finite (compare [24], p.43 and [6]). □

Define the dimension (rank) of $s \in M$ by

(3.6) $\dim(s) := \operatorname{Sup} n$ where n runs over all $n \in \mathbb{N}_0$ such that there is a product representation $s = s(1)\ldots s(n)$, $s(i) \in M$, where s(i) is not an isomorphism and the sup is taken in $\mathbb{N}_0 \cup \{\infty\}$. Obviously, dim(s)=0 if and only if s is an isomorphism.

(3.7)(Finite dimension): For all $s \in M$ the dimension dim(s) is finite. □

For the equivalence relation we need

(3.8) Isomorphy implies equivalence, i.e. if $s(2) = h(2)s(1)h(1)^{-1}$ with $s(i) \in M$ and isomorphisms h(i) then $s(1) \sim s(2)$. □

(3.9) The class $Iso(\underline{K}) \subset M$ of isomorphisms in $\underline{K}$ is $\sim$-saturated, i.e. if $s(1) \sim s(2)$ and s(1) is an isomorphism, then so is s(2). □

With $Iso(T) := Iso(\underline{K})/\sim$ one obtains a decomposition

(3.10) $\quad T = Iso(T) \,\dot{\cup}\, T(1) \quad , \quad T(1) = \{\bar{s};\ s \in M \text{ is not an isomorphism}\}$

We consider $Ob(\underline{K}) \subset Iso(\underline{K})$ with $X = id_X$.

(3.11) The maps domain (dom) and codomain (cod) are $\sim$-invariant, i.e. for $s: X \to Y$ in M the types $dom(\bar{s}) := \overline{id_X}$ and $cod(\bar{s}) = \overline{id_Y}$ in T are well-defined. □

The main axiom is (compare [6], pp.184)

(3.12) (Section coefficients): For types $t, t(1), t(2) \in T$ and $s \in M, \bar{s} = t$, the number $G(t; t(1)t(2)) := {}^{\#}\{[s(1),s(2)] \in S_2[M];\ \overline{s(i)} = t(i), s = s(1)s(2)\}$ is independent of the choice of the representative s of t. □

Then $\dim(t) := \dim(s), \bar{s}=t$, is well-defined.

Let k be a commutative coefficient ring. With the above data and conditions consider the linear topological k-module k^T with the product topology and the standard basis $e(t), t \in T$. Define the multiplication $f_1 f_2$ by

(3.13) $(f_1f_2)(t) = \Sigma\ \{G(t;t_1t_2)f_1(t_1)f_2(t_2);\ t_1,t_2 \in T\}$;

k^T with this multiplication is denoted by k[[T]]. Define

(3.14) $k[[T]]\ (d) = \Pi\{ke(t);\ \dim(t) \geq d\},\quad d \geq 0.$

(3.15) Theorem: Assumptions and notations as above. Then

(i) k[[T]] with the filtration (3.14) is an AIA with multiplication constants G(t;t(1)t(2)) with respect to the standard basis.

(ii) The family $(e(t);\ t \in \mathrm{Iso}(T))$ is a complete set of orthogonal idempotents.

(iii) Identify $k^{\mathrm{Iso}(T)} = \Pi\ \{ke(t);\ t \in \mathrm{Iso}(T)\} \subset k[[T]]$. Then the k-algebra $k^{\mathrm{Iso}(T)}$ with the componentwise structure is a closed subalgebra of k[[T]], and the projection $k[[T]] \longrightarrow k^{\mathrm{Iso}(T)}$ induces a topological isomorphism $k[[T]]/k[[T]](1) \cong k^{\mathrm{Iso}(T)}$, and the topological decomposition $k[[T]] = k^{\mathrm{Iso}(T)} \hat{\otimes}\ k[[T]](1)$. In particular, by (1.10) $f \in k[[T]]$ is invertible if and only if $f(t) \in U(k)$ for all $t \in \mathrm{Iso}(T)$.

The algebra k[[T]] is called the incidence algebra of M, reduced modulo ~ ([8], [24]). □

The higher multiplication constants G(t;1(1)...t(r)) are defined as in (1.16). They admit the following combinatorial interpretation. Define the class $S_r(M)$ of r-simplices and $S_r[M]$ as (3.3) and (3.4).

(3.16) Corollary: For $t, t(1), \ldots, t(r) \in T$, $r \geq 1$, and $s \in M$, $\bar{s}=t$,

$G(t;t(1)\ldots t(r)) = \#\{[s(1)\ldots s(r)] \in S_r[M];\ s(1)\ldots s(r) = s, \overline{s(i)} = t(i)\}$. □

A more suggestive interpretation is the following one.

For $X \in \mathrm{Ob}(\underline{K})$ consider the preordered class $M(X) = \{s \in M;\ \mathrm{dom}(s) = X\}$ with $s(1) \leq s(2)$ iff s(1)=ss(2) for some s. This s is then unique and in M and denoted by $s(1)s(2)^{-1}$.

For the equivalence relation $\approx$, induced by this preorder, i.e.

$s(1) \approx s(2) \Leftrightarrow s(1) \leq s(2)$ and $s(2) \leq s(1) \Leftrightarrow s(1)s(2)^{-1}$ is an isomorphism,

(3.17) the set $M[X] := M(X)/\approx\ \ni [s]$

of "factors of X in M" is ordered by the induced order and then locally finite by (3.5). The greatest element of M[X] is $1 := [id_X]$. If $s: X \longrightarrow Y$ is in M a decreasing sequence

(3.18) $1 = [id_X] \geq [f(1)] \geq \ldots \geq [f(r)] = [s]$

of length r in M[X] is called a normal series of [s] with r factors $\overline{f(i)f(i-1)^{-1}}$, $i = 1, \ldots, r$, in T.

(3.19) Theorem: Let $t, t(1), \ldots, t(r) \in T$ be types, and $s: Y \leftarrow X$ a representative of t. Then the map

$(X(r)=Y \xleftarrow{s(r)} X(r-1) \xleftarrow{s(r-1)} \ldots X(1) \xleftarrow{s(1)} X(0)=X) \longrightarrow$

$1 = [id_X] \geq [s(1)] \geq [s(2)s(1)] \geq \ldots \geq [s]$

induces a bijection from the set $\{[s(r), \ldots, s(1)];\ s(r)\ldots s(1)=s, \overline{s(i)}=t(i)\}$ onto the set of normal series of [s] with r factors $t(1),\dots,t(r)$. Hence $G(t;t(r),\dots,t(1))$ is the number of these normal series. The number $GS(t;m), 0 \neq m \in \mathbb{N}_0^{(T(1))}$, from (1.26) is then the number of all normal series of [s] of type m, i.e. with $m(t)$, $t \in T(1)$, factors of type t. □

(3.20) Corollary: Let $r = 2$. Then

$$G(t;t(2)t(1)) = \#\{[f] \in M[X]; [s] \leq [f], \bar{f} = t(1), \overline{sf^{-1}} = t(2)\}. \quad \square$$

The incidence algebra k[[M[X]]] of the locally finite ordered set M[X] is of course a special case of the above construction. It is the k-module of functions

$f: \{(i,j); i \leq j \text{ in } M[X]\} \to k$, $(i,j) \mapsto f(i,j)$ with the multiplication

$(fg)(i,j) = \Sigma\, f(i,k)g(k,j); i \leq k \leq j\}$. Then

(3.21) Theorem: (Connection with local incidence algebras)
Situation from above. The map

$$\text{can} : k[[T]] \longrightarrow k[[M[X]]],\ f \longmapsto \tilde{f},\ \text{given by}$$

$$\tilde{f}([s(1)],[s(2)]) = f(\overline{s(1)s(2)^{-1}}),\ s(1) \leq s(2) \text{ in } M(x)\ ,$$

is a continuous k-algebra automomorphism. It preserves the ζ-function and thus the Möbius function. In particular $\mu_{M[X]}([s],1) = \mu_T(\bar{s})$ where $s : X \longrightarrow Y$ in M. □

The preceding theorem shows that for the calculation of the Möbius function alone the incidence algebra of an ordered set is sufficient. It can be interpreted as the connection between standard and large incidence algebras in [8].

B. Second cohomology. Multiplication constants as 2-cocycles.

Let $\underline{K}$ and M satisfy (3.1) to (3.7), and let $\underline{L}$ be a small category whose morphisms are epimorphisms and whose isomorphisms are identities only. As standard examples one has

(3.23) A locally finite ordered set, considered as a category, and

(3.24) A locally finite cancellation monoid without invertible elements except 1.

Let $F : \underline{K} \to \underline{L}$ be a functor such that $F(M) = \text{Morph}(\underline{L})$ is the class of all morphisms of $\underline{L}$ and s is an isomorphism if ond only if Fs is an isomorphism, i.e. an identity. A generalization to $F(M) \subseteq \text{Morph}(\underline{L})$ is possible [9]. Define $s(1) \sim s(2)$, $s(i) \in M$, iff $Fs(1) = Fs(2)$. Then F induces a bijection

(3.25) $\quad F : M/\!\sim\ = T \cong \text{Morph}(\underline{L})$. Identify

(3.26) $\quad M/\!\sim\ = T = \text{Morph}(\underline{L}),\ \bar{s} = Fs$.

As is easily seen the conditions (3.8/9/10) are satisfied. Assume however that also (3.12) is fulfilled. That F is a functor implies that $G(t;t(1)t(2)) = 0$ unless $t = t(1)t(2)$. Hence define

(3.27) $z(t(1),t(2)) = G(t(1)t(2);t(1),t(2))$, $(t(1),t(2)) \in S_2(\underline{L})$
where $S_2(\underline{L}) := S_2(\mathrm{Morph}(\underline{L}))$ as in (3.3).

(3.28) Theorem: Assumptions as above.
(i) The function $z : S_2(\underline{L}) \longrightarrow \mathbb{Z}$ defined above is a normalized 2-cocycle.
This means
(a) $z(t_1,t_2) = 1$ if $t_1 = \mathrm{id}$ or $t_2 = \mathrm{id}$.
(b) For $(t_0,t_1,t_2) \in S_3(\underline{L})$ the cocycle condition
$z(t_1,t_2)z(t_0,t_1t_2) = z(t_0\, t_1,t_2)z(t_0,t_1)$ holds.
The incidence algebra $k[[T]]$ is given as $k^{\mathrm{Morph}(\underline{L})}$ ($T=\mathrm{Morph}(\underline{L})$) with the multiplication
(3.29) $(f_1f_2)(t) = \Sigma\{z(t_1,t_2)f_1(t_1)f_2(t_2)\ ;\ t_1t_2=t\}$.
(ii) If on the other hand z is any normalized 2-cocycle $z:S_2(\underline{L}) \longrightarrow k$ with values in k one defines $k[[\underline{L},z]] = k^{\mathrm{Morph}(\underline{L})}$ with the multiplication (3.29) and obtains an associative topological algebra.
(iii) If in (ii) z_1,z_2 are two cohomologous cocycles, i.e. if there is a normalized 1-chain $c:S_1(\underline{L}) = \mathrm{Morph}(\underline{L}) \longrightarrow U(k)$, $c(\mathrm{id}) = 1$, with values in the group $U(k)$ such that $z_2 = z_1(dc)$ where $(dc)(t_1,t_2) = c(t_2)^{-1}c(t_1,t_2)c(t_1)^{-1}$, then the map
(3.30) $k[[\underline{L},z_1]] \longrightarrow k[[\underline{L},z_2]]$, $f \longmapsto fc$ $(fc)(t) = f(t)c(t)$
is a topological algebra isomorphism. In particular $k[[\underline{L}]] := k[[\underline{L},1]] = k[[\underline{L},dc]]$.

□

The preceding theorem points to the interest of the monoid $H^2_{\mathrm{norm}}(\underline{L},\mathbb{Z}^{\times})$ of normalized $\mathbb{Z}^{\times} = (\mathbb{Z},\cdot)$-valued 2-cocycles on $\underline{L}$ modulo cohomology. Even $H^2_{\mathrm{norm}}(\mathbb{N}_0,\mathbb{Z}^{\times})$ is a very difficult object. A detailed study of this and a partial combinatorial interpretation is made in [9].
Assume now in addition that $z > 0$, i.e. that the numbers $z(t(1),t(2))$ are non-zero. Then z is a 2-cocycle with values in the group $U(\mathbb{Q}) = \mathbb{Q} - \{0\}$, and the second cohomology group $H^2(\underline{L},U(\mathbb{Q}))$ becomes interesting. This is known in many cases. Examples:

(3.31) $\underline{L} = \mathbb{N}_0$: Then $H^2(\mathbb{N}_0$, abelian group$) = 0$ ([3],pp.192). The z from (3.27) is cohomologous to 1,i.e. $z = dc$, and we obtain

$$\mathbb{Q}[[T]] = \mathbb{Q}[[\mathbb{N}_0,\ dc]] \cong \mathbb{Q}[[X]],\quad f \longrightarrow \Sigma\{\frac{f(t)}{c(t)}X^t\ ;\ t \geq 0\}.$$

This is the theory of algebras of full binomial type in [8], pp. 122. □

(3.32) L = free monoid on a set I of letters: Again $H^2(\underline{L}$ abelian group$) = 0$ by [3], pp.192 , and $\mathbb{Q}[[T]] \cong \widehat{\mathrm{Ass}}(I)$ (1.20).

□

(3.33) $\underline{L} = \mathbb{N}_0^{(I)}$, $|I| \geq 2$: The Koszul complex calculations of [3],pp. 192, show that z is cohomologous to a bilinear, alternating map $b:\ \mathbb{N}_0^{(I)} \times \mathbb{N}_0^{(I)} \longrightarrow U(\mathbb{Q})$, and thus $\mathbb{Q}[[T]] \cong \mathbb{Q}[[\mathbb{N}_0^{(I)},b]]$. The latter algebra is a skew power series albegra. For I the set of all prime numbers $\mathbb{N}_0^{(1)}$ is isomorphic to $\mathbb{N}$ with multiplication and we get the theory of algebras of full Dirichlet type in [8], pp.116. □

(3.34) $\underline{L}$ = a countable directed set: Then again $H^2(\underline{L}$, abelian group) = 0 by [22], and then $\mathbb{Q}[[T]]$ = $\mathbb{Q}$-algebra of upper triangular $\underline{L} \times \underline{L}$-matrices. This generalizes [8], pp.127, from $\mathbb{N}_0$ to arbitrary $\underline{L}$, e.g. $\underline{L} = \mathbb{N}_0^{(I)}$, I countable. □

(3.35) Remark: The most general construction of the type (3.28) (ii), is a generalization of the crossed product construction in Brauer theory (see [21],p.242). Let $A:\underline{L} \longrightarrow \underline{Al}$ be any functor (suggestive: $\underline{L}$ operates on A by algebra homomorphisms) and let z be a normalized 2-cocycle with values in A , i.e.

$$z(t(1),t(2)) \in A(\mathrm{cod}\ t(1)) \text{ for } (t(1),t(2)) \in S_2(\underline{L}).$$

Then

$$A[[\underline{L},z]] := \Pi\{A(\mathrm{cod}\, t);\ t \in T = \mathrm{Morph}(\underline{L})\} \ni f = (f(t)\ ;\ t \in T)$$

with the multiplication

$$(fg)(t) = \Sigma\ \{z(u,v)f(u)\,{}^u g(v); uv = t\}\ ,\ {}^u g(v) = (Au)(g(v))$$

is an associative topological algebra. It is the most general skew power series algebra

C. "Sheaf-like" categories.

We introduce now those "sheaf-like" categories $\underline{K}$ for which k[[T]] becomes a bialgebra, generalizing the Faa di Bruno coalgebra [24] and most of the other examples there. Suppose that $\underline{K}$ satisfies the following conditions:

(3.36) (Finite limits): $\underline{K}$ admits arbitrary finite limits (lim). □

(3.37) (Finite coproducts): $\underline{K}$ admits finite coproducts. □

(3.38) (Homomorphism theorem): For any $f:X \longrightarrow Y$ in $\underline{K}$ there is an exact sequence

$$R(f) = X \underset{Y}{\times} X \underset{\mathrm{proj}_2}{\overset{\mathrm{proj}_1}{\rightrightarrows}} X \xrightarrow{\mathrm{can}} X/R(f)$$, and the induced morphism f_{ind} with

$f = f_{ind} \circ \mathrm{can}$ is a monomorphism. In particular, R (can) = R(f) as subobjects of $X \times X$, i.e. the equivalence relation R(f) is effective (see below). If X and Y are sets then $R(f) = \{(x(1),x(2)) \in X \times X;\ f(x(1)) = f(x(2))\}$. □

A morphism $f:X \longrightarrow Y$ is called an effective (= regular) epimorphism ([14], pp.180), if f_{ind} is an isomorphism, and an equivalence relation $R \subset X \times X$ is called effective if X/R exists and R = R(can) as subobjects of $X \times X$. Let M be the class of effective equivalence relations on X considered as subobjects of $X \times X$. Then with (3.17) we obtain the following

(3.39) Corollary: The map

$$\mathrm{Rel}(X) \longrightarrow M[X],\quad R \longrightarrow [X \longrightarrow X/R]$$

is an order antiisomorphism. In (3.21) one can thus replace M[X] by Rel(X). □

In the category Setf of finite sets Rel(X) is the ordered set of partitions of X.

(3.40)(Universality): Finite coproducts are universal (see[14],p.243, and [12], p.156), and disjoint, i.e. (i) if $f:X\longrightarrow Y$ is a morphism and $Y = \coprod Y(i)$ a finite coproduct, then $X \cong \coprod f^{-1}(Y(i))$, $f^{-1}(Y(i)) := X \times_Y Y(i)$, and (ii) if $X \cong X \amalg X$ then $X = 0$. Then the canonical morphisms $Y(i)\longrightarrow Y$ are monomorphisms. □

Finally we require finiteness for strict monomorphisms . A monomorphism $s:Y\longrightarrow X$ in K is called strict ("echt" in [12],p.16) if any commutative diagram (without the dotted h)

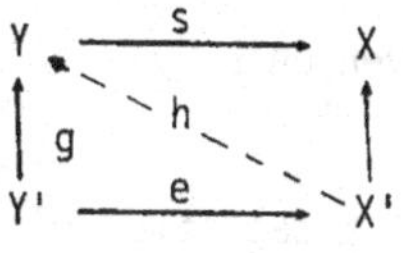

with an epimorphism e can be commutatively completed by an h which is then unique. For $X \in Ob(K)$ define $Sus(X) := \{s:Y\longrightarrow X$ strict monomorphism$\} / \approx$ with $s(1) \approx s(2)$ iff $s(2)h = s(1)$, h an isomorphism. Then Sus(X) is ordered like M[X] in (3.17).

(3.41) (Finiteness of strict monomorphisms): For $X \subset Ob(K)$ the set Sus(X) is finite. □

(3.42) The collection of conditions (3.36) to (3.41) is denoted by (Epi). □

If the above axioms are satisfied then also the sets Rel(X), and the Hom-sets Hom(X,Y) are finite. For enumeration problems (3.41) is a trivial requirement.

(3.43) Monomorphic situation: For the study of classes M of monomorphisms,i.e.subobjects, we do not dualize which is not interesting, but change only axioms (3.38) and (3.40) to

$(3.38)^{op}$ (Cohomomorphism theorem): Any f in K admits a factorization f = se with a strict monomorphism s and an epimorphism e. □

$(3.40)^{op}$ (3.40) plus: Epimorphisms are universal. □

In the monomorphic situation we denote the whole collection by (Mono).

(3.44) Main examples: The conditions (Epi) and (Mono) except the finiteness condition are satisfied in toposes = categories of sheaves ([14], pp.299). Actually, the theorem of Giraud ([14],p.303,[12],p.156, and [16],p.17) says that toposes are characterized by simple axioms of which the universality of coproducts is the most distinguishing. Many combinatorially interesting examples are given in [14],pp.311, see also §4 below. The conditions are also satisfied in elementary toposes [16]. The combinatorial standard example is

(3.44') I a category with finitely many objects and K := (I,Setf) = category of all functors from I to finite sets. For $X \in K$ the ordered set Sus(X) of subobjects of X is the Brouwer lattice af all subfunctors of X (compare [16],p.137).
Combinatorially standard examples are the following:
(a) $I = \mathbb{N}_0$ = the free monoid with one generator. The objects of $K = (\mathbb{N}_0,Setf)$ are pairs (M,s) of a finite set M with an endomorphism s.
(b) $I = \mathbb{Z}$ = the free group with one generator. As in (a) one obtains pairs (M,s) with a permutation s.

(c) I = the finite monoid on one generator g with the relation $g^m = g^n$, $1 \leq m \leq n < \infty$. The endomorphism s from (a) has to satisfy the same relation $s^m = s^n$. If $m = 1$, $n = 2$ this means that $s = s^2$ is an idempotent. □

However, the conditions (Mono) for the monomorphic situation are satisfied for combinatorially interesting categories which are no toposes. Special examples from the literature are the categories of finite topological spaces, ordered sets, graphs, (rooted) forests etc. The category of finite ordered sets does not satisfy the homomorphism theorem.

The next result shows how to derive new relevant categories from given ones by "induction".

(3.45) Induced categories: Let $F: \underline{K} \longrightarrow \underline{L}$ be a functor and assume that $\underline{L}$ satisfies (Mono) and that

$$F: \underline{K}(X(1),X(2)) \longrightarrow \underline{L}(FX(1), FX(2)) \ , \quad X(i) \in \underline{K},$$

is injective. Identify $f = Ff$, $F = \text{inj}$ for f in $\underline{K}$. An F-structure on $Y \in \underline{L}$ is an $X \in \underline{K}$ with $FX = Y$. Assume that F induces the following structures from $\underline{L}$ to $\underline{K}$:

(3.46) If $Y = \amalg F(X(i))$, $X(i) \in \underline{K}$, $Y \in \underline{L}$, is a finite coproduct with canonical morphisms $u(i) : F(X(i)) \longrightarrow Y$ then there is a unique F-structure X on Y such that $X = \amalg X(i)$ with canonical morphisms $u(i)$. □

$(3.46)^{op}$ The analogue of (3.46) for finite products. □

(3.47) If $Y' \xrightarrow{s} F(X)$, $Y' \in \underline{L}, X \in \underline{K}$, is a strict monomorphism in $\underline{L}$ there is a unique F-structure X' on Y' such that s is strict in $\underline{K}$, too and such that for all X'' in $\underline{K}$ $\underline{K}(X'',X') \cong \{f \in \underline{K}(X'',X);\ f = sg \text{ in } \underline{L}\} : g \longrightarrow sg$. □

Under these conditions also $\underline{K}$ satisfies the conditions (Mono). The combinatorial standard example is $\underline{L}$ = Setf = category of finite sets and $\underline{K}$ a category of sets with structure. A different approach is contained in [18] (catêgorie des espèces).

D. Krull-Schmidt categories and exponential formulas.

Krull-Schmidt categories are the suitable algebraic notion to distinguish between arbitrary and "connected" structures in combinatorics. They are abundant in this area because the finiteness conditions necessary for the proof of a "Krull-Schmidt"-theorem are trivially satisfied in enumeration problems.

An object X of a category $\underline{K}$ with finite coproducts is called indecomposable (in combinatorics often : connected) if $X \neq 0$ and if $X \cong X(1) \amalg X(2)$ implies $X(1)$ or $X(2) = 0$. A partition of X is a finite subset $\underset{\sim}{P} = \{[s:Y(s) \longrightarrow X]\}$ of $\text{Sus}(X)$ (see 3.41) such that $Y(s) \neq 0$ and such that the canonical morphism

$$(s;[s] \in \underset{\sim}{P}) : \amalg \{Y(s) ; [s] \in \underset{\sim}{P}\} \longrightarrow X$$

is an isomorphism. A partition is called a Krull-Schmidt(KS-)partition if the Y(s) are idecomposable.

(3.48) Definition: A category $\underline{K}$ is called a Krull-Schmidt(KS-)category if

(i) Finite coproducts in $\underline{K}$ exist and are disjoint (compare 3.40), and the canonical morphisms from the summands into the sum are strict monomorphisms, and if

(ii) each $X \in \underline{K}$ has a KS-partition, unique up to isomorphism.

(3.49) Example from section C: If a category $\underline{K}$ satisfies (Epi) or (Mono) from (3.42) respectively (3.43) then $\underline{K}$ is a KS-category with unique KS-partitions, i.e. the KS-partition of each object X is really unique, and not only unique up to isomorphism. □

The uniqueness of KS-partitions is typical for categories with universality of coproducts such as sheaves, ordered sets etc., but false for groups, modules etc. Often the dual situation with products instead of coproduct decompositions is considered (compare,for instance,[4],p.96).

If $\underline{K}$ is a skeletal-small KS-category and if $\underline{P}$ is a system of representatives of the isomorphism classes of indecomposable objects of $\underline{K}$ then the map

(3.50) $\mathbb{N}_0^{(\underline{P})} \longrightarrow \mathrm{Ob}(\underline{K})/\cong, \; n \longrightarrow X(n) := \amalg\{n(P)P; P \in \underline{P}\}$

is a monoid isomorphism where $n(P)P = P \amalg .. \amalg P$, n(P) times, and the addition on the right is induces from $\amalg$.

(3.51) Interpretation with connected components: In the situation of(3.44)we define for a functor $X \in (I,\underline{Setf})$ the oriented graph G(X) with vertex set $|X| := \cup\{X(i); i \in I\}$ and oriented edges (x,y), where $x \in X(i)$, $y \in X(j)$ and there is $\alpha : i \longrightarrow j$ with $(X\alpha)(x) = y$. Let $K \subset |X|$ be a connected component of the underlying non-oriented graph and define $X(K) \subset X$ by $X(K)(i) = K \cap X(i)$. Then X(K) is indecomposable and

$$X = \dot{\cup}\{X(K); K \text{ connected component of } |X|\}$$

is the unique KS-partition of X. □

If $\underline{K}$ is a KS-category and if $\underset{\sim}{P} = \{[s:Y(s) \longrightarrow X]\}$ is a partition of X, the type of $\underset{\sim}{P}$ is the vector $m = (m(n)) \in \mathbb{N}_0^{(T(1))}$, $T(1) := \mathbb{N}_0^{(\underline{P})} - \{0\}$, where

$$m(n) = \#\{[s] \in \underset{\sim}{P}; Y(s) \cong \amalg\{n(P)P; P \in \underline{P}\}\}.$$

From (1.33) one derives

(3.52) Theorem (Exponential formula for KS-categories):
Let $\underline{K}$ be a KS-category with finite automorphism groups. Notations as above. For $n \in T(1) := \mathbb{N}_0^{(\underline{P})} - \{0\}$ define $a(n) = |\mathrm{Aut}(\amalg n(P)P)|$. For $m \in \mathbb{N}_0^{(T(1))}$ let P(n,m) be the number of partitions of $\amalg n(P)P$ of type m. Then:

(i)

$$(3.53) \quad P(n,m) = \begin{cases} 0 & \text{if } \mathrm{weight}(m) := \Sigma\{m(t)t; t \in T(1)\} \neq n \\ \dfrac{a(n)}{a^m m!} & \text{if } \mathrm{weight}(m) = n. \end{cases}$$

(ii) If $X(P), P\in \underline{P}$, $T(n)$, $n\in T(1)$, and Y are indeterminates then one has in $\mathbb{Q}[Y,T][[X]]$

$\exp(Y\Sigma\{T(n)a(n)^{-1}X^n; n\in T(1)\}) = 1+\Sigma_n \Sigma_{k=1}^{|n|} B_{n,k}(T)Y^k a(n)^{-1}X^n$

with "partial Bell-Polynomial"

$B_{n,k}(T) = \Sigma\{P(n,m)T^m;\ \text{weight}(m) = n, |m| = k\}$. □

A comparable result in a different situation was derived by Stanley in [27], Theorem 3.2. The $M(n)$ from [27] coincide with the $P(n,n)$ from (3.53), i.e. are the numbers of KS-partitions of $X(n) = \amalg\, n(P)P$. For categories $\underline{K}$ with unique KS-partitions the $P(n,m)$ are given by the formula, (which is well-known in case $|\underline{P}| = 1$)

$$P(n,m) = n!\ [\Pi\{(t!)^{m(t)}m(t)!; t\in T(1)\}]^{-1}$$

and independent of $\underline{K}$. The typical case $\underline{K} = \underline{\mathrm{Setf}}^{(\underline{P})}$, $\underline{P}$ any set, is treated in [10].

(3.54) Corollary: In the situation of Theorem (3.52) assume moreover that $\underline{K}$ admits unique KS-partitions. Let $\underline{H}$ be another KS-category with unique KS-partitions and $F: \underline{H}\longrightarrow\underline{K}$ a functor which is injective (faithful) on morphisms and has the following "induction" property for direct sums (compare (3.46)):

F-structures on direct summands determine a unique F-structure on the direct sum, and, in turn, this F-structure on the direct sum uniquely determines the F-structures on the direct summands.

All these assumptions are satisfied in the "induction" situation of (3.45). For $n\in T(1) := \mathbb{N}_0^{(\underline{P})} - \{0\}$ let $f(n)$ be the number of indecomposable F-structures on $X(n) := \amalg\, n(P)P$. For $m\in\mathbb{N}_0^{(T(1))}$ and $k\in\mathbb{N}$ let $g(n,m,k)$ be the number of all F-structures Y on $X(n)$ such that the KS-partition $\underset{\sim}{P}$ of Y in $\underline{H}$ has k direct summands and such that the partition $F\underset{\sim}{P}$ of $FY = X(n)$ in $\underline{K}$ has type m. Then

(3.55) $\exp(Y\Sigma_n f(n)T(n)a(n)^{-1}X^n) =$

$$= 1 + \Sigma_n \Sigma_k B_{n,k}(fT)Y^k a(n)^{-1}X^n$$

where $B_{n,k}(fT) := \Sigma\{g(n,m,k)T^m;\ \text{weight}(m) = n, |m| = k\}$.

In particular, if $g(n)$ denotes the number of all F-structures on $X(n)$ we obtain from (3.56), by putting $Y = T(n) = 1$, the formula

(3.55') $\exp(\Sigma_n f(n)a(n)^{-1}X^n) = 1 + \Sigma_n g(n)a(n)^{-1}X^n$. □

This result is the typical connection between connected and arbitrary structures in exponential formulas (compare [28]) and extended to completely new situations in Theorem (3.74). Special examples, well-known from the literature (see, for instance, [28]), are the following. A nontrivial new example is given in §4, A, (4.12).

Examples: (i) If $\underline{K}$ is the category of finite sets and $\underline{H}$ that of finite graphs and if F is the functor mapping a graph to its vertex set, then $\underline{P} = \{\{1\}\}$, $T(1) = \mathbb{N}$, and $X(n) = \{1,\dots,n\}$. Since the number of all graphs with vertex set $\{1,\dots,n\}$ is $2^{\binom{n}{2}}$ one obtains

$$\exp\left(\Sigma_{n=1}^{\infty} f(n)\frac{X^n}{n!}\right) = 1+\Sigma_{n=1}^{\infty} 2^{\binom{n}{2}}\frac{X^n}{n!} \quad ,$$

where $f(n)$ is the number of connected graphs on $\{1,\dots,n\}$.

(ii) $\underline{K}$ as in (i). Let $\underline{H}$ be the category of finite sets with an operation of the cyclic group of order d, i.e. with a permutation s satisfying $s^d = 1$, and F the underlying functor. An easy calculation and (3.55') imply

$$\exp\left(\Sigma_{n|d} \frac{X^n}{n}\right) = 1 + \Sigma_{n=1}^{\infty} g(n)\frac{X^n}{n!} \quad ,$$

where $g(n)$ denotes the number of all solutions of $s^d = 1$ in S_n. □

E. Sheaf-like categories and unipotent groups

In this section we assume that $\underline{K}$ is a category which satisfies the assumptions (Epi) from (3.42). We show that these assumptions give rise to a natural construction of a combinatorial incidence algebra as in section A and a derived unipotent group. If $\underline{K}$ is the category of finite sets the corresponding affine bialgebra is the "Faa di Bruno" bialgebra from [24], pp.36.

By (3.49) $\underline{K}$ is a KS-category. Let $\underline{P}$ be a system of representatives of indecomposable objects of $\underline{K}$ modulo isomorphism, and let M be the class of effective epimorphisms of $\underline{K}$. We consider M as a full subcategory of the category Morph($\underline{K}$) of all morphisms of $\underline{K}$. As a functor category this inherits the properties of $\underline{K}$ and is thus itself a KS-category; M is closed under finite direct sums and taking direct summands in Morph($\underline{K}$) As equivalence relation on M we take isomorphy in Morph($\underline{K}$), and obtain $T := M/\cong$. These data ($\underline{K}$,M,~) satisfy the assumptions made in Theorem (3.15), and we obtain the incidence algebra k [[T]] with standard basis $e(t)$, $t \in T$. As for any KS-category (compare (3.50)), Morph($\underline{K}$)/$\cong$ is a free abelian monoid with addition induced from direct sum and the indecomposables as basic, and $T = M/\cong$ is a free submonoid of Morph($\underline{K}$)/$\cong$. The universality of ⨿ (3.40) implies that an effective epimorphism $f: X \longrightarrow Y$ in M ($\subset$Morph($\underline{K}$)) is indecomposable if and only if Y is indecomposable. On the other hand, if Y is indecomposable identify $Y = id_Y$ for $Y \in Ob(\underline{K})$. Let S denote a system of representatives of the isomorphism classes of effective indecomposable epimorphisms. With the above remarks one obtains without loss of generality a decomposition

(3.56) $S = S' \cup S''$, $S' = \{P = id_P; P \in \underline{P}\}$, $S'' = \{s \in S;\ s \text{ is not an isomorphism}\}$.

Then $U := \{\bar{s}, s \in S\}$ is a basis of T, and this basis also admits a decomposition

(3.57) $U = U' \dot{\cup} U''$, $U' = \{id_P; P \in \underline{P}\}$, $U'' = \{\bar{s}; s \in S''\}$.
This decomposition of the basis U of T gives a direct sum decomposition of abelian monoids

(3.58) $T := T' \oplus T''$, where $T'' := \oplus \{\mathbb{N}_0 u''; u'' \in U''\}$ and

$$T' := \mathrm{Isom}(T) = \oplus\{\mathbb{N}_0 u'; u' \in U'\} = \oplus\{\mathbb{N}_0 \overline{id_P}; P \in \underline{P}\} = \oplus\{\mathbb{N}_0 \bar{P}; P \in \underline{P}\} = \mathrm{Ob}(\underline{K})/\cong,$$

where $T' = \mathrm{Ob}(\underline{K})/\cong$ is identified through $\overline{id_Y} = \bar{Y}$, $Y \in \mathrm{Ob}(\underline{K})$. Hence $\mathrm{Ob}(\underline{K})/\cong$ from (3.50) is a free direct summand of $T = M/\cong$.
The monoid structure of T induces a topological coalgebra structure on k[[T]] by

(3.59) $\Delta : k[[T]] \longrightarrow k[[T]] \hat{\otimes} k[[T]]$, $e(t) \longrightarrow \Sigma\{e(t_1) \otimes e(t_2); t_1 + t_2 = t\}$
$\varepsilon : k[[T]] \longrightarrow k$, $e(t) \longrightarrow \delta_{t0}$.

Again the universality of $\amalg$ (3.40) implies the compatability of the algebra and coalgebra structure, and hence

(3.60) Theorem: Assumptions and definitions as above. Then $H := k[[T]]$ is a topological bialgebra, and (3.15), an abstract incidence algebra. □

Rota ([24],p.13) considers the dual situation. Let $A := k[T]$ be the dual (abstract) bialgebra of H with the k-basis $x(t)$, $t \in T$, multiplication
$x(t_1)x(t_2) = x(t_1 + t_2)$, $1_A = x(0)$, comultiplication

(3.61) $\Delta : A \to A \otimes A$, $x(t) \to \Sigma\{G(t;t(1)t(2))x(t(1)) \otimes x(t(2)); t(i) \in T\}$

(3.62) and counit $\varepsilon : A \longrightarrow k$, $x(t) \longrightarrow 1$ if $t \in T'$, 0 otherwise, the duality being given by

$$k[[T]] \times k[T] \longrightarrow k, \quad (e(t), x(t')) \longrightarrow \delta(t,t').$$

By definition, k[T] is the monoid algebra of T, i.e. the polynomial algebra in indeterminates $x(u)$, $u \in U$. For the category $\underline{K}$ = Setf of finite sets Rota ([24],pp.36) calls k[T] the Faa di Bruno bialgebra. In this case

$$S' = \{id: \{1\} \longrightarrow \{1\}\}, \quad S'' = \{const: \{1,\ldots,n\} \longrightarrow \{1\}; n \geq 2\},$$

i.e. $U = \mathbb{N}_0$, $U'' = \mathbb{N}$ by identification.
The algebra $A = k[T]$ defines the k-free affine monoid G by

(3.63) $G(R) := \mathrm{Al}(k[T],R) \cong \mathrm{Mon}(T,R)$, $g \to f$, $g(x(t)) = f(t)$,
where Mon(T,R) denotes the set of multiplicative ([24],p.40) functions f, i.e. $f(0)=1$ and $f(t(1)+t(2)) = f(t(1))f(t(2))$. If we identify in (3.63) and use the identification of (2.22) and (2.24) we obtain

(3.64) $G(R) = \mathrm{Mon}(T,R) \subset R[[T]] : f = \Sigma f(t)e(t)$,
i.e. $G(R) = \mathrm{Mon}(T,R)$ is a multiplicative submonoid of R[[T]]. Of course, since T is free on $u \in U$, f is already determined by the f(u). Contrary to the impression from [24] and other sources, since G is represented by k[T], the study of G(R) for all R is equivalent to that of k[T] or k[[T]] and not simpler.

In general, G is not a group. However, a group can easily be derived from G with preservation of the essential properties. Compare the remark in [24], p.42. The decomposition $T = T' \oplus T''$ (3.58) and a little calculation imply that

(3.65) $\quad A' := k[T'] \subset A = k[T] = k[T' \oplus T''] = k[T'] \otimes k[T'']$

is a subbialgebra and A is A'-free, hence flat, hence there is a faithfully flat epimorphism

(3.66) $\quad G = Al(k[T],-) \xrightarrow{Res} G' = Al(k[T'],-)$

of k-free affine monoids where Res is the restriction. Let

(3.67) $\quad G'' := \ker(Res) = Res^{-1}(1)$

be the kernel of Res. Then

(3.68) $\quad G''(R) = \{f \in Mon(T,R); f(T') = 1\} = Mon(T'',R) \subset R[[T]]$,

where an $f'' \in Mon(T'',R)$ is identified with an $f \in Mon(T,R)$ by

$$f = \Sigma_t f(t)e(t) = \Sigma_{t''} f''(t'')(\Sigma_{t'} e(t' + t'')).$$

The cartesian diagram (3.69)

$$(3.69)\quad \begin{array}{ccc} G & \xrightarrow{Res} & G' \\ \cup & & \cup \\ G'' & \longrightarrow & 1 \end{array} \qquad (3.70)\quad \begin{array}{ccc} k[T] & \supset & k[T'] \\ \downarrow k[proj] & & \downarrow \varepsilon \\ k[T''] & \supset & k \end{array} \qquad (3.71)\quad \begin{array}{ccc} k[[T]] & \xrightarrow{can} & k[[T']] \\ \uparrow k\,[[proj]] & & \uparrow \eta \\ k[[T'']] & \longrightarrow & k \end{array}$$

induces the cocartesian diagram (3.70) of the affine algebras, where $k[T'']$ is the affine algebra of G" and proj : $T \longrightarrow T''$ denotes the projection in (3.58). Dualization implies the commutative diagram of topological algebras (3.71) where the upper map can is that of (3.15) (iii), and the left k [[proj]] is given by $f'' \longrightarrow f''$ proj. The standard basis of $k[[T'']]$ is mapped onto the elements

$$(3.72)\left\{\begin{array}{l} e''(t'') := \Sigma\{e(t' + t''); t' \in T'\} \in k[[T]], \text{ thus} \\ k[[T'']] = \Pi\{ke''(t''), t'' \in T''\} \text{ by identification.} \end{array}\right.$$

We finally identify by (3.58)

(3.73) $\quad T = \mathbb{N}_0^{(U)}, T' = Iso(T) = \mathbb{N}_0^{(U')} = \mathbb{N}_0^{(\underline{P})}, T'' = \mathbb{N}_0^{(U'')}$

and write n instead of t. Let $x = (x(u); u \in U)$ and $y = (x(u); u \in U'')$. Then $k[T] = k[x]$, $k[T''] = k[y]$.

(3.74) <u>Theorem</u>: Conditions (Epi) from (3.42), $T := \{\text{effective epimorphisms}\}/\cong\, = \mathbb{N}_0^{(U)}$ Notations from above and § 2 . Then

(i) $G'' = G \cap \Lambda_{k[[T]]}$, and this is a closed, unipotent, k-free subgroup of $\Lambda_{k[[T]]}$. The affine algebra of G" is $k[T''] = k[y]$ with the universal element

(3.75) $\quad 1 + \Sigma\{y^n e''(n); n \neq 0\} \in G''(k[y]) \subset k[y][[T]]$.

The covariant topological algebra of G" is

$$k[[T'']] = \Pi\{ke''(n); n \in T'' = \mathbb{N}_0^{(U'')}\} \subset k[[T]].$$

(ii) The unit of $k[[T'']]$ is $e''(0) = 1_{k[[T]]}$, the coalgebra structure comes from the additive structure of $T'' = \mathbb{N}_0^{(U'')}$, and the multiplication constants $G''(n;n_1n_2)(n,n_1,n_2 \in T'')$ of $k[[T'']]$ with respect to the basis $e''(n)$, $n \in T''$, are

$$(3.76)\quad G''(n;n_1n_2) = G\binom{n}{n_1 + cod(n) - cod(n_1),\ n_2 + dom(n) - dom(n_2)}$$

if $cod(n_1) \leq cod(n)$, $dom(n_2) \leq dom(n)$ in the ordered monoid $T' = \mathbb{N}_0^{(P)}$, and 0 otherwise. Here $G\binom{n}{n_1 n_2}$ are of course the section coefficients of $k[[T]]$ with respect to the standard basis $e(n)$, $e \in T$.

(iii) The antipode $S:k[y] \longrightarrow k[y]$ of G'', i.e. the inverse of the universal element, is (1.31) given by

$$(3.77)\quad S(y^p) = \Sigma\ \{a(p,n)y^n;\ n \neq 0 \text{ in } T''\}\ ,\ 0 \neq p \text{ in } T'' = \mathbb{N}_0^{(U'')}\ , \text{ where}$$

$$a(p,n) = \Sigma\ \{(-1)^{|m|}GS''(p;m);\ weight(m) = n\},$$

m runs over the elements of $\mathbb{N}_0^{(T'')}$, $weight(m) = \Sigma\{m(n)n;n \in T''\}$ and $GS''(p;m)$ is derived from G'' as GS from G in (1.26). In particular, $S(x(u)) = \Sigma\{a(u,n)y^n;\ 0 \neq n \in T''\}$, where $u \in U'' \subset \mathbb{N}_0^{(U'')}$ is the u-th standard basis vector.

(iv) The Lie algebra $Lie(G'') \subset k[[T'']] \subset k[[T]]$ of G'' (2.39) has the topological k-basis $e''(u)$, $u \in U''$, and the Lie bracket is

$$(3.78)\quad [e''(u_1),e''(u_2)] = \Sigma\ \{[G''(u;u_1u_2) - G''(u;u_2u_1)]\ e''(u);u \in U''\} \text{ for } u_1,u_2 \in U''.$$

Thus Lie(G'') is determined by the indecomposable effective epimorphisms.

(v) (Exponential formula) If $\mathbb{Q} \subseteq k$ the exponential isomorphism

$$exp : R \hat{\otimes} Lie(G'') \longrightarrow G''(R), \text{ functorial in } R,$$

(from (2.42) again, as in (3.55)) gives a connection between indecomposable items on the left and all items on the right. □

The ζ-functions is the multiplicative function $\zeta:T \longrightarrow k$, $\zeta(t) = 1$, i.e. $\zeta(x(u)) = 1$ for all $u \in U$. This function is obviously contained in $G''(k)$, Hence $\mu = \zeta^{-1}$, the Möbius function, is also in $G''(k)$. But $\zeta = \zeta \circ id_{k[T]}$, $\mu = \zeta^{-1} = \zeta \circ S$, hence μ can be obtained by setting $x(u) = 1$ in (3.77). For $n \in T = \mathbb{N}_0^{(U)}$ let n'' be its component in $T'' = \mathbb{N}_0^{(U'')}$.

Then $n \in T(1) := T - Iso(T)$ if and only if $n'' \neq 0$.

(3.79) Corollary: Let $\mu \in k[[T]]$, $\mu = \Sigma\,\mu(n)e(n)$, be the Möbius function in $k[[T]]$. Then

$$\mu(n) = \begin{cases} 1 & \text{if } n \in Iso(T) = \mathbb{N}_0^{(P)}, \text{i.e. } n'' = 0 \\ \Sigma\{(-1)^{|m|}GS''(n'';m);m \in \mathbb{N}_0^{(T'')} - \{0\}\} & \text{if } n'' \neq 0. \end{cases}$$ □

(3.80) Remark: In the situation (Mono) of (3.43) one obtains a corresponding theory for the class of strict monomorphisms instead of effective epimorphisms. In this case the ordered set Sus(X) of strict subobjects of X is a distributive lattice, but in general not a Boolean algebra. However, if Sus(X) is a Boolean algebra for all

$X \in Ob(\underline{K})$, an equivalence relation on M satisfying the conditions (3.8)ff is given by

$$s_1 : Y_1 \longrightarrow X_1 \sim s_2 : Y_2 \longrightarrow X_2 \text{ iff } Y_1' \cong Y_2' ,$$

where $s_i' : Y_i' \longrightarrow X_i$ denotes the complement of $s_i : Y_i \longrightarrow X_i$ in $Sus(X_i)$.
Many examples in [24] are of this type.

(3.81) Example: The Butcher group
Let $\underline{K}$ be the category of finite rooted forests, $\underline{P}$ a system of representatives of finite rooted trees modulo isomorphism and M the class of (strict) monomorphisms. We define an equivalence relation on M satisfying the conditions (3.8)ff by

$$s_1 : Y_1 \longrightarrow X_1 \sim s_2 : Y_2 \longrightarrow X_2 \quad \text{iff } X_1 - s_1(Y_1) \cong X_2 - s_2(Y_2),$$

where $X_i - s_i(Y_i)$ is the rooted forest obtained from X_i by removing $s_i(Y_i)$.
The types are just the isomorphism classes of rooted forests, hence $T = \mathbb{N}_0^{(\underline{P})}$, and $k[[T]]$ is a topological bialgebra. The unipotent group $Mon(T,R)$ is isomorphic to the Butcher group - known from the theory of Ringe-Kutta-methods in numerical mathematics (see [31])- and admits a power series representation by Butcher series.

(3.82) Remark: The example ([24],pp.89) of finite ordered sets with the direct product decomposition and its specializations (matroids etc.) do not directly fall into the two cases (Epi) or (Mono) of this paragraph. The theory can, however, easily be adapted to this case, in particular, to obtain an analogue of Theorem (3.74).

§ 4 . Examples

A. Enumerations of effective epimorphisms and strict monomorphisms

Let $\underline{K}$ be a category satisfying the conditions (Epi) from (3.42) (or (Mono) from (3.43)), let M be the class of effective epimorphisms (or strict monomorphisms) of $\underline{K}$ and let $\underline{P}$ be a representative system of indecomposable objects in $\underline{K}$ modulo isomorphism. By (3.50), $Ob(\underline{K})/\cong$ is a monoid with elements $[X] := X$ modulo isomorphism and addition $[X(1)] + [X(2)] = [X(1) \amalg X(2)]$, and isomorphic to $\mathbb{N}_0^{(\underline{P})}$. We identify

(4.1) $\quad \mathbb{N}_0^{(\underline{P})} = Ob(\underline{K})/\cong \ , \ n = [\amalg\, n(P)P; P \in \underline{P}]$.

Moreover, if I is any index set we identify $i = (0\ldots010\ldots0)$, the entry 1 in the i-th position, and obtain I as the standard basis of $\mathbb{N}_0^{(I)}$. In particular,

(4.2) $\quad \underline{P} \subset \mathbb{N}_0^{(\underline{P})} = Ob(\underline{K})/\cong \ , \ P = [P] = (0\ldots010\ldots0)$, 1 at p-th place. By (3.50), $M/\cong$ is a free abelian monoid with the $\bar{f} := f$ modulo isomorphism, $f: X \longrightarrow Y$ indecomposable, as basis. Here, $f \in M$ is indecomposable if and only if Y is indecomposable. For such an f we define the type of f by

(4.3) $\quad type(f) = ([X],[Y]) \in \mathbb{N}_0^{(\underline{P})} \times \underline{P} \subset \mathbb{N}_0^{(\underline{P})} \times \mathbb{N}_0^{(\underline{P})}$, where we use the identifications (4.1) and (4.2). Define the set of indecomposable types by

(4.4) $\underline{Q}$:= {type(f);f ∈ M indecomposable} $\subset \mathbb{N}_0^{(\underline{P})} \times \mathbb{N}_0^{(\underline{P})}$.

We consider $\underline{Q} \subset \mathbb{N}_0^{(\underline{Q})}$ as standard basis as in (4.2), and call $T := \mathbb{N}^{(\underline{Q})}$ the free monoid of types. Extending linearly type(f) from the basis $\underline{Q}$ to $\mathbb{N}_0^{(\underline{Q})} = T$ we obtain the surjective monoid homomorphism

(4.5) $\text{type} : M/\cong \longrightarrow T = \mathbb{N}_0^{(\underline{Q})}$, $\bar{f} \longrightarrow \text{type}(f) = \Sigma\{\text{type}(f)(n,P)(n,P);(n,P) \in \underline{Q}\}$.

If here $f:X \longrightarrow Y$ is in M and $Y = \amalg_i Y_i$ is the KS-partition of Y, then $f = \amalg_i f_i, f_i := f|f^{-1}(Y_i) : f^{-1}(Y_i) \longrightarrow Y_i$, is the KS-partition of f, and

(4.6) $\text{type}(f)(n,P) = \#\{i;f^{-1}(Y_i) \cong \amalg_{p'} n(P')P', Y_i \cong P\}$.

We define type(R) := type($X \xrightarrow{\text{can}} X/R$) for an effective equivalence relation R on X and type ([s]) := type(s) for a strict subobject [s]. The injection $\underline{Q} \subset \mathbb{N}_0^{(\underline{P})} \times \mathbb{N}_0^{(\underline{P})}$ extends linearly to

(4.7) (weight, absolute value) : $\mathbb{N}_0^{(\underline{Q})} \longrightarrow \mathbb{N}_0^{(\underline{P})} \times \mathbb{N}_0^{(\underline{P})}$, $t \longrightarrow (\text{weight}(t), |t|)$,
where $\text{weight}(t) = \Sigma\{t(n,P)n;(n,P) \in \underline{Q}\}$ and $|t| = \Sigma\{t(n,P)P ; (n,P) \in \underline{Q}\}$
Setting

(4.8) $\underline{Q}(P) := \{n \in \mathbb{N}_0^{(\underline{P})}; (n,P) \in \underline{Q}\}$, $\underline{Q}(n) := \{P \in \underline{P};(n,P) \in \underline{Q}\}$

we obtain $|t|(P') = \Sigma\{t(n,P');n \in \underline{Q}(P')\}$.

The composition of the maps (4.5) and (4.7) is given by

$$(4.9) \qquad \begin{array}{l} M/\cong \xrightarrow{\text{type}} \mathbb{N}_0^{(\underline{Q})} \xrightarrow{(\text{weight},|\,|)} \mathbb{N}_0^{(\underline{P})} \times \mathbb{N}_0^{(\underline{P})} \\ \bar{f} \longrightarrow ([\text{dom}(f)],[\text{cod}(f)]) \; , \end{array}$$

i.e. if $f:X \longrightarrow Y$ is in M and has type t = type(f), then
$X \cong \amalg n(P)P$ and $Y \cong \amalg m(P)P$ with n := weight(t) and m := |t|.

(4.10) Standard example:
Let $\underline{K}$ = Setf be the category of finite sets. Then $\underline{P}$ = {{1}} and Setf/$\cong = \mathbb{N}_0$, [X] = |X|

(i) A system of representatives of indecomposable effective epimorphisms modulo isomorphism are the constant maps

$$\{1,\dots,n\} \longrightarrow \{1\},\ n \geq 1, \quad \text{hence } \underline{Q} = \{(n,1);\ n \in \mathbb{N}\} \cong \mathbb{N}.$$

Then (weight, | |) : $\mathbb{N}_0^{(\mathbb{N})} \longrightarrow \mathbb{N}_0 \times \mathbb{N}_0$, $t \longrightarrow (\Sigma\{t(n)n;n \geq 1\}, \Sigma\{t(n);n \geq 1\})$.

We identify an equivalence relation R on X with the partition X/R of X. Then the type $t \in \mathbb{N}_0^{(\mathbb{N})}$ of R is given by
t(k) = number of blocks of X/R with k elements, k ≥ 1.
Here weight(t) = |X| and |t| = |X/R|. The number of partitions of type t of a set with n elements, n = weight(t), is

$$R(t) := n!\,\Big(\prod_{k\in\mathbb{N}} (k!)^{t(k)} t(k)!\Big)^{-1}.$$

(ii) A system of representatives of indecomposable strict monomorphisms modulo isomorphism are the constant maps $\emptyset \longrightarrow \{1\}$ and $\{1\} \longrightarrow \{1\}$, hence $\underline{Q} = \{(0,1),(1,1)\} \cong \{0,1\}$. Then $(\text{weight}, |\ |) : \mathbb{N}_0^{\{0,1\}} \longrightarrow \mathbb{N}_0 \times \mathbb{N}_0$, $t \longrightarrow (t(1), t(0)+t(1))$.
We identify a subobject $[s : X \longrightarrow Y]$ of Y with the subset $s(X) \subset Y$.
Then the type $t \in \mathbb{N}_0^{\{0,1\}}$ of $[s]$ is given by $t(0) = |Y| - |X|$, $t(1) = |X|$.
Here $\text{weight}(t) = |X|$ and $|t| = |Y|$. The number of subsets of type t of a set with n elements, $n = \text{weight}(t)$, $m = |t|$, is $S(t) := \binom{m}{n}$.

We generalize this to

(4.11) <u>Theorem</u>: Situation as above. Let $t \in T = \mathbb{N}_0^{(\underline{Q})}$ be a type with $\text{weight}(t) = n$ and $|t| = m$. Then

$$M(t) := {}^{\#}\{f \in M; f: \amalg_P n(P)P \longrightarrow \amalg_{P'} m(P')P' \text{ has type } t\} =$$

$$= \Big(\prod_{P\in\underline{P}} n(P)!m(P)!\Big) \prod_{(n',P')\in\underline{Q}} \Big(M(n',P')\big(\prod_{P\in\underline{P}} n'(P)!\big)^{-1}\Big)^{t(n',P')} (t(n',P')!)^{-1},$$

where $M(n',P') = {}^{\#}\{f \in M;\ \amalg_P n'(P)P \longrightarrow P'\}$.
If M is the class of effective epimorphisms of $\underline{K}$, then
$\text{Aut}\,(\amalg_{P'} m(P')P')$ acts freely on $\{f \in M; f : \amalg_P n(P)P \longrightarrow \amalg_{P'} m(P')P'\}$ and we get

$$R(t) := {}^{\#}\{R \in \text{Rel}\,(\amalg_P n(P)P);\ \text{type}\,(R) = t\} = M(t)a(m)^{-1}.$$

If M is the class of strict monomorphisms of K, then
$\text{Aut}\,(\amalg_P n(P)P)$ acts freely on $\{f \in M; f: \amalg_P n(P)P \longrightarrow \amalg_{P'} m(P')P'\}$ and we get

$$S(t) := {}^{\#}\{S \in \text{Sus}\,(\amalg_{P'} m(P')P');\ \text{type}(S) = t\} = M(t)a(n)^{-1}. \qquad \square$$

The numbers $R(t)$ or $S(t)$ are often closely connected with the section coefficients of the AIA $H := k[[T]]$ from (3.60)(see, for instance, (4.18)).

(4.12) <u>An application of the exponential formula</u>:
We consider M as a full subcategory of the category $\text{Morph}(\underline{K})$ of all morphisms of $\underline{K}$. An element $f \in M$ is indecomposable if and only if $\text{cod}\,(f)$ is indecomposable. Since the functor

$$F : M \longrightarrow \underline{K} \times \underline{K},\ (f: X \longrightarrow Y) \longrightarrow (X,Y)$$

satisfies all conditions of (3.54), (3.55') implies

$$(4.13)\quad \exp\Big(\sum_{(n,P)\in\underline{Q}} M(n,P)\,\frac{X^n}{a(n)}\,\frac{Y(P)}{a(P)}\Big) = \sum_{n,m\in\mathbb{N}_0^{(\underline{P})}} M(n,m)\,\frac{X^n}{a(n)}\,\frac{Y^m}{a(m)}$$

and

$$(4.14)\quad \prod_{P\in\underline{P}}\Big(\sum_{n\in\underline{Q}(P)} M(n,P)\,\frac{X^n}{a(n)}\Big)^{m(P)} = \sum_{n\in\mathbb{N}_0^{(\underline{P})}} M(n,m)\,\frac{X^n}{a(n)},$$

where $M(n,m) = \#\{f \in M; \amalg_P n(P)P \longrightarrow \amalg_{P'} m(P')P'\}$ and $X(P), Y(P), P \in \underline{P}$ are indeterminates.

(4.13) and (4.14) are not only the generatung functions for the $M(n,m)$, but also for the numbers of effective equivalence relations and strict subobjects.
If $\underline{K}$ is the category of finite sets, (4.13) and (4.14) reduce to well-known results about Bell numbers, Stirling numbers of the second kind and binomial numbers.
For instance,

$$\exp(\exp X - 1) = \sum_{n=0}^{\infty} B(n) \frac{X^n}{n!}, \quad \frac{(\exp X - 1)^m}{m!} = \sum_{n=0}^{\infty} S(n,m) \frac{X^n}{n!}$$

and $(1+X)^m = \sum_{n=0}^{\infty} \binom{m}{n} X^n$.

Our next example is more complicated, but still very similar to finite sets.

(4.15) Equivariant partitions of G-sets: Let G be a finite group and $K := G\text{-}\underline{\mathrm{Setf}}$ the presheaf category of finite G-sets, i.e. sets with an operation of G from the left. The epimorphisms of $\underline{K}$ are the G-homogeneous surjections and effective. If X is a G-set an equivalence relation $R \in \mathrm{Rel}(X)$ is simply an equivalence relation $R \subset X \times X$ such that $(x,y) \in R$, $g \in G$ implies $(gx,gy) \in R$, or equivalently, that G permutes the blocks of X/R. In this case we call X/R an equivariant partition of X. Let S be a system of representatives of conjugacy classes of subgroups of G, ordered by $U \leq V$ if there is a $g \in G$ so that $g^{-1}Ug \subset V$. Then $\underline{P} = \{G/U; U \in S\}$ is a representative system of indecomposable G-sets modulo isomorphism. For $U,V \in S$ we define

$$(U,V) := {}^{\#}\{g \in G; g^{-1}Ug \subset V\}$$

and w.l.o.g. we set $\underline{Q} = \{(n,U) \in (\mathbb{N}_0^S - \{0\}) \times S;\ \mathrm{supp}\, n \leq U$, i.e. for all $V \in \mathrm{supp}\, n$: $V \leq$ where $\mathrm{supp}\, n := \{V \in S; n(S) \neq 0\}$, and $T = \mathbb{N}_0^{(\underline{Q})}$ .From (4.13) we get the generating functions of

$$S(n,m) := {}^{\#}\{R \in \mathrm{Rel}(\amalg_U n(U)G/U); (\amalg_U n(U)G/U)/R \cong \amalg_V m(V)G/V\} \text{ and}$$

$B(n) := {}^{\#}(\mathrm{Rel}(\amalg_U n(U)G/U))$

$$(4.16) \quad \sum_{n,m \in \mathbb{N}_0^S} S(n,m) \frac{X^n}{a(n)} Y^m = \exp\{ \sum_{U \in S} [\exp(\sum_{V \leq U} \frac{(V,U)}{|U|} \frac{X(V)}{a(V)} -1] \frac{Y(U)}{a(U)} \}$$

and

$$(4.17) \quad \sum_{n \in \mathbb{N}_0^S} B(n) \frac{X^n}{a(n)} = \exp\{ \sum_{U \in S} \frac{|U|}{(U,U)} [\exp(\sum_{V \leq U} \frac{(V,U)}{|U|} \frac{X(V)}{a(V)}) -1] \},$$

where $a(n) = \prod_{U \in S} \left(\frac{(U,U)}{|U|}\right)^{n(U)} n(U)!$.

By differentiating we derive the recursive relations

$$B(n+U) = \sum_{\substack{k,l \in \mathbb{N}^S \\ k+l = n^{o}}} \binom{n}{k} B(k) \{ \sum_{\substack{V \geq U \\ \mathrm{supp}\, l \leq V}} \frac{(U,V)}{(V,V)} \prod_{W \leq V} \frac{(W,V)}{|V|}^{l(W)} \}$$

and

$$S(n+u, m) = \sum_{V \geq U} \{ m(V) \frac{(U,V)}{|V|} S(n,m) + \frac{(U,V)}{(V,V)} S(n,m-V) \} .$$

The number of G-homogeneous maps from $\amalg_U n(U)G/U$ onto $\amalg_V m(V)G/V$ is

$$M(n,m) = S(n,m)a(m) = \sum_{\substack{k \in \Pi[0,m(U)] \\ U}} \prod_U (-1)^{m(U)-k(U)} \binom{m(U)}{k(U)} \left(\sum_{V \geq U} k(V) \frac{(U,V)}{|V|} \right)^{n(U)} .$$

Applying our results to finite cyclic groups we can solve the following combinotorial problem:

Let X be a finite set and s a permutation of X.

How many partitions P of X are there such that s permutes the blocks of P?

Let a be the least common multiple of the lengths of cycles of s and $G := \mathbb{Z}/\mathbb{Z}a$ be the cyclic group of order a. Then G acts on X by

$$\bar{k} \circ x := s^k(x), \text{ where } \bar{k} \in G \text{ and } x \in X.$$

Let $S := \{d\mathbb{Z}/\mathbb{Z}a ; d \text{ divides } a\}$ be the set of all subgroups of G and define $n \in \mathbb{N}_0^S$ by

$$n(d\,\mathbb{Z}/\,\mathbb{Z}a) := \text{number of cycles of s of length d.}$$

Then the number of partitions P of X such that s permutes the blocks of P is just B(n) from above.

B. Two unipotent groups in combinatorics

(4.18) A generalization of the Faa di Bruno bialgebra: Let now G = A be a finite abelian group in (4.15). Then $S = P(A)$ is the set of all subgroups of A, i.e. $\underline{Q} \subset \mathbb{N}_0^{P(A)} \times P(A) \subset \mathbb{N}_0^{P(A)} \times \mathbb{N}_0^{P(A)}$. In this case, as for the special case Setf = 1-Setf, the map (4.6) type: $M/\cong \longrightarrow \mathbb{N}_0^{(\underline{Q})}$ is an isomorphism, i.e. two effective epimorphisms f and g are isomorphic if and only if type(f) = type(g). We identify $M/\cong\, = T = \mathbb{N}_0^{(\underline{Q})}$. According to (3.15) and (3.60), k[[T]] is an AIA and a topological bialgebra. Using (4.11), we can calculate the elementary section coefficients:

(4.19) Theorem: Let (n,U) be an element of $\underline{Q} \subset \mathbb{N}_0^{P(A)} \times P(A)$.

For types $t_1, t_2 \in T$, $G((n,U);t_1,t_2) = 0$ unless

$t_2(m,V) = 0$ for all $(m,V) \in \underline{Q}$ with $V \not\subseteq U$, weight(t_2) = n and $t_1 = (|t_2|, U)$.

If $t \in T$ satisfies $t(m,V) = 0$ for all $(m,V) \in \underline{Q}$ with $V \not\subseteq U$ and the weigth of t is n, then $G((n,U);(|t|,U),t) =$

$$= \left(\prod_{H \subseteq U} n(H)! \right) \prod_{\substack{(m,V) \in \underline{Q} \\ V \subseteq U}} \left\{ \left(\frac{|U|}{|V|} \prod_{H \subseteq V} \left[\left(\frac{|V|}{|U|} \right)^{m(H)} n(H)! \right] \right)^{t(m,V)} t(m,V)! \right\}^{-1} \qquad \square$$

For sets this is the Faa di Bruno coefficient

$$G(n;|t|,t) = n! \left(\prod_{k=1}^{\infty} (k!)^{t(k)} t(k)! \right)^{-1}, \text{ where } t \in \mathbb{N}_0^{(\mathbb{N})} \text{ has weight } n.$$

The following theorem shows the structure of the monoid Mon(T,k) of multiplicative functions in k[[T]] and generalizes Theorem 5.1 of Doubilet, Rota and Stanley in [8].

(4.20) Theorem: Let Mon := $\prod_{U \in P(A)} \{a \in k[[X(H); H \in P(U)]]; a(0) = 0\}$ be a monoid of vectors of formal power series with componentwise substitution composition and unit $(X(U))_{U \in P(A)}$. Then the map

$$F: \mathrm{Mon}(T,k) \longrightarrow \mathrm{Mon}$$

$$f \longmapsto (\Sigma \{f(m,U)\frac{X^m}{a_U(m)} ;\ m \in \mathbb{N}_0^{P(A)} - \{0\} \text{ with supp } m \subseteq U\})_{U \in P(A)},$$

where $a_U(m) = |\mathrm{Aut}(\amalg_{H \subseteq U} m(H) U/H| = \prod_{H \subseteq U} (\frac{|U|}{|H|})^{m(H)} m(H)!$,

is an isomorphism of monoids. □

$$F(\zeta) = (\exp[\sum_{H \subseteq U} \frac{|H|}{|U|} X(H)] - 1)_{U \in P(A)} \quad \text{implies}$$

$$F(\mu) = F(\zeta)^{-1} = (\sum_{H \subseteq U} \mu_{P(A)}(H,U) \frac{|H|}{|U|} \log(1+X(H)))_{U \in P(A)},$$

where $\mu_{P(A)}$ is the Möbius function in P(A). If we denote the set of all prime numbers by P, then

$$\mu_{P(A)}(H,U) = \prod_{p \in P} (-1)^{o(p)} p^{o(p)(o(p)-1)/2}$$

for $U/H \cong \prod_{p \in P} (\mathbb{Z}/\mathbb{Z}p)^{o(p)}$ and some $o \in \mathbb{N}_0^{(P)}$, where $\mathbb{Z}/\mathbb{Z}p$ is the cyclic group of order p, and $\mu_{P(A)}(H,U) = 0$ otherwise.

(4.21) <u>Proposition</u>: Let (n,U) be an element of <u>Q</u>. Then

$$\mu(n,U) = (-1)^{k-1}(k-1)! \prod_{p \in P} (-1)^{o(p)} p^{o(p)(o(p)+2k-3)/2}$$

if $n = kH$ with $k \in \mathbb{N}$, $H \subseteq U$ and $U/H \cong \prod_{p \in P} (\mathbb{Z}/\mathbb{Z}p)^{o(p)}$ for some $o \in \mathbb{N}_0^{(P)}$

and $\mu(n,U) = 0$ otherwise. □

Applying $F(fg) = F(f)(F(g))$ for $f,g \in \mathrm{Mon}(T,k)$ to the ζ-function, we finally get the generating functions of B(n) and S(n,m), i.e. (4.16) and (4.17) for G abelian.

(4.22) <u>Representations of ordered sets</u>:

Let O be a finite ordered set.

A representation of O is an order preserving map $F: O \longrightarrow \mathrm{Pot}(X)$, where Pot(X) is the ordered set of all subsets of a finite set X.

A morphism from one representation $F: O \longrightarrow \mathrm{Pot}(X)$ to another representation $G: O \longrightarrow \mathrm{Pot}(Y)$ is a map $f: X \longrightarrow Y$ so that $f(F(o)) \subseteq G(o)$ for all $o \in O$.

Let <u>K</u> be the category of representations of O, let M be the class of all monomorphisms of <u>K</u>, i.e. all morphisms that are one-to-one, and take isomorphy as equivalence relation on M. Then $T := M/\cong$ is the free monoid of types and $H := k[[T]]$ a topological bialgebra.

Here $G(R) := \mathrm{Mon}(T,R)$ is algebraic and has a faithful linear representation by triangular matrices. In the special case, where O is the empty set, i.e. <u>K</u> = <u>Setf</u>, G(R) is just the monoid of affine maps of the line R.

References

[1] S.Anantharaman, Schémas en groupes sur une base de dimension 1, Bull.Soc.Math.France, Mem.33 (1973), 5-79.

[2] H.Bass - E.H.Connell, Polynomial automorphisms and the Jacobian conjecture, to appear.

[3] H.Cartan - S.Eilenberg, Homological Algebra, Princeton University Press 1956.

[4] P.M. Cohn, Universal Algebra, Harper & Row 1965.

[5] L. Comtet, Advanced Combinatorics, Reidel 1974.

[6] M. Content - F.Lemay - P.Leroux, Catégories de Möbius et fonctorialités: un cadre général pour l'inversion de Möbius, J.Comb.Th.A 28 (1980), 169-190.

[7] M.Demazure - P.Gabriel, Groupes Algébriques, North Holland 1970.

[8] P.Doubilet - G.C.Rota - R.P.Stanley, The idea of generating function, in: Finite Operator Calculus, Academic Press 1975.

[9] A.Dür, Inzidenzalgebren zu Cozyklen mit nichtleerer Nullstellenmenge, Diplomarbeit, Innsbruck 1981.

[10] D.Foata - M.P.Schützenberger, Théorie Géométrique des Polynômes Eulériens, Springer Lecture Notes (SLN) 138, 1970.

[11] P.Gabriel, Étude Infinitésimale des Schémas en Groupes, in: SLN 151, 1962/63/64.

[12] P.Gabriel - F.Ulmer, Lokal präsentierbare Kategorien, SLN 221, 1971

[13] P.Gabriel - M.Zisman, Calculus of Fractions and Homotopy Theory, Springer Ergebnisse 35, 1967.

[14] A.Grothendieck - J.V.Verdier, Theorie des Topes et ..., SLN 269, 1963/64.

[15] M.Hazewinkel, Formal Groups and Applications, Academis Press 1978.

[16] P.T.Johnstone, Topos Theory, Academic Press 1977.

[17] S.A.Joni, Lagrange inversion in higher dimensions and umbral operators, Lin.Multil.Alg.6 (1978), 111-122.

[18] A.Joyal, Une théorie combinatoire des séries formelles, Adv.Math. 42 (1981), 1-82.

[19] I.G.Macdonald, Symmetric Functions and Hall Polynomials, Clarendon 1979.

[20] D.Mumford, Lectures on Curves on an Algebraic Surface, Ann.Math.Stud. 59, 1966.

[21] I.Reiner, Maximal orders, Academic Press 1975.

[22] J.E.Roos, Sur les derives de lim. Applications, C.R.Acad.Sc.Paris 254, (1961), 3702-3704.

[23] G.C.Rota - Kahaner - Odlyzko, Finite Operators calculus, in: Finite Operator Calculus, Academic Press 1975.

[24] G.C.Rota, Coalgebras and Bialgebras in: Combinatorics (Notes by S.A.Joni), Umbral Calculus Conference, Oklahoma 1978.

[25] J.P.Serre, Lie Algebras and Lie Groups, W.A.Benjamin 1965.

[26] J.P.Serre, Groupes de Grothendieck des Schémas en Groupes Réductifs Déployés, in: Publ.Math.34,(IHES 1968), 37-52.

[27] R.P.Stanley, Exponential structures, Stud.Appl.Math. 59, (1978), 73-82

[28] R.P.Stanley, Generating Functions, in: Stud. in Comb., MAA Studies 17, 1978.

[29] M.E.Sweedler, Hopf Algebras and Combinatorics, Umbral Calculus Conference, Oklahoma 1978.

[30] B.J.Veisfeiler - I.V.Dolgacev, Unipotent group schemes over integral rings, Math.USSR Izv.8, (1974), 761-800.

[31] E.Hairer - G.Wanner, On the Butcher group and general multi-value methods, Computing 13, (1974), 1-15.

On Irreducible Translation Planes in Odd Characteristic, I

David Foulser
Department of Mathematics, University of Illinois at Chicago Circle, Chicago, Illinois 60680, U.S.A.

Geoffrey Mason
Department of Mathematics, University of California, Santa Cruz, California 95064, U.S.A.

Michael Walker
Mathematisches Institut der Universität, Auf der Morgenstelle 10, D-7400 Tübingen, West Germany

§ 1 Introduction

Suppose that Π is a (finite, affine) translation plane, O is some fixed but arbitrary point of Π, and G is a group of collineations of Π which fixes O and commutes with the kernel of Π. Then it is well known that there is a field K and a KG-module V such that the points of Π correspond to the elements of V (with O corresponding to the zero vector of V), and the lines of Π containing O correspond to a G-invariant spread of V. It is because of this correspondence that the theory of group representations becomes important for the study of translation planes.

It is useful to alter our perspective and start with a KG-module V. We then ask under what conditions does V support a G-invariant spread? This is a vast question, and certainly not one which we shall consider here in anything like its full generality. For example, if G is fixed, there are obviously an infinite number of choices for V that can be made. However, taking our cue from representation theory, it would seem to make sense to start with modules which are "small". This does not mean that we bound the dimension of V (though it might), instead we interpret small to mean irreducible. This certainly makes sense from a representation-theoretic point of view, though it is admittedly not quite clear how important such a hypothesis is geometrically. On the other hand, since very little is known about these questions, it seems worthwhile to persue anything which can provide a systematic approach.

As for our choice of G, historical reasons and known examples suggest that taking G to be of Lie-type might prove fruitful. For more technical teasons, we limit ourselves to quasisimple groups of type A_ℓ, D_ℓ, E_6, E_7 or E_8. These are the groups associated to a root system with a simply-laced diagram (i.e., one with no double bonds), and of course we have

$$A_\ell(q) = SL_{\ell+1}(q) \quad ; \quad D_\ell(q) = O^+_{2\ell}(q)'$$

Finally, a word about the choice of K. The representation theory of G (of Lie-type, as above) falls into two categories, distinguished according as to whether the characteristic of K does or does not divide the order of G. Furthermore, if G is associated with the field GF(q), where $q = p^d$ with p a prime, then the first category bifurcates into the cases char$K \neq p$ and char$K = p$. Again, historical precedent and known examples suggest that the latter case is of high interest, and for technical reasons we take $K = GF(q)$.

Thus we are led to consider the following situation. We take G to be a quasisimple group (that is, $G = G'$ and $G/Z(G)$ is simple) of Lie-type

$$A_\ell(\ell \geq 1),\ D_\ell(\ell \geq 4),\ E_6,\ E_7 \text{ or } E_8$$

defined over the Galois field GF(q), where $q = p^d$ with p a prime. Then we let V be an irreducible GF(q)G-module, which we assume supports a G-invariant spread? Observe that our hypotheses mean that the kernel of the translation plane represented by spread includes the field GF(q) over which G is defined.

Anyone familiar with [4] will recognize that the foregoing is a restatement of the philosophy of that paper. Moreover, it was suggested there that the following should be true.

Conjecture <u>Let G be a group of the type described above, and let V be an irreducible GF(q)G-module. If V supports a spread, then</u> $G \cong SL_2(q)$.

In fact, the conjecture was proved in [4] for the case $p = 2$, and the present report is an initial attempt to extend the ideas of that paper to the case p odd. There are two main difficulties which appear to arise. The relative plethora of irreducible GF(q)G-modules in odd characteristic, and the fact that the notation of a dispersive module, introduced in [4], seems to have no direct analogue in odd characteristic.

Partial compensation for the second of these difficulties is to be found in a theorem due to the first author, which is restated below (Lemma 3.2) in a form convenient for us. In order to study the irreducible GF(q)G-modules, it appears useful to employ the theory of weights in a systematic way. Indeed, though we cannot as yet establish the conjecture in complete generality, we thought that it would be of interest to see how the above ideas, in particular those involving Lie theory, can be used to gain a strong hold on the possible modules V which can occur. The results themselves are stated in the next section.

The authors wish to acknowledge the support and hospitality of the University of California at Santa Cruz during the winter of 1982. The second author was also supported by a grant from the N.S.F.

§ 2 Statement of results

As mentioned above, our present results are quite modest. They are recorded here because we believe that they may play a role in a conclusive proof of the conjecture of § 1, moreover they have the merit of showing how the Lie theory and results from [1] on planar collineation groups help the analysis in a particularly clear way.

In order to state our results we need to recall some facts from [4]. First, it was shown there [4, 7.1] that the conjecture is true in general if it can be shown that $G \not\cong SL_3(q)$. So it suffices to assume, by way of contradiction, that the following holds:

$$G \cong SL_3(q),\ \text{where } q = p^d \text{ with } p \text{ an odd prime,}$$

and V is an irreducible GF(q)G-module which supports a spread.

We assume this hypothesis for the remainder of the section.

Next, we recall (cf. [4, § 5]) that the Steinberg tensor product theorem tells us that V can be represented as a twisted tensor product

$$V = V_1^{\sigma_1} \otimes \ldots\ldots \otimes V_d^{\sigma_d}$$

where each V_i is one of the p^2 so-called basic GF(q)G-modules, and $V_i^{\sigma_i}$ is the "twisted" version corresponding to the field automorphism $\sigma_i \in \mathrm{Gal}(GF(q))$. In [3] the basic modules are constructed explicitly, but all we need here is the well-known fact that there is a distinguished basic module, the so called <u>basic Steinberg module</u>, of dimension p^3 over GF(q).

We can now state our principle result.

<u>Theorem 2.1</u> <u>At least one of the factors</u> V_i <u>occuring in the representation of</u> V <u>as a twisted tensor product is the basic Steinberg module</u>.

There are several consequences of this result. For example we have

Corollary 2.2 $q \nmid p$

To see why the corollary is true, assume that $q = p$. Thus $d = 1$, so that V is itself a twisted basic module. By the theorem we see that V is a twisted version of the basic Steinberg module; in particular, $\dim V = p^3$ is odd. However, V is supposed to support a spread, and so must have even dimension.

As explained above, this corollary implies the validity of the conjecture when q is a prime.

Corollary 2.3 *Let* V, G *and* $q = p^d$ *be as in the conjecture. If* V *supports a spread and if* $d = 1$, *then* $G \cong SL_2(p)$.

The theorem is proved as follows. In § 3 we show that if $\alpha \in G \cong SL_3(q)$ is a non-identity root element (a transvection), then the fixed structure of α in its action on Π is a pth-root subplane. In particular, V is a free $\langle\alpha\rangle$-module. Then in §4, we show how this observation and the representation theory allow us to arrive at the conclusion of the theorem.

§ 3 The action of root elements on Π

If Π is an affine plane admitting a collineation α, we denote by $F(\alpha)$ the fixed structure of α. If A is a group of collineations of Π, then $F(A)$ is the fixed structure of A.

Now assume that V is a vector space over a (finite) field of characteristic p, which supports a spread giving rise to a translation plane Π. If α is a collineation of order p in the linear translation complement of Π, we shall call α *uniform* provided the following holds:

If a matrix representing the action of α on V is put in Jordan canonical form then each Jordan block has the same size.

The importance of this concept stems from the following results established in [1]. (The second is restated in a form convenient for our present purposes).

Lemma 3.1 [1, 3.1] *If* α *is a collineation of* Π *of order* p *which is planar (that is,* $F(\alpha)$ *is a subplane), then* α *is uniform.*

Lemma 3.2 [1,4.3 and 4.5] *Suppose that* α *and* β *are a pair of commuting planar collineations of* Π *of order* p. *Assume that* $F(\alpha) \nmid W$

and $F(\beta) \nsubseteq W$, where $W = F(\alpha) \cap F(\beta)$, and the restriction of α to $F(\beta)$ is uniform. Then at least one of $F(\alpha)$ and $F(\beta)$ is a pth-root subplane of Π .

In the sequel we shall only apply Lemma 3.2 when both α and β are conjugate inside a larger group of collineations. Thus, in our applications of the Lemma, we shall always conclude that both $F(\alpha)$ and $F(\beta)$ are p^{th}-root subplanes.

We now begin the proof of Theorem 2.1, so let G and V be as in § 2. That is, $G \cong SL_3(q)$, where $q = p^d$ with p an odd prime, and V is an irreducible GF(q)G-module, which supports a spread giving rise to a translation plane Π .

We need to recall some facts concerning the structure of G. Fix a p-Sylow subgroup P of G. Then $|P| = q^3$, and P is the product $P = Q\tilde{Q}$ of a pair of elementary abelian p-groups Q and $\tilde{Q}$ each of order q^2. Geometrically Q can be viewed as the group of all translations with axis a distinguished line ℓ_∞ of the desarguesian projective plane of order q, and $\tilde{Q}$ is the group of dual translations with distinguished centre $[\infty]$ on ℓ_∞. The groups $N = N_G(Q)$ and $\tilde{N} = N_G(\tilde{Q})$ are the maximal parabolic subgroups of G containing P; that is, the stabilizers in G of ℓ_∞ and $[\infty]$ respectively. The group Q is complemented in N by its Levi-factor L, and $L' \cong SL_2(q)$. Of course, L is just the translation complement in N of the desarguesian affine plane with ℓ_∞ the line at infinity. Moreover, if $Z = Q \cap \tilde{Q}$, so that $Z = Z(P)$, then Z has just q+1 distinct conjugates under the action of L. These are the root subgroups contained in Q; that is, the elation groups corresponding to the distinct centres on ℓ_∞ . Similar comments apply to $\tilde{N}$, and we may choose our notation so that $\tilde{Q} = \langle Z, \tilde{Q} \cap L \rangle$. Then Z and $\tilde{Q} \cap L$ are conjugate under the action of the Levi-factor of $\tilde{N}$.

In [4] it was shown that Q or $\tilde{Q}$ is planar on Π . We choose our notation so that F(Q) is a subplane of Π . Certain other facts relevant to our present discussion were also established, and these are summarized in the next lemma.

Lemma 3.3 [4, 5.5 and 7.1]. The fixed structure F(Q) is a subplane, and the Levi-factor L acts irreducibly on F(Q). Moreover, F(P) is 1-dimensional.

The fact that L is irreducible on F(Q) and that dim F(P) = 1 are just special cases of Smith's theorem [5] .

We need to interpolate another result from [1, 3.1] (see also [2, 2.2]).

Lemma 3.4 Suppose that $F(Q) \leq W \leq V$ and that W is a subplane of Π. Then dim $F(Q)$ divides dim W.

We are now ready to establish the main result of this section.

Proposition 3.5 If $\alpha \in Q$ with $\alpha \neq 1$, then the following hold:
(1) $F(\alpha)$ is a pth-root subplane;
(2) $F(Q)$ is a proper subplane of $F(\alpha)$.

Proof: First assume that there exists $1 \neq \beta \in Q$ with $F(\beta) \neq F(\alpha)$. Since $F(Q)$ is a subplane (Lemma 3.3), both $F(\alpha)$ and $F(\beta)$ are subplanes and α is planar on $F(\beta)$. Therefore α is uniform on $F(\beta)$ by Lemma 3.1. Moreover, $\langle\alpha\rangle$ is conjugate to $\langle\beta\rangle$ because L is transitive on the subgroups of order p in Q. In particular, $|F(\alpha)| = |F(\beta)|$ and thus $F(\alpha) \neq F(\beta)$ implies $F(\alpha) \neq F(\alpha) \cap F(\beta) \neq F(\beta)$. Now (1) follows from Lemma 3.2 and (2) is obvious.

If the above assumption is false, then $F(\beta) = F(Q)$ for all $\beta \in Q-\{1\}$. We argue that this leads to a contradiction. In accordance with the notation introduced prior to Lemma 3.3, let $1 \neq \gamma \in \tilde{Q} \cap L$. Since $\tilde{Q} \cap L$ is conjugate to Z, we have $F(\tilde{Q} \cap L) = F(\gamma)$, and hence

$$F(P) = F(Q) \cap F(\tilde{Q} \cap L) = F(\alpha) \cap F(\gamma).$$

Now choose γ to commute with α. Since $F(P)$ is 1-dimensional by Lemma 3.3, the restriction of α to $F(\gamma)$ is uniform (indeed, there is a unique Jordan block). Moreover, $F(\alpha) \not\subseteq F(\gamma) \not\subseteq F(\alpha)$ because $\dim F(\alpha) = \dim F(\gamma) \geq 2$. Thus $F(\alpha)$ is a pth-root subplane by Lemma 3.2 and the fact that α and γ are conjugate in G. As $F(\alpha) = F(Q)$ by hypothesis, it follows that

$$\dim V = p \dim F(Q) \qquad (*)$$

On the other hand, L acts irreducibly on $F(Q)$ by Lemma 3.3, and $L' \cong SL_2(q)$ has a unique involution t which is (of course) central in L. Therefore t has a unique eigenvalue in its action on $F(Q)$. Thus the eigenspace W corresponding to this eigenvalue is a Baer subplane of Π which contains $F(Q)$. By Lemma 3.4 we have

$$\dim F(Q) \mid (\dim W = 1/2 \dim V).$$

This is not compatible with (*), and the desired contradiction is reached.

§ 4 Some representation theory

Proposition 3.5(1) implies that if $\alpha \neq 1$ is a root element in G, then V is free as an $\langle\alpha\rangle$ -module; that is to say, each Jordan block of a matrix representing α has size p. The following result is elementary, and the reader is left to provide a proof.

Lemma 4.1 If X and Y are $\langle\alpha\rangle$-modules, then $X \otimes Y$ is free if, and only if, X or Y is free.

Applying this observation to the decomposition

$$V = V_1^{\sigma_1} \otimes \dots \otimes V_d^{\sigma_d}$$

provided by Steinberg's tensor product theorem, we deduce that at least one of the tensor factors $V_i^{\sigma_i}$ is a free $\langle\alpha\rangle$-module. It is now clear that Theorem 2.1 is a consequence of the following result.

Proposition 4.2 Let B be a basic GF(q)G-module, where $G \cong SL_3(q)$, and let $\alpha \neq 1$ be a root element in G. Then B is a free $\langle\alpha\rangle$-module if, and only if, B is the basic Steinberg module.

Note that Proposition 4.2 is purely a problem of representation theory. It is possible to establish the result by directly examining the basic GF(q)G-modules, a description of which appear in [3] . We shall, however, use a little of the theory of weights in our approach. Not only is an explicit knowledge of the relevant modules not required, but we anticipate that our arguments will prove useful (when suitably generalized) in a more general context.

Proof of Proposition 4.2:

First, if G_0 is the canonical subgroup of G isomorphic to $SL_3(p)$ then we may assume that $\alpha \in G_0$. Now assume that $B \cong St$, the basic Steinberg module. Then

$$B|G_0 \cong B_0 \otimes_{GF(p)} GF(q) ,$$

where B_0 is the Steinberg module for G_0 . But it is well-known that B_0 is projective as G_0-module, so B_0 is a free $\langle\alpha\rangle$-module. Hence the above isomorphism shows that B is a free $\langle\alpha\rangle$-module.

This proves one half of 4.2. The remaining assertion, that if B is a free $\langle\alpha\rangle$-module, then $B \cong St$, requires some preparation. Let us denote by $\{\alpha_1, \alpha_2, \alpha_1 + \alpha_2\}$ the positive roots of the root system attached to G. Furthermore we take λ_1, λ_2 to be the fundamental dominant weights satisfying $(\lambda_i, \alpha_j) = \delta_{ij}$. As B is a basic module it has a highest weight given by $\lambda = a_1\lambda_1 + a_2\lambda_2$, for some $0 \leq a_i \leq p-1$.

We must show that each $a_i = p-1$.

It is convenient to adjust the nomenclature for certain subgroups of G introduced prior to Lemma 3.3. Let X_{α_1}, X_{α_2}, and $X_{\alpha_1+\alpha_2}$ be the root subgroups of P corresponding to α_1, α_2 and $\alpha_1+\alpha_2$. Then we may take

$$Q = X_{\alpha_1} X_{\alpha_1+\alpha_2} \quad \text{and} \quad \tilde{Q} = X_{\alpha_2} X_{\alpha_1+\alpha_2} \quad ;$$

$$L_2 = L' = \langle X_{\alpha_2}, X_{-\alpha_2} \rangle \quad \text{and} \quad L_1 = \tilde{L}' = \langle X_{\alpha_1}, X_{-\alpha_1} \rangle \quad .$$

If $H = \langle h_{\alpha_1}, h_{\alpha_2} \rangle$ is the usual Cartan subgroup of $N_G(P)$, we need to compute $C_H(L_i)$. In fact it is a simple computation involving the Chevalley commutator formulae that

$$C_H(L_2) = \langle h_{\alpha_1}^2 h_{\alpha_2} \rangle \quad \text{and} \quad C_H(L_1) = \langle h_{\alpha_1} h_{\alpha_2}^2 \rangle$$

Now set $B_2 = C_B(Q)$ and $B_1 = C_B(\tilde{Q})$. Either by [5] , or by direct computation, we find that B_i is an irreducible module for L_i . In fact, from [5] one knows the following.

<u>Lemma 4.3</u> <u>If B_γ is the weight-space of</u> B <u>corresponding to weight</u> γ , <u>then</u> $B_i = \Sigma \oplus B_\mu$, <u>the sum running over those weights</u> μ <u>satisfying</u> $\lambda - \mu = k\alpha_i$, <u>for some</u> $k \in \mathbb{N}$.

Also this implies

<u>Lemma 4.4</u> <u>As</u> L_i<u>-module,</u> B_i <u>has highest weight</u> $a_i\lambda_i$.

Now the irreducibility of B_i as L_i-module tells us that a generator h_i of $H_i = C_H(L_i)$ has a unique eigenvalue in its action on B_i . Let this eigenvalue be σ_i .

<u>Lemma 4.5</u> <u>The full</u> σ_i<u>-eigenspace of</u> h_i <u>acting on</u> B <u>is precisely</u> B_i .

<u>Proof</u>: One knows that B decomposes into a sum of weight spaces B_μ under the action of H, where μ is a weight satisfying

$$\lambda - \mu = \text{a sum of positive roots}$$

Let μ be such a weight with the property that $h_i B_\mu = \sigma_i B_\mu$. Now, as B_λ is a weight space contained in B_i, we have $h_i B_\lambda = \sigma_i B_\mu$. Taking the case i = 1 (the case i = 2 is the same), we have $h_1 = h_{\alpha_1} h_{\alpha_2}^2$, and now the foregoing equations together with the definition of a weight-space yield

$$(\lambda-\mu)(\alpha_1 + 2\alpha_2) = 0$$

But if $\lambda-\mu = k\alpha_1 + \ell\alpha_2$ with $k,\ell > 0$, it follows from $(\alpha_i,\alpha_i) = 2$ and $(\alpha_1,\alpha_2) = -1$ that $3\ell = 0$, so $\ell = 0$, whence $\lambda - \mu = k\alpha_1$ and $B_\mu \subsetneq B_1$ by Lemma 4.3. This completes the proof of Lemma 4.5.

Corollary 4.6 The eigenspace B_i has a complement in B which is invariant under L_i.

Now we can complete the proof that B $\cong$ St if B is free as $\langle\alpha\rangle$-module. Since α is conjugate to the p-elements in each L_i, and as B_i is an L_i-summand of B by Corollary 4.6, it follows that the root elements of L_i are free on D_i. As D_i has highest weight $a_i\lambda_i$ by Lemma 4.4, it has dimension $a_i + 1 \leqslant p$. But B_i has dimension divisible by p since it admits a free p-element, so we get

$$a_i = p - 1, \text{ for } i = 1 \text{ and } 2$$

As explained at the beginning of the proof, this suffices to prove the Proposition.

§ 5 Concluding remarks

We have been able to establish the conjecture in several other cases, in addition to those covered by Corollary 2.3 and [4]. For example when $p = 3$, or when d is even and $p \equiv 2 \pmod 3$.

References

1. D. Foulser, Planar collineations of order p in translation planes of order p^r; Geometriae Dedicata 5 (1976), 393-409.
2. R. Liebler, Combinatorial representation theory and translation planes; in "Finite Geometries", ed. Kallaher and Long, Marcell Decker (to appear).
3. J.C. Mark, On the modular representations of the group GLH(3,p); University of Toronto Thesis, 1939.
4. G. Mason, Irreducible translation planes and representations of Chevalley groups in characteristic 2; in "Finite Geometries", ed. Kallaher and Long, Marcell Decker (to appear).
5. S. Smith, Irreducible modules and parabolic subgroups; J. Algebra (to appear).

The equivalence classes of the Vasil'ev codes of length 15

Ferdinand Hergert
Technische Hochschule Darmstadt
Fachbereich Mathematik - AG 1
D-6100 Darmstadt, W.Germany

In this paper we determine the equivalence classes of the perfect Vasil-ev codes of length 15: There exist 19 non equivalent Vasil'ev codes (including the Hamming code). If we restrict the equivalence transformations to permutations of coordinates we get 64 different Vasil'ev codes.

In 1962 Vasil'ev [4] constructed a class of nonlinear perfect single-error-correcting binary codes. His construction works as follows:
Let C be a perfect single-error-correcting code of length n, not necessarily linear. Let $g: C \to \mathbb{F}_2$ be any mapping with $g(\underline{0})=0$. Set $\pi(\underline{v})=0$ or 1 depending on whether $wt(\underline{v})$ (wt denotes the Hamming weight) is even or odd. Then $V := \{(\underline{u}|\underline{u}+\underline{v}|\pi(\underline{u})+g(\underline{v})) \mid \underline{u}\in\mathbb{F}_2^n, \underline{v}\in C\}$ is a perfect single-error-correcting code of length 2n+1, which is nonlinear if g is nonlinear.

The smallest nonlinear Vasil'ev code has length 15, it is constructed from the Hamming code of length 7. Since this Hamming code has 15 nonzero codewords, the above construction yields a total of 2^{15} different Vasil'ev codes of length 15.

Two binary codes C and $\tilde{C}$ are called <u>equivalent</u> if one can be obtained from the other by permuting the coordinates and adding a constant vector, i.e. $\tilde{C} = \{\pi(\underline{c})+\underline{a} \mid \underline{c}\in C\} = \pi(C)+\underline{a}$, where $\underline{a}\in\mathbb{F}_2^n$. In this paper we want to determine the number of nonequivalent Vasil'ev codes of length 15.

To do so, we shall use some notations introduced in [1]. We give the basic definitions here.

By an <u>(n,k)-code</u> we mean a <u>binary</u> code of length n, which is <u>systematic</u> with respect to the first k coordinates. I.e. for every $\underline{x}=(x_1,x_2,..,x_k)\in\mathbb{F}_2^k$ there exists a unique element $c(\underline{x})=(x_1,x_2,..,x_k,x_{k+1},..,x_n)\in C$. W.l.o.g. we shall always assume that $\underline{0}=(0,0,...,0)\in C$. Since the entries $x_{k+1},..,x_n$ are uniquely determined by $x_1,...,x_k$ we may write C in the form

$$C = \{ (x_1,...,x_k,f_1(\underline{x}),...,f_r(\underline{x})) \mid \underline{x}=(x_1,...,x_k)\in\mathbb{F}_2^k \} \text{, where}$$

$f_1,...,f_r$ $(r:=n-k)$ are mappings $f_i:\mathbb{F}_2^k \to \mathbb{F}_2$ with $f_i(\underline{0})=0$.
Since every mapping $f:\mathbb{F}_2^k \to \mathbb{F}_2$ can be uniquely written as a (reduced) polynomial in variables $x_1,...,x_k$ over the two element field $\mathbb{F}_2$, we shall refer to $f_1,...,f_r$ as the <u>redundancy polynomials</u> of C.

Every $\mathbb{F}_2$-polynomial in variables $x_1,\dots,x_k$ formally looks like a (square free) real polynomial, thus we can form the formal <u>derivative</u> f_{x_i} of f with respect to the variable x_i, which again can be viewed as an $\mathbb{F}_2$-polynomial.

The <u>Jacobian</u> of an (n,k)-code C with redundancy polynomials $f_1,\dots,f_r$ is defined as the r×n matrix (r=n-k)

$$\text{jac } C = \begin{bmatrix} f_{1x_1} & f_{1x_2} & \dots & f_{1x_k} \\ f_{2x_1} & f_{2x_2} & \dots & f_{2x_k} \\ \vdots & \vdots & & \vdots \\ f_{rx_1} & f_{rx_2} & \dots & f_{rx_k} \end{bmatrix} .$$

Note that the entries of the Jacobian are polynomials.
In case C is a linear (n,k)-code with parity check matrix $(H|I_r)$, jacC=H. Since we assume $\underline{0}\in C$ the Jacobian uniquely determines the code C.

We shall use the Jacobian to characterize the Vasil'ev codes. Since every perfect single-error-correcting code of length 7 is (equivalent to) the linear Hamming code $H_3 := \{(v_1,v_2,v_3,v_4,h_1(\underline{v}),h_2(\underline{v}),h_3(\underline{v})) \mid \underline{v}\in\mathbb{F}_2^4\}$ with the (linear) redundancy polynomials $h_1(\underline{v})=v_2+v_3+v_4$, $h_2(\underline{v})=v_1+v_3+v_4$ and $h_3(\underline{v})=v_1+v_2+v_4$, every Vasil'ev code of length 15 is composed of H_3 and a mapping $g:H_3 \to \mathbb{F}_2$. But since H_3 is systematic in its first 4 coordinates, g is already determined by v_1,v_2,v_3,v_4 so that we may think of g as a mapping $g:\mathbb{F}_2^4 \to \mathbb{F}_2$, i.e. as an $\mathbb{F}_2$-polynomial in variables $v_1,\dots,v_4$.
It is clear that by permuting the coordinates we can transform any given Vasil'ev code in a kind of normal form. We describe this normal form via the Jacobian (for a proof see [2]).

<u>1.Lemma:</u> Every Vasil'ev code V of length 15 is equivalent to a Vasil'ev code

$$\tilde{V} = \{(x_1,x_2,x_3,x_4,y_1,y_2,y_3,z_1,z_2,z_3,z_4,f_1(\underline{x},\underline{y},\underline{z}),\dots,f_4(\underline{x},\underline{y},\underline{z})) \mid (\underline{x},\underline{y},\underline{z})\in\mathbb{F}_2^{11}\}$$

in <u>normal</u> <u>form</u>. We say $\tilde{V}$ has normal form (or $\tilde{V}$ is a <u>normal</u> <u>code</u>) if jac$\tilde{V}$ looks as follows:

$$\begin{matrix} f_1 \\ f_2 \\ f_3 \\ f_4 \\ \\ \end{matrix} \begin{bmatrix} 0 & 1 & 1 & 1 & 1 & 0 & 0 & 0 & 1 & 1 & 1 \\ 1 & 0 & 1 & 1 & 0 & 1 & 0 & 1 & 0 & 1 & 1 \\ 1 & 1 & 0 & 1 & 0 & 0 & 1 & 1 & 1 & 0 & 1 \\ 1 & 1 & 1 & 1 & 1 & 1 & 1 & 0 & 0 & 0 & 0 \\ +g_{x_1} & +g_{x_2} & +g_{x_3} & +g_{x_4} & & & & +g_{z_1} & +g_{z_2} & +g_{z_3} & +g_{z_4} \end{bmatrix}$$

Where $g(v_1,v_2,v_3,v_4)\in G := \{g:\mathbb{F}_2^4 \to \mathbb{F}_2 \mid g(\underline{0})=0, \nabla g(\underline{0}) := (g_{v_1}(\underline{0}),\dots,g_{v_2}(\underline{0}))=\underline{0}\}$ and with the substitution $v_i := x_i+z_i$.

Lemma 1 essentially says that $V = \{(\underline{u}|\underline{u}+\underline{v}|\pi(\underline{u})+g(\underline{v})) \mid \underline{u}\in\mathbb{F}_2^7, \underline{v}\in H_3\}$ has normal form if the linear code $L = \{(\underline{u}|\underline{u}+\underline{v}|\pi(\underline{u})) \mid \underline{u}\in\mathbb{F}_2^7, \underline{v}\in H_3\}$, which is equivalent to the linear Hamming code of length 15, has the following parity check matrix:

$$\underbrace{\begin{bmatrix} 0&1&1&1&1&0&0&0&1&1&1 \\ 1&0&1&1&0&1&0&1&0&1&1 \\ 1&1&0&1&0&0&1&1&1&0&1 \\ 1&1&1&1&1&1&1&0&0&0&0 \end{bmatrix}}_{\text{jac } H_4}\begin{bmatrix} 1&0&0&0 \\ 0&1&0&0 \\ 0&0&1&0 \\ 0&0&0&1 \end{bmatrix}.$$

We denote this distinguished Hamming code by H_4.

A second remark on lemma 1: In the original definition we had to consider all 2^{15} functions $g:\mathbb{F}_2^4 \to \mathbb{F}_2$ with $g(\underline{0})=0$. Now the above lemma tells us that if the polynomial g contains linear terms v_i (i.e. $\nabla g(\underline{0}) \neq (0,0,0,0)$) we can obtain an equivalent normal form with $\tilde{g}\in G$. This can be done by exchanging the variables (columns of jacV) $x_i \leftrightarrow z_i$ for every linear term v_i in g. This yields the polynomial $\tilde{g}:\mathbb{F}_2^4 \to \mathbb{F}_2$ with $\tilde{g}(\underline{0})=0$, $\nabla\tilde{g}(\underline{0})=\underline{0}$ and $g(\underline{v})+\tilde{g}(\underline{v})$ is linear.

Since a normal code is determined by the function $g\in G$ we call g the <u>characterizing function</u> of the normal code and denote the code by V_g.

To determine the equivalence classes of the Vasil'ev codes we may now restrict ourselves on codes in normal form. The following lemma characterizes the permutations of the coordinates π that transform a normal code into a normal code.

2.Lemma: Let V_g be a nonlinear normal code. Then $\tilde{V}=\pi(V_g)$ is again a normal code if and only if π is an automorphism of H_4, which fixes the last coordinate ($\pi(15)=15$).

<u>Proof:</u> If the last coordinate is fixed, it is clear (by definition of normal code) that π must be an automorphism of H_4. So we only have to show that the permutation π' fixes the last coordinate.
Since deleting the last coordinate of a normal form gives a linear code (shortened Hamming code), we are through if we can proof that deleting any other coordinate of V_g yields a nonlinear code. But this follows immediately since V_g has minimum distance 3:
Every shortened code contains again all 2^{11} (shortened) codewords and therefore the last coordinate makes it nonlinear □

Let us define two codes C, $\tilde{C}$ to be p-equivalent if there exists a permutation π of the coordinates such that $\pi(C)=\tilde{C}$ (so this is a more restrictive definition of equivalence then the usual one).

We denote by Π the set of all permutations described in lemma 2. Thus we have V_g p-eqivalent to V_h if and only if there exists $\pi\in\Pi$ with $\pi(V_g)=V_h$. Furthermore lemma 2 says that Π is a group:
Since the coordinates of the Hamming code H_4 can be naturally interpreted as the points of the 3 dimensional projective space PG(3,2) the automorphismgroup of H_4 is isomorphic to the automorphism group of PG(3,2) and the subgroup Π fixing the last coordinate is isomorphic to the stabilizer of a point. So we get:

3.Corollary: The p-equivalence classes of Vasil'ev codes in normal form are the orbits of the group Π acting on the normal codes. Furthermore Π is isomorphic to the group GL(3,2) of all invertible 3×3-matrices over $\mathbb{F}_2$.

Since there is a one-to-one correspondence between the normal codes and the set of characterizing functions G, the group Π acts naturally on G via $g^{\pi}=h$ if $\pi(V_g)=V_h$. So to find the equivalence classes of the Vasil'ev codes we may as well determine the orbits on G under this group.

We shall see that the group Π acting on G has a representation as a group $\bar{\Pi}$ of linear transformations on the arguments of the polynomials in G, i.e. $g^{\pi}(v_1,v_2,v_3,v_4) = g(\bar{\pi}(v_1),\bar{\pi}(v_2),\bar{\pi}(v_3),\bar{\pi}(v_4))$, where $\bar{\pi}$ is an automorphism of the 4-dimensional $\mathbb{F}_2$-vectorspace V with base $\langle\langle v_1,v_2,v_3,v_4\rangle\rangle$.

We indicate here, how this representation of Π can be found.

It is easy to check that the following 3 permutations of the coordinates are automorphisms of H_4 fixing the last coordinate:

π_1 : $(x_1x_2)(z_1z_2)(f_1f_2)(y_1y_2)$ $\quad$ π_2 : $(x_1x_3)(z_1z_3)(f_1f_3)(y_1y_3)$
π_3 : $(x_2x_3)(z_2z_3)(f_2f_3)(y_2y_3)$.

Since $v_i=x_i+z_i$ (see lemma1) an exchange $x_i\leftrightarrow x_j$, $z_i\leftrightarrow z_j$ just means an exchange $v_i\leftrightarrow v_j$ of the variables of the characterizing function g. The other transpositions $(f_if_j),(y_iy_j)$ do not affect g.
We write $\bar{\pi}_i$ as a 4×4-matrix to describe the linear mapping on V:

$$\bar{\pi}_1 : \begin{bmatrix}0&1&0&0\\1&0&0&0\\0&0&1&0\\0&0&0&1\end{bmatrix}\begin{matrix}v_1\\v_2\\v_3\\v_4\end{matrix} \qquad \bar{\pi}_2 : \begin{bmatrix}0&0&1&0\\0&1&0&0\\1&0&0&0\\0&0&0&1\end{bmatrix} \qquad \bar{\pi}_3 : \begin{bmatrix}1&0&0&0\\0&0&1&0\\0&1&0&0\\0&0&0&1\end{bmatrix}$$

Let us finally discuss a more complicate automorphism of H_4:
The permutation π_4: $(z_1f_2)(x_1y_2)(x_3x_4)(z_3z_4)$ is an element of Π. Looking at jac H_4 we find that $f_2(\underline{x},\underline{y},\underline{z})=v_1+v_3+v_4+y_2=x_1+z_1+x_3+z_3+x_4+z_4+y_2$. So we have $z_1=x_1+f_2+v_3+v_4+y_2$. The transposition of the coordinates (z_1f_2) means that we have to substitute z_1 by the above expression, this yields

$$v_1 = x_1+z_1 \to x_1+(x_1+f_2+v_3+v_4+y_2) = f_2+v_3+v_4+y_2.$$

Furthermore, because f_2 now takes the position of z_1, we rename f_2 into
and so we get $\qquad v_1 \to z_1+v_3+v_4+y_2$.

Because of the transposition (x_1y_2) y_2 is renamed into x_1, so

$$v_1 \to z_1+v_3+v_4+x_1 = v_1+v_3+v_4 \, ,$$

and the transpositions $(x_3x_4)(z_3z_4)$ result in renaming $v_3 \leftrightarrow v_4$.
All together we have

$$\bar{\pi}_4: \begin{bmatrix} 1&0&0&0\\ 0&1&0&0\\ 1&0&0&1\\ 1&0&1&0 \end{bmatrix} \begin{matrix} v_1\\ v_2\\ v_3\\ v_4 \end{matrix}$$

An example: $g^{\pi_4}(v_1,v_2,v_3,v_4) = g(v_1+v_3+v_4,v_2,v_4,v_3)$.

So it is not difficult to find a generating set for the group $\bar{\Pi}$ (which is a subgroup of GL(4,2) isomorphic to GL(3,2)). The following table gives such a generating set:

$\bar{\pi}_1 \begin{bmatrix} 0&1&0&0\\ 1&0&0&0\\ 0&0&1&0\\ 0&0&0&1 \end{bmatrix}$ or $\begin{matrix} v_1\to v_2\\ v_2\to v_1\\ v_3\to v_3\\ v_4\to v_4 \end{matrix}$	$\bar{\pi}_2 \begin{bmatrix} 0&0&1&0\\ 0&1&0&0\\ 1&0&0&0\\ 0&0&0&1 \end{bmatrix}$ or $\begin{matrix} v_1\to v_3\\ v_2\to v_2\\ v_3\to v_1\\ v_4\to v_4 \end{matrix}$
$\bar{\pi}_3 \begin{bmatrix} 1&0&0&0\\ 0&0&1&0\\ 0&1&0&0\\ 0&0&0&1 \end{bmatrix}$ or $\begin{matrix} v_1\to v_1\\ v_2\to v_3\\ v_3\to v_2\\ v_4\to v_4 \end{matrix}$	$\bar{\pi}_4 \begin{bmatrix} 1&0&0&0\\ 0&1&0&0\\ 1&0&0&1\\ 1&0&1&0 \end{bmatrix}$ or $\begin{matrix} v_1\to v_1+v_3+v_4\\ v_2\to v_2\\ v_3\to v_4\\ v_4\to v_3 \end{matrix}$
$\bar{\pi}_5 \begin{bmatrix} 0&1&0&1\\ 1&0&0&1\\ 0&0&1&0\\ 0&0&0&1 \end{bmatrix}$ or $\begin{matrix} v_1\to v_2\\ v_2\to v_1\\ v_3\to v_3\\ v_4\to v_1+v_2+v_4 \end{matrix}$	(figure 1)

It is readily checked that $v_1+v_2+v_3$ (or $(1,1,1,0)^T$) is a fixpoint of every element of $\bar{\Pi}=\langle\bar{\pi}_1,\bar{\pi}_2,\bar{\pi}_3,\bar{\pi}_4,\bar{\pi}_5\rangle$. Therefore taking the new base $w_1:=v_1+v_2+v_3$, $w_2:=v_2$, $w_3:=v_3$, $w_4:=v_4$ the following lemma is immediate:

4. Lemma: Let $\bar{\Pi} := \left\{ \left(\begin{array}{c|ccc} 1 & 0 & 0 & 0 \\ \hline 0 & & & \\ 0 & & M & \\ 0 & & & \end{array}\right) \in GL(4,2) \;\middle|\; M \in GL(3,2) \right\}$ be the subgroup of $GL(4,2)$ which acts on the vector space V with base $\langle\langle w_1, w_2, w_3, w_4 \rangle\rangle$. Then

$$V_{g(w_1,w_2,w_3,w_4)} \text{ is p-equivalent to } V_{h(w_1,w_2,w_3,w_4)}$$

if and only if there exists $\bar{\pi} \in \bar{\Pi}$ with

$$g(\bar{\pi}(w_1), \bar{\pi}(w_2), \bar{\pi}(w_3), \bar{\pi}(w_4)) \equiv h(w_1,w_2,w_3,w_4) \ .$$

By $g \equiv h$ we mean that g and h differ only by some affine term $\varepsilon_o + \sum_1^4 \varepsilon_i w_i$, $\varepsilon_i \in \{0,1\}$.

Finally, we want to describe the orbits of Π on G. The set of all polynomials $g: \mathbb{F}_2^4 \to \mathbb{F}_2$ without constant term $(g(\underline{0})=0)$ and linear terms $(\nabla g(\underline{0}) = \underline{0})$ is itself an $\mathbb{F}_2$-vector space. As a base we have for example all monomials of degree ≥ 2, $\langle\langle w_1w_2, w_1w_3, \ldots, w_1w_2w_3, \ldots, w_2w_3w_4, w_1w_2w_3w_4 \rangle\rangle$. We choose $\langle\langle b_1, b_2, \ldots\ldots, b_{11} \rangle\rangle$ (given below) as a base of G.

$b_1 := w_1(w_2w_3w_4 + w_3w_4 + w_2w_4 + w_2w_3 + w_2 + w_3 + w_4)$

$b_2 := w_2w_3w_4 + w_3w_4 + w_2w_4$

$b_3 := w_1w_3w_4$ $\quad b_4 := w_1w_2w_4$ $\quad b_5 := w_1w_2w_3$

$b_6 := w_3w_4$ $\quad b_7 := w_2w_4$ $\quad b_8 := w_2w_3$

$b_9 := w_1w_2$ $\quad b_{10} := w_1w_3$ $\quad b_{11} := w_1w_4$.

(figure 2)

Now we have the following situation: For $\bar{\pi}_M = \left(\begin{array}{c|ccc} 1 & 0 & 0 & 0 \\ \hline 0 & & & \\ 0 & & M & \\ 0 & & & \end{array}\right) \in \bar{\Pi}$ we get the matrix

$$\tilde{\pi}_M := \left(\begin{array}{c|c|c|c|c} 1 & & & & \\ \hline & 1 & & \multicolumn{2}{c}{0} \\ \hline & & \hat{M} & & \\ \hline & 0 & & \hat{M} & \\ \hline & & R_M & 0 & M \end{array}\right) \begin{array}{l} b_1 \\ b_2 \\ b_3 \ldots b_5 \\ b_6 \ldots b_8 \\ b_9 \ldots b_{11} \end{array}$$

where $\hat{}$ denotes the outer automorphism $\hat{} : \begin{cases} GL(3,2) \to GL(3,2) \\ M \mapsto \hat{M} := (M^{-1})^T \end{cases}$

and R_M is some 3×3-matrix determined by M. We denote this group of 11×11-matrices by $\tilde{\Pi}$.

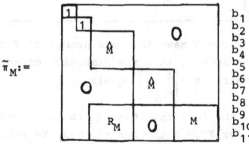

Now the p-equivalence classes of the Vasil'ev codes correspond to the orbits of the $\mathbb{F}_2$-vector space $G = \langle\langle b_1, b_2, \dots, b_{11} \rangle\rangle$ under $\tilde{\Pi}$ and we inten to use the Burnside lemma to count those orbits. To accomplish that we need some information about GL(3,2).

5. Lemma: In the group GL(3,2) of all invertible 3×3-matrices over $\mathbb{F}_2$ for every matrix $M \in GL(3,2)$ the dimension of the fixpoint space d_M depends only on the order of M. Furthermore the elements of order 4 and the elements of order 2 are all conjugate in GL(3,2).

number of elements	order	d_M
1	1	3
21	2	2
56	3	1
42	4	1
48	7	O

Σ 168

Lemma 5 can be obtained rather easily via the Sylow theorems and using the fact that GL(3,2) is the automorphism group of the projective plane PG(2,2).

To use the Burnside lemma we need to know the dimensions $d_{\tilde{\pi}}$ of the fixpoint spaces of the elements $\tilde{\pi} \in \tilde{\Pi}$.

6. Lemma: Let $M \in GL(3,2)$, d_M the dimension of the fixpoint space of M. Then the fixpoint space of $\tilde{\pi}_M \in \tilde{\Pi}$ has dimension $d_{\tilde{\pi}_M} = 3 \cdot d_M + 2$.

$\tilde{\pi}_M =$
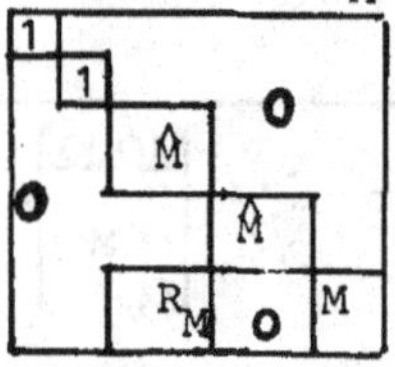

Proof: Since ord $\hat{M}$ = ord M, M and $\hat{M}$ have the same number of fixpoints (lemma 5). So all we have to show is that the dimension of the fixpoint space of the 6×6-submatrix $M' = \left[\begin{array}{c|c} \hat{M} & O \\ \hline R_M & M \end{array}\right]$ equals $2 \cdot d_M$.

This is trivial for M being the identity (then R_M is the O-matrix) and it follows for orders 3 or 7 by a theorem of Maschke since there (ord M, char $\mathbb{F}_2$)=1.
Because all elements of order 4 and order 2 are conjugate (lemma 5), we have to check only two matrices, e.g.

$$\mathrm{ord}\begin{pmatrix}1&1&0\\0&1&1\\0&0&1\end{pmatrix} = 4 \qquad M' = \left(\begin{array}{ccc|ccc}1&0&0&&&\\1&1&0&&0&\\1&1&1&&&\\\hline 0&0&1&1&1&0\\1&0&0&0&1&1\\0&0&0&0&0&1\end{array}\right),\ d_{M'} = 2.$$

$$\mathrm{ord}\begin{pmatrix}1&0&1\\0&1&0\\0&0&1\end{pmatrix} = 2 \qquad M' = \left(\begin{array}{ccc|ccc}1&0&0&&&\\0&1&0&&0&\\1&0&1&&&\\\hline 0&1&0&1&0&1\\0&0&0&0&1&0\\0&0&0&0&0&1\end{array}\right),\ d_{M'} = 4.$$

and the assertion is true in both cases □

Now we can formulate our first main result:

7.Theorem: There exist 64 p-equivalence classes of Vasil'ev codes of length 15.

Proof: The number of orbits b of G under $\tilde{\Pi}$ is

$$b = \frac{1}{|\tilde{\Pi}|} \sum_{\tilde{\pi}\in\tilde{\Pi}} 2^{d_{\tilde{\pi}}} \qquad \text{(Burnside lemma)}$$

So using 5,6 we have

$$b = \frac{1}{168}\ (1\cdot 2^{11}+21\cdot 2^{8}+56\cdot 2^{5}+42\cdot 2^{5}+48\cdot 2^{2}) = 64 \qquad □$$

Since we know the group $\tilde{\Pi}$ quite well, it is now possible to determine representatives of the equivalence classes explicitely. The arguments are a little lengthy, so we give only the result:

8.Theorem: The following polynomials (given as vectors of the 11-dimensional $\mathbb{F}_2$-vector space G with base $\langle\langle b_1,\dots,b_{11}\rangle\rangle$) are representatives of the p-equivalence classes of the Vasil'ev codes:

α	β	0	0	0	0	0	0	0	0	0	1
α	β	0	0	0	0	0	1	0	0	0	7
α	β	0	0	0	0	0	0	0	0	1	7
α	β	0	0	0	0	0	1	1	0	0	21
α	β	0	0	0	1	0	0	1	0	0	28
α	β	1	0	0	0	0	0	0	1	1	7
α	β	1	0	0	1	0	0	0	1	1	7
α	β	1	0	0	0	1	0	0	1	1	42
α	β	1	0	0	0	0	0	0	1	0	21
α	β	1	0	0	1	0	0	0	1	0	21
α	β	1	0	0	0	1	0	0	1	0	42
α	β	1	0	0	0	0	1	0	1	0	84
α	β	1	0	0	0	0	0	1	1	1	28
α	β	1	0	0	1	0	0	1	1	1	28
α	β	1	0	0	1	1	0	1	1	1	84
α	β	1	0	0	0	1	0	1	1	1	84

Where $\alpha,\beta\in\{0,1\}$. The second column gives the number of elements in the orbit.

An example, how to use the list: The base $<<b_1,\ldots,b_{11}>>$ is given in figure 2. So the first coordinate stands for b_1, the last denotes b_{11}. The vector (0,0,1,0,0,0,1,0,0,1,1) (8th row) denotes the function

$g(w_1,w_2,w_3,w_4)=w_1w_3w_4+w_2w_4+w_1w_3+w_1w_4$ or using the original variables v_i:

$$g(v_1,v_2,v_3,v_4)=(v_1+v_2+v_3)v_3v_4+v_2v_4+(v_1+v_2+v_3)v_3+(v_1+v_2+v_3)v_4=$$
$$=(v_1+v_2)v_3v_4+(v_1+v_2)v_3+v_1v_4+v_3\equiv(v_1+v_2)v_3v_4+(v_1+v_2)v_3+v_1v_4.$$

Having determined the p-equivalence classes, we shall now find the equivalence classes of the Vasil'ev codes.

Besides permutation of coordinates, we now have a second type of equivalence transformations, the addition of a constant vector $\tilde{C}=C+a$.

Given an (n,k)-code $C = \{(x_1,\ldots,x_k,f_1(\underline{x}),\ldots,f_r(\underline{x})) \mid \underline{x}\in\mathbb{F}_2^k\}$.
For $M\subseteq\{1,2,\ldots,k\}$ we denote by $c_i(\varepsilon_M)$ the codeword in C_i (to be defined below) $c_i(x_1,x_2,\ldots,x_k)\in C_i$ where $x_i=1$ if $i\in M$ and $x_i=0$ otherwise. If we want to form the code $\tilde{C}=C+c(\varepsilon_M)$ with $M=\{i_0,i_1,\ldots,i_{l-1}\}$, it is clear that we can do this in l successive steps namely

$$C_{j+1} := C_j + c_j(\varepsilon_{\{i_j\}}) \quad j=0,1,\ldots,l-1,$$

where $C_0=C$. We then get $C_l=\tilde{C}$.

So we only need to consider addition of codewords, which have exactely one nonzero coordinate in their systematic part, i.e. $c(\varepsilon_{\{i\}})$ $i\in\{1,\ldots,k\}$

Given a Vasil'ev code $V=\{(\underline{x},\underline{y},\underline{z},f_1(\underline{x},\underline{y},\underline{z}),\ldots,f_4(\underline{x},\underline{y},\underline{z})) \mid (\underline{x},\underline{y},\underline{z})\in\mathbb{F}_2^{11}\}$.
By $c(x_i)$ we mean the codeword with all systematic places zero, except th coordinate x_i. Accordingly we define $c(y_i),c(z_i)$. Since every normal cod V is linear with respect to y_i i=1,2,3 (see lemma 1) , adding the codeword $c(y_i)$ does not change the code. Furthermore it is easy to check tha adding a codeword $c(x_i)$ results in the same normal form as adding the codeword $c(z_i)$. So we have to consider only 4 new transformations:

$$\psi_i : V \mapsto V+c(z_i) \quad i=1,2,3,4 .$$

It is not hard to recognize that adding $c(z_i)$ to V_g results in substituting the variable v_i by v_i+1 in the characterizing function $g(v_1,\ldots,v_4$

For example take $g(v_1,v_2,v_3,v_4) = v_1v_2+v_1v_3v_4$. Then

$$g^{\psi_1}(v_1,v_2,v_3,v_4) \equiv g(v_1+1,v_2,v_3,v_4) =$$
$$= (v_1+1)v_2+(v_1+1)v_3v_4 = v_1v_2+v_2+v_1v_3v_4+v_3v_4 \equiv$$
$$\equiv v_1v_2+v_1v_3v_4+v_3v_4 = h(v_1,v_2,v_3,v_4)\in G .$$

So $V_{g(\underline{v})}+c(z_1) = V_{h(\underline{v})}$.

Using the substitution $w_1=v_1+v_2+v_3$, $w_2=v_2$, $w_3=v_3$, $w_4=v_4$ again, we may assume the same 4 fundamental transformations $\psi_i : w_i \to w_i+1$ as a generating set for addition of codewords.

Together with the transformations
$\bar{\pi}_1$: $(w_1 \to w_1,\ w_2 \to w_1+w_2+w_3,\ w_3 \to w_3,\ w_4 \to w_4)$ to
$\bar{\pi}_5$: $(w_1 \to w_1,\ w_2 \to w_1+w_2+w_3,\ w_3 \to w_3,\ w_4 \to w_1+w_3+w_4)$ from figure 1 we then have a generating set for the group of equivalence transformations on the Vasil'ev codes (in normal form).

9.Lemma: Let $\bar{\Psi}$ be the group generated by $\bar{\pi}_1, \bar{\pi}_2, \bar{\pi}_3, \bar{\pi}_4, \bar{\pi}_5, \bar{\psi}_1, \bar{\psi}_2, \bar{\psi}_3, \bar{\psi}_4$.
Then

$$V_{g(w_1,w_2,w_3,w_4)} \text{ is equivalent to } V_{h(w_1,w_2,w_3,w_4)}$$

if and only if there exists $\psi \in \bar{\Psi}$ with

$$g(\psi(w_1),\psi(w_2),\psi(w_3),\psi(w_4)) \equiv h(w_1,w_2,w_3,w_4) .$$

Unfortunately this group $\bar{\Psi}$ is not as nice as our $\bar{\Pi}$, which was isomorphic to GL(3,2). So the following theorem was obtained by computing the orbits in G under $\bar{\Psi}$ using a computer.

10.Theorem: There exist 19 equivalence classes of Vasil'ev codes of length 15. The following polynomials (given as vectors of the $\mathbb{F}_2$-vector space G with base $\langle\langle b_1,\dots,b_{11}\rangle\rangle$) are representatives of the equivalence classes :

0 0 0 0 0 0 0 0 0 0 0	1	0 1 0 0 0 0 0 0 0 0 1	56
0 0 0 0 0 0 0 1 0 0 0	7	0 1 1 0 0 0 0 0 0 1 0	168
0 0 0 0 0 0 0 0 0 0 1	7	0 1 1 0 0 0 0 0 1 1 1	224
0 0 0 0 0 0 0 1 1 0 0	21	0 1 1 0 0 0 0 0 0 1 1	56
0 0 0 0 0 1 0 0 1 0 0	28	1 0 0 0 0 0 0 0 0 0 0	16
0 0 1 0 0 0 0 0 0 1 0	56	1 0 0 0 0 0 0 0 0 0 1	112
0 0 1 0 0 0 0 0 1 1 1	56	1 0 0 0 0 0 0 1 0 0 0	112
0 0 1 0 0 0 0 1 0 1 0	168	1 0 0 0 0 1 0 0 1 0 0	448
0 0 1 0 0 0 1 0 1 1 1	168	1 0 0 0 0 0 0 1 1 0 0	336
0 1 0 0 0 0 0 0 0 0 0	8		

The second column gives the number of elements in the orbit.

Although we have now classified all Vasil'ev codes of lenth 15, we are far from a complete classification of all perfect codes of this length, since there exists a great number of codes, which are not equivalent to any Vasil'ev code. In [1] three examples of length 15 are constructed.

References

[1] H.Bauer, B.Ganter, F.Hergert: Algebraic techniques for nonlinear codes. TH Darmstadt, Preprint-Nr. 609 (1981).

[2] F.Hergert: Beiträge zur Theorie nichtlinearer Fehler-korrigierender Codes. Diplomarbeit TH Darmstadt (1980) (unpublished).

[3] F.J.MacWilliams, N.J.A.Sloane: The theory of error-correcting codes. North-Holland Publ. Comp. (1978).

[4] J.L.Vasil'ev: On nongroup close-packed codes. Probl.Kibernet., 8 (1962).

On the new projective planes of R. Figueroa

by

Christoph Hering and Hans-Jörg Schaeffer

We define a <u>proper projective plane</u> to be a projective plane whose automorphism group does not fix any point or line. Until recently only 2 types of finite proper projective planes had been known: The classical planes and the Hughes planes constructed by Hughes in 1957 (resp., in the smallest case, by Veblen and Wedderburn in 1907). Recently a very interesting third class has been discovered by Figueroa [1], who obtained a plane of order q^3 for each prime power q such that $q \not\equiv 1 \pmod 3$. We present here a slight modification of Figueroa's construction, which works for all prime powers. Also, we investigate the correlation groups of these planes.

Let q be a prime power, K a field of order q^3, $(\mathfrak{P},\mathfrak{L})$ the classical projective plane over K and $(\bar{\mathfrak{P}},\bar{\mathfrak{L}})$ a subplane of $(\mathfrak{P},\mathfrak{L})$ of order q. Define $\mathfrak{L}_i = \{\ell \in \mathfrak{L} \mid |\ell \cap \bar{\mathfrak{P}}| = i\}$ and $\mathfrak{P}_i = \{P \in \mathfrak{P} \mid |[P] \cap \bar{\mathfrak{L}}| = i\}$ for $i = 0, 1$. (Here $[P] = \{\ell \in \mathfrak{L} \mid P \in \ell\}$.) Clearly $\mathfrak{L} = \bar{\mathfrak{L}} \cup \mathfrak{L}_1 \cup \mathfrak{L}_0$ and $\mathfrak{P} = \bar{\mathfrak{P}} \cup \mathfrak{P}_1 \cup \mathfrak{P}_0$. There is a group $G \cong PGL(3,q)$ of automorphisms of $(\mathfrak{P},\mathfrak{L})$ fixing $(\bar{\mathfrak{P}},\bar{\mathfrak{L}})$ which is generated by perspectivities. Let m be any permutation of $\mathfrak{P}_0 \cup \mathfrak{L}_0$ interchanging $\mathfrak{P}_0$ and $\mathfrak{L}_0$ such that $X^{gm} = X^{mg}$ for all $X \in \mathfrak{P}_0 \cup \mathfrak{L}_0$ and $g \in G$.

<u>Lemma.</u> <u>Let</u> $X, Y \in \mathfrak{P}_0$ <u>and</u> $X \neq Y$. <u>Then</u> $XY \in \mathfrak{L}_1$ <u>if and only if</u> $X^m \cap Y^m \in \mathfrak{P}_1$.

<u>Proof.</u> Assume that $XY \in \mathfrak{L}_1$ and let $XY \cap \bar{\mathfrak{P}} = \{P\}$ (where XY denotes the line joining X and Y). The group $G(P)$ consisting of all perspectivities in G with center P has order $q^2(q-1)$ and acts semiregularly on $XY \cap \mathfrak{P}_0$. Thus $G(P)$ is transitive on $XY \cap \mathfrak{P}_0$, and there exists $\alpha \in G(P)$ such that $X^\alpha = Y$. Let a be the axis of α. Then $a \cap X^m \in X^m \cap X^{m\alpha} = X^m \cap Y^m$ and clearly $a \cap X^m \in \mathfrak{P}_1$. The dual argument finishes our proof.

We now introduce the following replacement: Denote

$$\ell^* = (\ell \cap \mathfrak{P}_1) \cup ([\ell^m] \cap \mathfrak{L}_0)^m \qquad \text{for } \ell \in \mathfrak{L}_0$$

and

$$\mathfrak{L}^* = \bar{\mathfrak{L}} \cup \mathfrak{L}_1 \cup \mathfrak{L}_0^*$$

and consider the incidence geometry $(\mathfrak{P},\mathfrak{L}^*)$. Clearly

$$|\ell| = q^3+1 \text{ for } \ell \in \mathfrak{L}^* \text{ and } |\{\ell \in \mathfrak{L}^* \mid P \in \ell\}| = q^3+1 \text{ for } P \in \mathfrak{P}. \qquad (*)$$

Let $\ell, k \in \mathfrak{L}_0$ and $\ell \neq k$. Assume at first that there exists

$S \in \ell^* \cap k^* \cap \mathfrak{P}_0$. Then $\ell^m, k^m \in S^{m^{-1}}$ so that $S^{m^{-1}} = \ell^m k^m$ and $S = (\ell^m k^m)^m$ is uniquely determined. Hence $|\ell^* \cap k^* \cap \mathfrak{P}_0| = 1$. As $\ell^m k^m = S^{m^{-1}} \in \mathfrak{L}_0$, we have $\ell \cap k \in \mathfrak{P}_0$ by the dual of our Lemma, so that $\ell \cap k \cap \mathfrak{P}_1 = \emptyset$ and $|\ell^* \cap k^*| = 1$. Assume now $\ell^* \cap k^* \cap \mathfrak{P}_0 = \emptyset$. Then $\ell^m k^m \in \mathfrak{L}_1$ and, again by our Lemma, $\ell \cap k \in \mathfrak{P}_1$. Thus once more $|\ell^* \cap k^*| = 1$.

Let $t \in \mathfrak{L}_1$ and $\ell \in \mathfrak{L}_0$. Suppose that $X, Y \in t \cap \ell^* \cap \mathfrak{P}_0$ and $X \neq Y$. Then $XY \in \mathfrak{L}_1$ while $X^{m^{-1}} \cap Y^{m^{-1}} = \ell^m \in \mathfrak{P}_0$, a contradiction. So $|t \cap \ell^* \cap \mathfrak{P}_0| \leq 1$ and therefore $|t \cap \ell^*| \leq 2$. Thus we have

<u>Theorem 1.</u> <u>Let</u> $k \neq \ell$. <u>If</u> $k, \ell \in \mathfrak{L}^*$, <u>then</u> $|k \cap \ell| \leq 2$. <u>If</u> $k, \ell \in \bar{\mathfrak{L}} \cup \mathfrak{L}_0^*$, <u>then</u> $|k \cap \ell| = 1$.

We now choose $\bar{\mathfrak{P}} = \{\langle(x,\bar{x},\bar{\bar{x}})\rangle \mid x \in K\backslash\{0\}\}$, where $\bar{x} = x^q$ for $x \in K$. Then G is induced by the group of matrices of the form

$$\begin{bmatrix} a & b & c \\ \bar{c} & \bar{a} & \bar{b} \\ \bar{\bar{b}} & \bar{\bar{c}} & \bar{\bar{a}} \end{bmatrix}$$

and determinant $\neq 0$, where $a, b, c \in K$. Let $S = \langle(1,0,0)\rangle$ and s be the line corresponding to the kernel of $(1,0,0)^t$. Then G_S is induced by matrices of the form

$$\begin{bmatrix} a & & \\ & \bar{a} & \\ & & \bar{\bar{a}} \end{bmatrix}$$

for $a \in K\backslash\{0\}$. In particular $[G : G_S] = q^3(q-1)^2(q+1) = |\mathfrak{P}_0|$, so that G is transitive on $\mathfrak{P}_0$. Also, $G_S = G_s$ so that there exists a permutation m such that $S^m = s$ and $s^m = S$.

<u>Theorem 2.</u> <u>Assume that</u> $S^m = s$ <u>and</u> $s^m = S$.

a) <u>If</u> $\ell \in [S] \cap \mathfrak{L}_0$ <u>and</u> $P \in s \cap \mathfrak{P}_1$, <u>then</u> $P\ell^m \in \mathfrak{L}_0$.

b) $(\mathfrak{P},\mathfrak{L}^*)$ <u>is a projective plane.</u>

c) $s \cap s^* = s \cap \mathfrak{P}_1 \cup \{\langle(0,0,1)\rangle , \langle 0,1,0\rangle\}$.

<u>Proof</u> a) Clearly $(1,1,1)^t \in \bar{\mathfrak{L}}$ so that $\langle(0,1,-1)\rangle \in s \cap \mathfrak{P}_1$. As $G_S = G_s$ is transitive on $s \cap \mathfrak{P}_1$, we can assume $P = \langle(0,1,-1)\rangle$. Because G is transitive on $\mathfrak{L}_0$, there exists an element $x \in G$ such that $s^x = \ell$. Here $S^{x^{-1}} \in \ell^{x^{-1}} = s$, so that x^{-1} is represented by a matrix

$$x^{-1} = \begin{bmatrix} 0 & b & c \\ \bar{c} & 0 & \bar{b} \\ \bar{\bar{b}} & \bar{\bar{c}} & 0 \end{bmatrix} .$$

Thus $P^{x^{-1}} = \langle(\bar{c}-\bar{\bar{b}}, -\bar{\bar{c}}, \bar{b})\rangle$ and $P^{x^{-1}}S = (0,\bar{b},\bar{\bar{c}})^t = Y(1,0,0)^t$, where

$$Y = \begin{bmatrix} 0 & c & b \\ \bar{b} & 0 & \bar{c} \\ \bar{\bar{c}} & \bar{\bar{b}} & 0 \end{bmatrix}.$$

As $\det Y = \det X^{-1} \neq 0$, it follows that $P^{x^{-1}}S \in \mathfrak{L}_0$ and hence $PS^x = Ps^{mx} = Ps^{xm} = P\ell^m \in \mathfrak{L}_0$.

b) If $t \in \mathfrak{L}_1$ and $t \cap s^* \cap \mathfrak{P}_0 \neq \emptyset$, then $t \cap s \cap \mathfrak{P}_1 = \emptyset$ by a). As G is transitive on $\mathfrak{L}_0$, this together with Theorem 1 implies that any two different lines in $\mathfrak{L}^*$ intersect in at most one point. By (*) this implies that $(\mathfrak{P},\mathfrak{L}^*)$ is a projective plane.

c) Let $\ell \in [S] \cap \mathfrak{L}_0$, and suppose $\ell^m \in s$. We define x and X as above. Then $S^x = \ell^m \in s$, so that

$$X = \begin{bmatrix} 0 & \beta & \gamma \\ \bar{\gamma} & 0 & \bar{\beta} \\ \bar{\bar{\beta}} & \bar{\bar{\gamma}} & 0 \end{bmatrix},$$

where $\beta, \gamma \in K$. This implies that $\gamma c = 0$ and $S^x = \langle(0,1,0)\rangle$ or $S^x = \langle(0,0,1)\rangle$.

<u>In the following we assume that</u> $S^m = s$ <u>and</u> $s^m = S$.

Suppose that there is an isomorphism Φ of $(\mathfrak{P},\mathfrak{L}^*)$ onto $(\mathfrak{P},\mathfrak{L})$, that is a permutation of $\mathfrak{P}$, mapping $\mathfrak{L}^*$ onto $\mathfrak{L}$. As $\bar{\mathfrak{P}}$ is a subplane of order q of the plane $(\mathfrak{P},\mathfrak{L})$, whose automorphism group is transitive on such planes, we can assume that $\bar{\mathfrak{P}}^\Phi = \bar{\mathfrak{P}}$. Also, $\mathrm{Aut}(\mathfrak{P},\mathfrak{L})$ induces all automorphisms of $(\bar{\mathfrak{P}},\bar{\mathfrak{L}})$ so that we actually can assume that Φ fixes $\bar{\mathfrak{P}}$ pointwise. If τ is a perspectivity in G, then $[\tau,\Phi]$ is an automorphism of $(\mathfrak{P},\mathfrak{L})$ leaving invariant all points in $\bar{\mathfrak{P}}$ and on each axis of τ. Thus $[G,\Phi] = 1$. In particular, $[G_S,\Phi] = 1$, so that S^Φ is one of the 3 fixed points of G_S. Now the pointwise stabilizer Z of $\mathrm{Aut}(\mathfrak{P},\mathfrak{L})$ on $\bar{\mathfrak{P}}$ is transitive on the set of fixed points of G_S. Therefore we can assume $S^\Phi = S$, and $[G,\Phi] = 1$ forces Φ to fix all points in $\mathfrak{P}_0$. As $|s^* \cap s \cap \mathfrak{P}_0| \geq 2$, we have $s^{*\Phi} = s$ and $s^* \cap \mathfrak{P}_0 = s \cap \mathfrak{P}_0$. But this implies $q = 2$ by Theorem 1 c).

Let A be the stabilizer of $\bar{\mathfrak{P}}$ in $\mathrm{Aut}(\mathfrak{P},\mathfrak{L})$ and C the stabilizer of $\bar{\mathfrak{P}} \cup \bar{\mathfrak{L}}$ in the group of correlations of $(\mathfrak{P},\mathfrak{L})$. Then $[C : A] = 2$ and $A/Z \cong P\Gamma L(3,q)$. The particular permutation m which we have chosen can be described in the following way: If $X \in \mathfrak{P}_0$, then the fixed point structure of G_X is a triangle, and X^m is the side of this triangle oposite to X. From this one easily derives, that C is compatible with m, i.e. that $P^{\xi m} = P^{m\xi}$ and $\ell^{\xi m} = \ell^{m\xi}$ for all $P \in \mathfrak{P}_0$, $\ell \in \mathfrak{L}_0$, and $\xi \in C$. Thus A leaves invariant $\mathfrak{L}^*$. Obviously $\mathrm{Aut}(\mathfrak{P},\mathfrak{L}^*)$ leaves invariant $\bar{\mathfrak{P}}$.

Also, the pointwise stabilizer Z^* of $\mathrm{Aut}(\mathfrak{P},\mathfrak{L}^*)$ on $\bar{\mathfrak{P}}$ centralizes G. Thus Z^* leaves invariant the set of fixed points of G_S, which implies $Z^* = Z$ and $\mathrm{Aut}(\mathfrak{P},\mathfrak{L}^*) = A$.

For $\xi \in C\setminus A$ we define a permutation ξ^* of $\mathfrak{P} \cup \mathfrak{L}^*$ by

$$X^{\xi^*} = X^{\xi} \qquad \text{for } X \in (\bar{\mathfrak{P}} \cup \mathfrak{P}_1)\cup(\bar{\mathfrak{L}} \cup \mathfrak{L}_1)$$
$$P^{\xi^*} = (P^{\xi})^* \qquad \text{for } P \in \mathfrak{P}_0$$
$$(\ell^*)^{\xi^*} = \ell^{\xi} \qquad \text{for } \ell \in \mathfrak{L}_0$$

Also, denote $C^* = A \cup (C\setminus A)^*$.
Let $P \in \mathfrak{P}_0$ and $\ell \in \mathfrak{L}_0$. If $P \in \ell^*$, then $\ell^m \in P^m$, $P^{\xi m} = P^{m\xi} \in \ell^{m\xi} = \ell^{\xi m}$ and hence $\ell^{*\xi^*} = \ell^{\xi} \in (P^{\xi})^* = P^{\xi^*}$. If on the other hand $\ell^{*\xi^*} \in P^{\xi^*}$, then $\ell^{\xi} \in (P^{\xi})^*$, $P^{m\xi} = P^{\xi m} \in \ell^{\xi m} = \ell^{m\xi}$, $\ell^m \in P^m$ and $P \in \ell^*$. This shows that ξ^* is a correlation of $(\mathfrak{P},\mathfrak{L}^*)$. Also, C^* is a group isomorphi to C and equal to the group of all correlations of $(\mathfrak{P},\mathfrak{L}^*)$.

References

[1] R. Figueroa: A family of not (V,ℓ)-transitive projective planes of order q^3, $q \not\equiv 1 \pmod 3$ and $q > 2$. To appear in Math. Zeitschrift.

Mathematisches Institut der Universität
Auf der Morgenstelle 10
74oo Tübingen

COUNTING SYMMETRY CLASSES OF FUNCTIONS BY WEIGHT AND AUTOMORPHISM GROUP

Adalbert Kerber and Karl-Josef Thürlings*

Lehrstuhl II für Mathematik, Univ. Bayreuth
Postfach 3008, 8580 Bayreuth, W.-Germany

Let us at first recall the basic problem of the theory of enumeration of symmetry classes of functions. If $\underline{m} := \{1,\ldots,m\}$ and $\underline{n} := \{1,\ldots,n\}$ denote two standard sets of orders m and n, then we denote by $\underline{m}^{\underline{n}}$ the set

$$\underline{m}^{\underline{n}} := \{f \mid f : \underline{n} \longrightarrow \underline{m}\}.$$

Any permutation group P acting on $\underline{n}$ induces an action on $\underline{m}^{\underline{n}}$, if we put for $\pi \in P$

$$\pi f := f \circ \pi^{-1}$$

(composition of mappings). This induced permutation group on $\underline{m}^{\underline{n}}$ is usually denoted by

$$E^P ,$$

and its orbits are called *symmetry classes* of functions $f \in \underline{m}^{\underline{n}}$.

The theory of enumeration deals with the count of such symmetry classes. If for $\pi \in P$ we denote by

* The authors would like to thank the Deutsche Forschungsgemeinschaft for fincancial support under contract Ke 201/8-1.

$$a_i(\pi)$$

the number of cyclic factors of length i of π, and if we denote by

$$c(\pi) := \sum_i a_i(\pi),$$

the number of cyclic factors of π, then by the Cauchy-Frobenius lemma, the number of all the symmetry classes is equal to

$$\frac{1}{|P|} \sum_{\pi \in P} m^{c(\pi)} ,$$

as it is well known and easy to see.

If for $f \in \underline{m}^{\underline{n}}$ we put

$$w(f) := (|f^{-1}[\{1\}]|,\ldots,|f^{-1}[\{m\}]|),$$

the *weight* of f, then a result of Pólya says that the number of symmetry classes of functions of given weight $(w_1,\ldots,w_m)$ is equal to the coefficient of

$$x^{(w)} := x_1^{w_1} \ldots x_m^{w_m}$$

in the polynomial

$$\frac{1}{|P|} \sum_{\pi \in P} \prod_{i=1}^{n} (x_1^i + \ldots + x_m^i)^{a_i(\pi)} .$$

An example is the number 2, being the number of graphs on 4 points with weight (4,2), i.e. with 2 edges (see e.g. [1], 5.1). These

graphs are

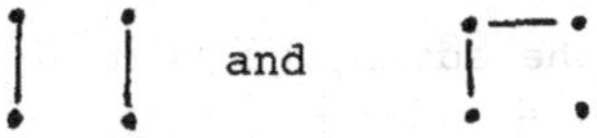

This picture shows that these graphs have different automorphism groups so that at least in this case the count of symmetry classes by weight and automorphism group is a refinement of the usual count by weight. It is therefore the aim of this paper to describe a method to solve this problem of counting symmetry classes of functions by weight and automorphism group.

1. The table of marks

Double cosets in symmetric groups are the link between the theories of enumeration of symmetry classes of functions and the theory of representations of symmetric groups. In particular the problem of evaluating a transversal of the symmetry classes of weight $w := (w_1,\dots,w_m)$ is equivalent to the problem of constructing a system of representatives of the double cosets

$$(S_{w_1} \oplus \dots \oplus S_{w_m})\ \sigma\ P \subseteq S_{\underline{n}}$$

(see [1], 5.1). Hence we consider first double cosets in an arbitrary finite group in order to present a general approach, afterwards we shall restrict attention to our special problem.

If U and V denote subgroups of a finite group G, then the double cosets $UgV \subseteq G$ are obviously the orbits of the following action of $U\times V$ on G:

$$(u,v)g := ugv^{-1}\ .$$

A seemingly different operation is in fact similar to this one (and has therefore also the double cosets as orbits) and although it looks complicated, it has the advantage that we immediately see how it can be generalized in a natural way. It is an action of $U\times V$ on the set $G\times G/\Delta(G\times G)$ of left cosets of $G\times G$ with respect to the diagonal $\Delta(G\times G) := \{(g,g) \mid g \in G\}$ and reads as follows:

1.1 $$(u,v)((g_1,g_2)\Delta(G\times G)) := (ug_1,vg_2)\Delta(G\times G).$$

Our first remark (which is easy to check) shows how one can see that the two operations are similar:

1.2 (i) *The bijection from the set* $G\times G/\Delta(G\times G)$ *of left cosets onto* G,

$$\varphi\colon G\times G/\Delta(G\times G) \longrightarrow G\colon\ (g_1,g_2)\Delta(G\times G) \longmapsto g_1g_2^{-1},$$

has the property

$$\varphi((u,v)(g_1,g_2)\Delta(G\times G)) = (u,v)\varphi((g_1,g_2)\Delta(G\times G)).$$

(ii) *The stabilizer of* $(g_1,g_2)\Delta(G\times G)$ *under the action* 1.1 *is equal to*

$$U\times V \cap (g_1,g_2)\Delta(G\times G)(g_1,g_2)^{-1}.$$

Hence the stabilizers of left cosets are uniquely determined by subgroups of $\Delta(G\times G)$ and therefore we may very well ask for the number of orbits of the action 1.1., the elements of which have stabilizers conjugate to $\Delta(W\times W)$, W being a subgroup of G.

In order to consider this question we consider the lattice

$$U(G)$$

of subgroups

$$U, U', U'', \ldots \leqslant G.$$

G acts on this lattice by conjugation:

$$g \longmapsto \binom{U}{gUg^{-1}}$$

in a way that the following holds:

1.3 $$\forall U, U', g (U \leqslant U' \Longleftrightarrow (gU) \leqslant (gU')).$$

The orbit of U under this action is the class

$$\tilde{U}$$

of subgroups conjugate to U.

It is not difficult to see that these orbits can be numbered in a way that the following holds:

1.4 $$[U' \in \tilde{U}_i, U'' \in \tilde{U}_j, U' \leqslant U''] \Longrightarrow i \leqslant j.$$

Having fixed such a numbering, we put (for $U_i \in \tilde{U}_i$, $1 \leqslant i \leqslant d$)

$$a_{ijk} := \text{no. of orbits of } U_i \times U_j \text{ under 1.1, the elements of which have stabilizers conjugate to } \Delta(U_k \times U_k).$$

Let these numbers form the matrices

$$A_i := (a_{ijk}),\ 1 \leqslant i \leqslant d,$$

d := no. of conjugacy classes of subgroups, and define

1.5
$$\omega_{ij} := |\{U_i h \mid h \in G,\ U_i h g^{-1} = U_i h, \text{ for all } g \in U_j\}|$$

$$= \frac{1}{|U_i|}\ |\{g \in G \mid U_j \leqslant g^{-1} U_i g\}|.$$

These ω_{ij} are the so-called *marks* which already Burnside introduced ([2], p. 236) and which form a table

$$\Omega = (\omega_{ij}),$$

the table of marks of G, which is uniquely determined up to permutation of rows and columns. It forms the main tool in the enumeration by automorphism group and it has the following properties (see [2],[3]):

<u>1.6</u> (i) *Ω is a lower triangular matrix. The main diagonal contains the indices of the U_i in their normalizers, the last row consists of 1's, the first column of the indices of the U_i in G:*

$$\Omega = \begin{pmatrix} |G| & \cdot & & & 0 \\ \vdots & & \ddots & & \\ |G:U_i| & & & |N_G(U_i):U_i| & \\ \vdots & & * & & \ddots \\ 1 & \cdots & \cdots & \cdots & 1 \end{pmatrix}$$

Hence in particular Ω is invertible.

(ii) *The entries of Ω satisfy the equations*

$$\omega_{ij}\,\omega_{kj} = \sum_{\nu} a_{ik\nu}\,\omega_{\nu j}\ ,$$

in other words:

$$\text{if } \omega_j := \begin{pmatrix} \omega_{1j} \\ \vdots \\ \omega_{dj} \end{pmatrix} \quad \text{then } A_i\,\omega_j = \omega_{ij}\,\omega_j,$$

i.e. the (linearly independent) vectors ω_j form a system of simultaneous eigenvectors of the matrices A_i to the eigenvalues ω_{ij}.

In order to draw conclusions for the enumeration theory of symmetry classes of functions, we notice that the definition of marks implies

<u>1.7</u> (i) *If we denote by ρ_i the (transitive) permutation representation of G induced by U_i, i.e.*

$$\rho_i := IU_i \uparrow G,$$

then the set $\{\rho_1,\ldots,\rho_d\}$ is just the set of all the essentially different transitive permutation representations, ρ_1 is the regular representation, ρ_d the identity representation (by 1.3).

(ii) *Each permutation representation δ is thus a linear combination of the ρ_i, say*

$$\delta = \sum_{i=1}^{d} d_i\,\rho_i, \quad d_i \in \mathbb{N}\,.$$

In particular, the mark μ_i^δ *of* U_i *in* δ *is*

$$\mu_i^\delta = \sum_k d_k \mu_i^{\rho_k} = \sum_k d_k \omega_{ki} .$$

Keeping this in mind we consider a finite G-set, i.e. a permutation representation $\delta : G \longrightarrow S_M$. Let G_m denote the stabilizer of $m \in M$ and put

$$y_j^M := |\{m \in M \mid U_j \leq G_m\}|, \ 1 \leq j \leq d.$$

If furthermore

x_j^M := no. of orbits of G on M, the elements of which have their stabilizers in $\tilde{U}_j$,

then the column vectors y^M and x^M consisting of these numbers satisfy

1.8 $$\boxed{x^M = {}^t\Omega^{-1} y^M .}$$

Thus $(\alpha_{ik}) := \Omega^{-1}$, the inverse of the table of marks, turns out to be the crucial matrix for the enumeration of orbits of G by automorphism group of elements. Its entries can be evaluated by Moebius inversion on U(G) as follows.

$$\delta_{ij} = \sum_\nu \omega_{i\nu} \alpha_{\nu j}$$

$$\underset{1.5}{=} \sum_\nu \frac{|N_G(U_\nu)|}{|U_i|} |\{U \leq U_i \mid U \in \tilde{U}_\nu\}| \alpha_{\nu j}.$$

Using the notation

$$\varphi_j(U) := |N_G(U)| \cdot \alpha_{ij}, \text{ if } U \in \tilde{U}_i ,$$

this shows that

$$\psi_j(U_i) := |U_i|\delta_{ij} = \sum_{U \leq U_i} \varphi_j(U),$$

or, equivalently (if μ_G denotes the Moebius function on $U(G)$),

$$\varphi_j(U_i) = \sum_{U \leq U_i} \mu_G(U,U_i)\psi_j(U).$$

This proves the equation

<u>1.9</u> $$\alpha_{ij} = \frac{|U_j|}{|N_G(U_i)|} \sum_{\substack{U \leq U_i \\ U \in \tilde{U}_j}} \mu_G(U,U_i),$$

which should be compared with the following reformulation of 1.5:

<u>1.10</u> $$\omega_{ij} = \frac{|N_G(U_j)|}{|U_i|} \sum_{\substack{U \leq U_i \\ U \in \tilde{U}_j}} 1.$$

2. Symmetry classes of functions

We would like to apply the results mentioned above to the G-set $\underline{m}^{\underline{n}}$, where $g \in G$ acts on $f \in \underline{m}^{\underline{n}}$ as follows:

$$gf := f \circ \rho(g)^{-1},$$

$\rho: G \longrightarrow S_{\underline{n}}$ being a permutation representation of G on $\underline{n}$.

In order to do this we recall from the introduction that

$$w(f) := (|f^{-1}[\{1\}]|,\dots,|f^{-1}[\{m\}]|)$$

denotes the weight of f. If $(w) := (w_1,\dots,w_m)$ denotes such a weight, i.e. if $w_i \in \mathbb{N}$, $\Sigma w_i = n$, then we denote by F(w) the set of functions with this particular weight:

$$F(w) := \{f \in \underline{m}^{\underline{n}} \mid W(f) = (w)\}.$$

This subset of $\underline{m}^{\underline{n}}$ is obviously a G-subset. The number of orbits of G on F(w) is (as we know from the weighted form of the Cauchy-Frobenius lemma (see e.g. [1], 5.1.15)) equal to the coefficient of

$$x^{(w)} := x_1^{w_1}\dots x_m^{w_m}$$

in the following polynomial arising from the cycle index $ZI(\rho[G])$ of $\rho[G]$:

$$ZI(\rho[G] \mid x_1+\dots+x_m) := \frac{1}{|G|}\sum_{g\in G}\prod_{i=1}^{n}\left(\prod_{r=1}^{m} x_r^i\right)^{a_i(\rho(g))}.$$

We want to refine this count by asking for the number of orbits of G on F(w) the elements of which have their stabilizers in the conjugacy class $\tilde{U}_i$ of subgroups of G. Denoting this desired number by

$$n^i(w)$$

we put

$$m^i(w) := |\{f \in F(w) \mid U_i \leqslant G_f\}|.$$

Then 1.8 yields

2.1 $$n^i(w) = \sum_\nu \alpha_{\nu i}\, m^\nu(w), \text{ if } (\alpha_{ik}) := \Omega^{-1}.$$

This gives us the following expression for the polynomial

$$\sum_{(w)} n^i(w) x^{(w)}$$

(which we want to display in some detail):

2.2 $$\sum_{(w)} n^i(w) x^{(w)} = \sum_\nu \alpha_{\nu i} \sum_{(w)} m^\nu(w) x^{(w)}.$$

We now try to refine the inner sum of the right hand side with the aid of the following numbers:

$$s_i := \text{no. of orbits of } \rho_i[U_i],$$

$$t_{ij} := \text{length of j-th orbit of } \rho_i[U_i].$$

2.3 Lemma: $$\sum_{(w)} m^\nu(w) x^{(w)} = \prod_{j=1}^{s_\nu} \left(\sum_{r=1}^{m} x_r^{t_{\nu j}} \right).$$

This follows immediately from the fact that U_i is contained in the stabilizer of $f \in \underline{m}^{\underline{n}}$ if and only if f is constant on the orbits of $\rho_i[U_i]$.

Let us now introduce the symbol "$\lesssim$" for "being conjugate to a subgroup of" and put

$$c_{il} := \begin{cases} \dfrac{|U_i|}{|U_l|}, & \text{if } U_l \lesssim U_i \\ 0, & \text{otherwise} \end{cases},$$

k_{il} := no. of orbits of $\rho_i[U_i]$, the elements of which have their stabilizers in $\tilde{U}_l$.

Then the following is true:

2.4 Lemma: $$\prod_{j=1}^{s_\nu} \left(\sum_{r=1} x_r^{t_{\nu j}} \right) = \prod_{l=1}^{d} \left(\sum_{r=1}^{m} x_r^{c_{\nu l}} \right)^{k_{\nu l}} .$$

In order to agree with this we need only to observe that in case $k_{\nu l} \neq 0$ the length of each one of these $k_{\nu l}$ orbits is equal to $c_{\nu l}$. Gathering up we have proved the following ([4], IV.8):

2.5 Theorem:

The number of symmetry classes of functions $f \in \underline{m}^{\underline{n}}$ of weight (w) the elements of which have their automorphism group in $\tilde{U}_i$ is equal to the coefficient of $x^{(w)}$ in the polynomial

$$\sum_{\nu} \alpha_{\nu i} \prod_{l=1}^{d} \left(\sum_{r=1}^{m} x_r^{c_{\nu l}} \right)^{k_{\nu l}} .$$

This solves the problem of counting symmetry classes by weight and automorphism group. Moreover the given form of the generating function for this problem clearly shows how far we can get with the knowledge of the isomorphism type G of the symmetry group $E^{\rho[G]}$ alone (for it yields both the numbers $\alpha_{\nu i}$ and the $c_{\nu l}$) and what depends on the particular permutation representation $\delta: G \longrightarrow S_{\underline{m}^{\underline{n}}}$ of G on $\underline{m}^{\underline{n}}$ (namely the $k_{\nu l}$).

This result of Plesken together with the expression 1.9 of the α_{ik} in terms of the Moebius function of the subgroup lattice U(G) also yields the results of Stockmeyer ([7]).

It is clear that vice versa this theorem implies Pólya's theorem which we should obtain from 2.5 by a summation over i. In order to show this we need only to remark that for each ν the following holds ([3]):

<u>2.6 Lemma:</u>

$$\sum_i \alpha_{\nu i} = \begin{cases} \phi(|U_\nu|)/|N_G(U_\nu)|, & \text{if } U_\nu \text{ is cyclic} \\ 0, & \text{otherwise,} \end{cases}$$

from which it follows that

$$\sum_{i,\nu} \alpha_{\nu i} \prod_l (\sum_r x_r^{c_{\nu l}})^{k_{\nu l}} = \frac{1}{|G|} \sum_{\substack{\nu \\ U_\nu \text{ cyclic}}} \prod_l (\sum_r x_r^{c_{\nu l}})^{k_{\nu l}} \frac{\phi(|U_\nu|)}{|N_G(U_\nu)|} |G|$$

$$= ZI(\rho[G] \mid x_1+\ldots+x_m),$$

i.e. Pólya's theorem!

Further results can be obtained by applying 1.9:

<u>2.7 Corollary:</u>

$$\sum_{U' \leq U} |U'| \mu_G(U',U) = \begin{cases} \phi(|U|), & \text{if } U \text{ is cyclic} \\ 0, & \text{otherwise} \end{cases}$$

<u>Proof:</u> $\sum_i \alpha_{\nu i} \underset{1.9}{=} \frac{1}{|N_G(U_\nu)|} \sum_{U \leq U_\nu} |U| \mu_G(U,U_\nu)$,

so that the statement follows from 2.6.

□

A further remark is implied by

$$\alpha_{\nu 1} \underset{1.9}{=} \frac{1}{|N_G(U_\nu)|} \mu_G(\{1_G\},U_\nu),$$

so that we get from 2.5:

2.8 Corollary: *The number of orbits of G on $\underline{m}^{\underline{n}}$, the elements of which have trivial stabilizer, is equal to*

$$\sum_{\nu} \frac{\mu_G(\{1_G\},U_\nu)}{|N_G(U_\nu)|} \cdot m^{\tau_\nu} ,$$

if τ_ν denotes the number of orbits of $\rho_\nu[U_\nu]$

3. Examples

Let us begin with the example we already mentioned in the introduction: the graphs on 4 points having 2 edges.

Defining a *labelled graph* on p points as a mapping f from the set

$$\underline{p}^{[2]}$$

of 2-element subsets of $\underline{p}$ (i.e. the set of $\binom{p}{2}$ pairs of points) into 2 := {0,1}, for short:

$$f \in 2^{\underline{p}^{[2]}} ,$$

we regard this set as an $S_{\underline{p}}$-set, and define a *graph* on p points as an orbit of $S_{\underline{p}}$.

Hence in our concrete example p := 4, we have to consider the $S_{\underline{4}}$-set

$$F(4,2) := \{f \in 2^{\underline{4}^{[2]}} \mid w(f) = (4,2)\}$$

$$= \{f \in 2^{\underline{4}^{[2]}} \mid |f^{-1}[\{1\}]| = 2\}.$$

In order to be prepared for an application of 1.8, we take the table of marks of $S_{\underline{4}}$ from a paper of H.O. Foulkes ([5]), who clarified the ideas of J.H. Redfield on enumeration of symmetry classes by automorphism group (see [6], which was in fact the first paper concerning this theory), and showed the connections to representation theory. This table reads as follows:

$$\begin{bmatrix} 24 & & & & & & & & & & \\ 12 & 2 & & & & & & & & & \\ 12 & 0 & 4 & & & & & & & & \\ 8 & 0 & 0 & 2 & & & 0 & & & & \\ 6 & 0 & 2 & 0 & 2 & & & & & & \\ 6 & 0 & 6 & 0 & 0 & 6 & & & & & \\ 6 & 2 & 2 & 0 & 0 & 0 & 2 & & & & \\ 4 & 2 & 0 & 1 & 0 & 0 & 0 & 1 & & & \\ 3 & 1 & 3 & 0 & 1 & 3 & 1 & 0 & 1 & & \\ 2 & 0 & 2 & 2 & 0 & 2 & 0 & 0 & 0 & 2 & \\ 1 & 1 & 1 & 1 & 1 & 1 & 1 & 1 & 1 & 1 & 1 \end{bmatrix}$$

The vector Y^M is equal to

$$\begin{pmatrix} 15 \\ 3 \\ 3 \\ 0 \\ 1 \\ 3 \\ 1 \\ 0 \\ 1 \\ 0 \\ 0 \end{pmatrix},$$

so that by 1.8 we get

$$X^M = \begin{pmatrix} 0 \\ 1 \\ 0 \\ 0 \\ 0 \\ 0 \\ 0 \\ 0 \\ 1 \\ 0 \\ 0 \end{pmatrix},$$

which means that the two graphs on 4 points and with weight (4,2) have automorphism groups conjugate to

$$U_2 := \{1,(12)\},$$

and

$$U_9 := \underline{D_4},$$

in accordance with the picture drawn above.

In our next example we take

$$\rho : G \longrightarrow S_G : g \longmapsto \binom{h}{gh},$$

the regular representation of G, from which we get $\underline{m}^G$ as a G-set. This time instead of using 1.8, we prefer to consider the polynomial given in 2.5. As

$$a_1(\rho(g)) = \begin{cases} |G|, \text{ if } g=1_G \\ 0 \text{ , otherwise,} \end{cases}$$

we have for $1 \leq i \leq d$

$$k_{i1} = \begin{cases} |G:U_i|, & \text{if } U_1 = \{1_G\} \\ 0 & \text{, otherwise,} \end{cases}$$

and hence

3.1 *The coefficient of* $x^{(w)}$ *in*

$$\sum_{\nu=1}^{d} \alpha_{\nu i} \left(\sum_{r=1}^{m} x_r^{|U_\nu|}\right)^{|G:U_\nu|}$$

is equal to the number of orbits of $E^{\rho[G]}$ *on* $\underline{m}^G$*,* ρ *being the regular representation of G, the elements of which are of weight (w) and have their automorphism group in* $\tilde{U}_i$*.*

In order to interprete this result, we notice that $U \leq G_f$, $f \in \underline{m}^G$, means that f is constant on the right cosets of G with respect to U. Hence G_f must be the subgroup of G which is maximal with respect to f being constant on its right cosets. In the case $m := 2$ we can identify the set $\underline{2}^G$ with the power set

$$\mathbb{P}(G) := \{M \mid M \subseteq G\}.$$

The stabilizer G_M of such a subset M (which has to be identified with the mapping f which satisfies $M = f^{-1}[\{1\}]$) is now the subgroup $U \leq G$ which is maximal in the sense that M is a union of right cosets of U in G. Thus 3.1 yields for this particular case

<u>3.2</u> *The coefficient of* $x_1^r x_2^{|G|-r}$ *in*

$$|G:U_i| \sum_{\nu=1}^{d} \alpha_{\nu i} \left(x_1^{|U_\nu|} + x_2^{|U_\nu|}\right)^{|G:U_\nu|}$$

is equal to the number of subsets $M \subseteq G$ *of order* r *such that the maximal subgroup of* G *for which* M *is a union of its right cosets, lies in* $\tilde{U}_i$.

A numerical example is provided by $G := S_{\underline{4}}$ and

$$U_i := \{1,(12),(34),(12)(34)\},$$

which yields the polynomial

$$6\left(\tfrac{1}{2}(x_1^4+x_2^4)^6 - \tfrac{1}{2}(x_1^8+x_2^8)^3\right)$$

$$= 18x_1^{20}x_2^4 + 36x_1^{10}x_2^8 + 60x_1^{12}x_2^{12} + 36x_1^8x_2^{16} + 18x_1^4x_2^{20} .$$

This means for example that there exist 18 subsets $M \subseteq S_{\underline{4}}$ of order 20 such that M is a union of right cosets of an $U \in \tilde{U}_i$ in a way that U is maximal with respect to this property of M.
(This number 18 agrees with the general formula

$$|G:U_i| \cdot |G: N_G(U_i)|$$

for the number of such subsets.)

Having considered the regular representation ρ of G which yields $\underline{m}^G$ as a G-set, we turn to the natural representation

$$\iota : S_n \longrightarrow S_n : \pi \longmapsto \pi$$

of $S_{\underline{n}}$, which yields $\underline{m}^{\underline{n}}$ as an $S_{\underline{n}}$-set. As $S_{\underline{n}}$ is n-fold transitive, the orbits of $S_{\underline{n}}$ on $\underline{m}^{\underline{n}}$ are just the subsets

$$F(w)$$

of functions of a given weight (w). The corresponding stabilizer of an $f \in F(w)$ is a Young subgroup

$$S_{(w)} = S_{w_1} \oplus \ldots \oplus S_{w_m} \leqslant S_{\underline{n}}.$$

Hence for each proper partition λ of n, i.e.

$$\lambda = (\lambda_1, \lambda_2, \ldots) \ , \ \lambda_1 \geqslant \lambda_2 \geqslant \ldots, \ \Sigma\, \lambda_i = n,$$

where we put

$$a_i^m(\lambda) := |\{j \mid \lambda_j = i\}| \quad , \quad 1 \leqslant i \leqslant n$$
$$a_o^m(\lambda) := m - \sum_{i=1}^{m} a_i^m(\lambda)$$

there are so many orbits the elements of which have a stabilizer conjugate to S_λ $(:= S_{\lambda_1} \oplus S_{\lambda_2} \oplus \ldots$ (see [1])):

3.3 $$x_i^{(m)} := \begin{cases} \dfrac{m!}{a_o^m(\lambda)!\,a_1^m(\lambda)!\,a_2^m(\lambda)!\ldots} & , \text{ if } S_\lambda \in \tilde{U}_i \\ 0 & , \text{ otherwise.} \end{cases}$$

These $x_i^{(m)}$ form the solution $x^{(m)}$ of the system of linear equations

3.4 $$^t\Omega x^{(m)} = y^{(m)} \ ,$$

which corresponds to this problem by 1.8. The coefficients of $y^{(m)}$ have the form

<u>3.5</u> $$y_i^{(m)} = m^{t_i}, \quad t_i := \text{no. of orbits of } U_i.$$

3.3/4/5 yield various relations in the matrix Ω^{-1}. For if U_i is not a conjugate of a Young subgroup, we obtain from 3.4:

<u>3.6</u> $$\sum_{\nu} \alpha_{\nu i}\, m^{t_\nu} = 0, \text{ for all } m \geqslant 1.$$

This implies that

$$\sum_{t=1}^{n} \Big(\sum_{\substack{\nu \\ t_\nu = t}} \alpha_{\nu i} \Big) x^t$$

is the zero polynomial, and hence the following holds:

<u>3.7</u> $$\sum_{\substack{\nu \\ t_\nu = t}} \alpha_{\nu i} = 0 ,$$

if the sum is taken over all the U_ν which have the same orbits as has U_i. Among these there is exactly one Young subgroup, say U_k. This group is maximal with respect to $\lesssim$, and so we obtain

<u>3.8 Theorem:</u>

If U_i is not a Young subgroup, but U_k is and has the same orbits as U_i, then

$$\sum_{\substack{\nu \\ U_i \lesssim U_\nu \lesssim U_k}} \alpha_{\nu i} = 0 .$$

For further results the reader is referred to [3].

REFERENCES

[1] G.D. James/A. Kerber: The representation theory of the symmetric group. Addison-Wesley 1981.

[2] W. Burnside: Theory of groups of finite order, 2nd ed. 1911 (Dover Publications 1955)

[3] A. Kerber/K.-J. Thürlings: Symmetrieklassen von Funktionen und ihre Abzählungstheorie (in preparation)

[4] W. Plesken: Counting with groups and rings (to appear)

[5] H.O. Foulkes: On Redfield's group reduction functions. Can. J. Math. 15 (1963), 272-284.

[6] J. H. Redfield: The theory of group reduced distributions. Amer. J. Math. 49 (1927), 433-455

[7] P.K. Stockmeyer: Enumeration of Graphs with prescribed auto morphism group, Dissertation, University of Michigan, 1971.

Quadruple Systems over $\mathbb{Z}_p$ Admitting the Affine Group

Egmont Köhler
Mathematisches Seminar
der Universität Hamburg
2000 Hamburg 12
Federal Republic of Germany

Summary

A necessary and sufficient condition for the existence of t-(p,k,λ) designs which are invariant under the affine group $A_p = \{x \to ax+b : a,b \in GF(p),\ a \neq 0\}$ is given. From this we derive sufficient criteria für the existence of A_p-invariant 3-$(p,4,\lambda)$ designs for all primes p. These designs are simple in the case $p \equiv 5 \pmod{12}$ and $\lambda = 2$. As a corollary to our considerations, we obtain some infinite series of simple 2-(p,r,λ) designs for all primes p and certain values of λ which are also invariant under A_p.

Definitions and Notations

For a set M and $r \in \mathbb{N}$ let $\binom{M}{r} = \{N \subseteq M : |N|=r\}$ and V(M) be the $\mathbb{Q}$-vectorspace with M as a basis. For $r_1,r_2,\dots,r_n \in \mathbb{N}$ and $m_1,m_2,\dots,m_n \in M$ the vector $B = \sum_{i=1}^{n} r_i m_i \in V(M)$ is called a <u>multiset</u>, and the r_i's are the multiplicities of the m_i's. (Also for $N = \{n_1,\dots,n_s\} \subseteq M$ we consider N as the vector $\sum_{i=1}^{s} n_i \in V(M)$.) Furthermore, we define $|B| = \sum_{i=1}^{n} r_i$ and we write $m \in B$ if $m \in M$ is a term of the sum in B.

Now let $1 < t < k < p \in \mathbb{N}$ and $\lambda \in \mathbb{N}$, where p is a prime. Then for $\mathcal{B} = \Sigma \beta_i B_i \in V(\binom{\mathbb{Z}_p}{k}), \beta_i \in \mathbb{N}$, and arbitrary $T \in \binom{\mathbb{Z}_p}{t}$ we define

$$\mathcal{B}_T = \Sigma \beta_i^T (B_i - T) \text{ by } \beta_i^T = \begin{cases} \beta_i, & \text{if } T \subseteq B \\ 0 & \text{otherwise.} \end{cases}$$

Such a multiset $\mathcal{B}$ is called a <u>cyclic t-design</u> over $\mathbb{Z}_p$ (in short $cS_\lambda(t,k,p)$) if the following two conditions hold:

(1) $|\mathcal{B}_T| = \lambda$ for all $T \in \binom{\mathbb{Z}_p}{t}$, and

(2) if $B = \{b_1,\dots,b_k\} \in \mathcal{B}$ and $c \in \mathbb{Z}_p$, then $c + B = \{c+b_1,\dots,c+b_k\} \in \mathcal{B}$.

(If $\mathcal{B}$ is a set the corresponding $cS_\lambda(t,k,p)$ is called <u>simple</u>.)

It is wellknown that the existence of a $cS_\lambda(t,k,p)$ implies that λ is a multiple of

$$\lambda_0 = \text{lcm}\ \left\{ \frac{\binom{k-i}{t-i}}{\gcd\ \{ \binom{p-i}{t-i}, \binom{k-i}{t-i} \}} : i = 0,1,\ldots,t \right\}.$$

Such a λ will be called putative (with respect to t,k, and p.).

Let $\mathcal{B}$ be a $cS_\lambda(t,k,p)$. From now on we will assume that for all $B = \{b_1,\ldots,b_k\} \in \mathcal{B}$ we have the following :

(1) the b_i's are represented by the numbers $0,1,\ldots,p-1$, and

(2) $b_1 < b_2 < \ldots < b_k$.

Let $2 < k < p \in \mathbb{N}$ be as before. We write $T_{k,p} = \{(a_1,\ldots,a_k): a_i \in \mathbb{Z}_p \setminus \{0\},\ a_k = -\sum_{i=1}^{k-1} a_i\}$. The components of the elements of $T_{k,p}$ are again represented as non-negative integers $a_i \in \{0,1,\ldots,p-1\}$.

Now for $A,B \in T_{k,p}$ we write $A \sim B$ iff A differs from B only by a cyclic permutation of the components. We denote the set of orbits of the equivalence relation $\sim$ in $T_{k,p}$ by $K_{k,p}$ and for $(a_1,\ldots,a_k) \in T_{k,p}$ we denote the orbit of $(a_1,\ldots,a_k)$ with respect to $\sim$ by $[a_1,\ldots,a_k]$. Such a $[a_1,\ldots,a_k] \in K_{k,p}$ is called a k-difference-cycle over $\mathbb{Z}_p$.

For $t \in \mathbb{N}$, $1 < t < k$, and $\kappa = [a_1,\ldots,a_k] \in K_{k,p}$ an element $\tau = [b_1,\ldots,b_t] \in K_{t,p}$ is called a t-subcycle of κ iff (1) and (2) hold, where

(1) for all $i \in \mathbb{Z}_t$ there exist $j,j^* \in \mathbb{Z}_k$ with $b_i = \sum_{h=j}^{j^*} a_h$, and

(2) if $b_i = \sum_{h=j}^{j^*} a_h$ and $b_{i+1} = \sum_{h=l}^{l^*} a_h$, then $l = j^* + 1$.

(Hence we assume the indices of the components of τ and κ to be elements of $\mathbb{Z}_t$ and $\mathbb{Z}_k$ respectively, represented as the numbers $0,1,\ldots,t-1$ and $0,1,\ldots,k-1$ respectively).

If τ is a t-subcycle of κ we write $\tau < \kappa$.

This construction yields exactly $\binom{k}{t}$ not necessarily mutually distinct t-subcycles of a given $\kappa \in K_{k,p}$. Now let κ_τ be the multiplicity of the appearence of $\tau \in K_{t,p}$ as a t-subcycle of $\kappa \in K_{k,p}$. Then we denote the multiset of all t-subcycles of κ by $\kappa^t = \sum_{\tau<\kappa} \kappa_\tau \cdot \tau$. Therefore, we have $\sum_{\tau<\kappa} \kappa_\tau = \binom{k}{t}$ and $\kappa^t \in V(K_{t,p})$.

Orbits of Difference-cycles

The following theorem is proved in [6].

Theorem 1. If there exist $\beta_1,\ldots,\beta_n \in \mathbb{N}$ and $S = \{\kappa_1,\ldots,\kappa_n\} \in K_{k,p}$ with $\sum_{i=1}^{n} \beta_i \kappa_i^t = \lambda \cdot K_{t,p}$, then there exists a $cS_\lambda(t,k,p)$.
(Clearly, this $cS_\lambda(t,k,p)$ is simple iff $\beta_1 = \beta_2 = \ldots = \beta_n = 1$.) □

Now let $S = \{\kappa_1,\ldots,\kappa_n\} \subseteq K_{k,p}$ with $\sum_{i=1}^{n} \beta_i \kappa_i^t = \lambda \cdot K_{t,p}$. The $cS_\lambda(t,k,p)$ which, by Theorem 1, belongs to this S is of the Form $\mathcal{B} = \sum_{i=1}^{n} \beta_i S_{\kappa_i} \in V(\binom{\mathbf{Z}_p}{k})$, where the S_{κ_i}'s are defined by the bijection $\Phi_k : S \to \{S_{\kappa_1},\ldots,S_{\kappa_n}\}$ with

$$S_{\kappa_j} = \Phi_k(\kappa_j) = \langle 0, a_1^{(j)}, a_1^{(j)} + a_2^{(j)}, \ldots, \sum_{h=1}^{k-1} a_h^{(j)} \rangle =$$
$$\{\{l, l + a_1^{(j)}, l + a_1^{(j)} + a_2^{(j)}, \ldots, l + \sum_{h=1}^{k-1} a_h^{(j)}\} : l \in \mathbf{Z}_p\}, \text{ where } \kappa_j = [a_1^{(j)},\ldots,a_k^{(j)}]$$

Then Φ_k^{-1} may be described as follows.

Consider an element $A_j \in S_{\kappa_j}$ containing $0 \in \mathbf{Z}_p$, say $A_j = \{0, b_1^{(j)},\ldots,b_{k-1}^{(j)}\}$ with $0 < b_1^{(j)} < \ldots < b_{k-1}^{(j)}$. Then

$$\Phi_k^{-1}(S_{\kappa_j}) = [b_1^{(j)}, b_2^{(j)} - b_1^{(j)}, \ldots, p - b_{k-1}^{(j)}] = [a_1^{(j)},\ldots, a_k^{(j)}] = \kappa_j .$$

We define a multiplication of an element $y \in \mathbf{Z} \setminus \{0\}$ with an element $\kappa = [a_1,\ldots,a_k] \in K_{k,p}$ such that $y \cdot \kappa \in K_{k,p}$ by

$$y \cdot \kappa = \Phi_k^{-1}(y \cdot S_\kappa), \text{ where}$$
$$y \cdot S_\kappa = y \cdot \langle 0, a_1, a_1 + a_2, \ldots, \sum_{l=1}^{k-1} a_l \rangle = \langle 0, ya_1, y(a_1+a_2), \ldots, y \sum_{l=1}^{k-1} a_l \rangle .$$

The above construction yields the following

Lemma 1. If $\kappa \in K_{k,p}$ and $y \in \mathbf{Z}_p \setminus \{0\}$, then for all $\tau \in K_{t,p}$ we have $\tau < \kappa$ if and only if $y\tau < y\kappa$. □

With respect to this multiplication the set $K_{k,p}$ consists of orbits and we write $\langle\kappa\rangle = \{y \cdot \kappa : y \in \mathbf{Z}_p \setminus \{0\}\}$ for $\kappa \in K_{k,p}$. Hence, if ε is a primitive root modulo p we can write $\langle\kappa\rangle = \{\varepsilon^i \cdot \kappa : i = 1,2,\ldots,p-1\}$. Therefore, $|\langle\kappa\rangle| \le p-1$ and $|\langle\kappa\rangle|$ is a divisor of p-1. Now let $\kappa \in K_{k,p}$ such that $a \cdot |\langle\kappa\rangle| = p-1$. Then, $\varepsilon^x \cdot \kappa = \varepsilon^y \cdot \kappa$ iff $x \equiv y \pmod{\frac{p-1}{a}}$.

Furthermore, let $\kappa^t = \sum_{h=1}^{r} x_h \tau_h$ with $x_h \in \mathbb{N}$ and $b_h \cdot |\langle\tau_h\rangle| = p-1$. Then we have $\sum_{i=1}^{p-1} (\varepsilon^i\kappa)^t = \sum_{j=1}^{r} x_j \sum_{i=1}^{p-1} (\varepsilon^i\tau_j)$. On the other hand $\sum_{i=1}^{p-1} (\varepsilon^i\kappa)^t = a \sum_{i=1}^{(p-1)/a} (\varepsilon^i\kappa)^t$ and $x_h \sum_{i=1}^{p-1} (\varepsilon^i\tau_h) = x_h b_h \sum_{i=1}^{(p-1)/b_h} (\varepsilon^i\tau_h) = x_h b_h <\tau_h>$. This yields

<u>Lemma 2.</u> For $\kappa^t = \sum_{h=1}^{r} x_h\tau_h$ the indices of the sum may be chosen in such a way that $\kappa^t = \sum_{j=1}^{r_1} x_j\tau_j + \sum_{j=r_1+1}^{r_2} x_j\tau_j + \ldots + \sum_{j=r_{f-1}+1}^{r_f} x_j\tau_j$ with

$$<\tau_1> = <\tau_2> = \ldots = <\tau_{r_1}>,$$
$$<\tau_{r_1+1}> = <\tau_{r_1+2}> = \ldots = <\tau_{r_2}>,$$
$$\vdots$$
$$<\tau_{r_{f-1}+1}> = <\tau_{r_{f-1}+2}> = \ldots = <\tau_{r_f}>, \text{ and } <\tau_{r_i}> \neq <\tau_{r_j}> \text{ if } i \neq j.$$

Furthermore, let $|<\kappa>| = \frac{p-1}{a}$ and $|<\tau_{r_n}>| = \frac{p-1}{b_n}$ for $n = 1,2,\ldots,f$.

Then, putting $r_o = 0$, the following holds:

$$\sum_{i=1}^{(p,1)/a} (\varepsilon^i\kappa)^t = \sum_{j=1}^{f} \left(\frac{b_j}{a} \sum_{i=r_{j-1}+1}^{r_j} x_i\right) \cdot <\tau_{r_j}>.$$ □

Instead of $\sum_{i=1}^{(p-1)/a} (\varepsilon^i\kappa)^t$ we write $<\kappa>^t$. As a consequence of Lemma 2 $<\kappa>^t$ is a multiset over $<K_{t,p}> = \{<\tau> | \tau \in K_{t,p}\}$.

Using Theorem 1 and Lemma 2, we obtain the following

<u>Theorem 2.</u> There exists a $cS_\lambda(t,k,p)$ which is invariant under the multiplication with elements from $\mathbb{Z}_p \setminus \{0\}$ if and only if there exists a multiset $\Gamma = \sum_{i=1}^{r} \beta_i <\kappa_i>, \beta_i \in \mathbb{N}$ of orbits of k-difference-cycles over $\mathbb{Z}_p$ with $\Gamma = \lambda \cdot <K_{t,p}>$.
(Clearly, if Γ is a set, the corresponding $cS_\lambda(t,k,p)$ is simple.) □

We denote the multiset $cS_\lambda(t,k,p)$ with the properties given in Theorem 2 by $aS_\lambda(t,k,p)$.

Now we state some properties of the sets $K_{k,p}$:

Firstly, since $(p,k) = 1$ we have $|K_{k,p}| = \binom{p}{k} \cdot p^{-1}$.

In order to obtain an important classification of the elements of $K_{k,p}$ we introduce the following definition.

Let $\kappa = [a_1,\ldots,a_k] \in K_{k,p}$. If $[a_1,\ldots,a_k] = [a_1,a_k,a_{k-1},\ldots,a_2]$ we call κ symmetric . We denote the set of all symmetric k-difference-cycles over $\mathbf{Z}_p$ by $sK_{k,p}$. Futhermore, we write $s^*K_{k,p} = K_{k,p} - sK_{k,p}$. By looking at the corresponding definitions, one can prove, using elementary tools, the following

Lemma 3. $\kappa \in sK_{k,p}$ if and only if $\kappa = (-1) \cdot \kappa$. □

Lemma 3 implies that the cardinality of the set of symmetric difference-cycles over $\mathbf{Z}_p$ is the same as the cardinality of the orbits in $\binom{\mathbf{Z}_p}{k}$ under the group $D_p = \{x \to ax+b : a \in \{+1,-1\}, b \in \mathbf{Z}_p\}$. We can, therefore, apply Polya's counting-theorem to determine this cardinality. But this requires some calculations which we omit. So we just state the following

Lemma 4. $|sK_{k,p}| = \binom{\frac{p-1}{2}}{\lfloor \frac{k}{2} \rfloor}$.

(If $\alpha \in \mathbb{R}$, then $\lfloor\alpha\rfloor = \max\{z \in \mathbf{Z} : z \le \alpha\}$.)

A proof of Lemma 4 can be found in [2]. □

Lemma 5.

(i) If $\kappa \in sK_{k,p}$ then $|\langle\kappa\rangle| \le \frac{p-1}{2}$,

(ii) If $\kappa \in K_{k,p}$ and $|\langle\kappa\rangle| = \frac{p-1}{2}$ then $\kappa \in sK_{k,p}$,

(iii) If $\kappa \in sK_{k,p}$ and $\kappa' \in \langle\kappa\rangle$ then $\kappa' \in sK_{k,p}$. □

Finally, we introduce the following notation. We write

$\langle sK_{k,p}\rangle = \{\langle\kappa\rangle : \kappa \in sK_{k,p}\}$ and $\langle s^*K_{k,p}\rangle = \{\langle\kappa\rangle : \kappa \in s^*K_{k,p}\}$.

The Sets $K_{3,p}$ and $K_{4,p}$

In the previous chapter we defined for $\kappa = [a_1,\ldots,a_k] \in K_{k,p}$ and $y \in \mathbf{Z}_p \setminus \{0\}$ the element $y \cdot \kappa \in K_{k,p}$ using the bijection Φ_k. Now we ask if it is possible to write $y \cdot \kappa$ without making use of Φ_k. In general, we have

$y[a_1,\dots,a_k] = \Phi_k^{-1}(<0, ya_1, y(a_1+a_2),\dots,y\cdot\sum_{l=1}^{k-1} a_l>)$. In order to apply Φ_k^{-1} the elements $0, ya_1, y(a_1+a_2), y\cdot\sum_{l=1}^{k-1} a_l$ have to be ordered regarded as nonnegative integers. (Without this ordering $[ya_1,\dots,ya_k]$ is sometimes not an element of $K_{k,p}$. We call this ordering the reduction of $[ya_1,\dots,ya_k]$.

Remark 1. For $\kappa=[a,b,-(a+b)]\in K_{3,p}$ and $y\in \mathbb{Z}_p\setminus\{0\}$ we have

(i) $y\kappa = [ya,yb,-y(a+b)]$, or

(ii) $y\kappa = [y(a+b),-yb,-ya]$.

Proof: It is $y\kappa = \Phi_3^{-1}(<0,ya,y(a+b)>) = [ya,yb,-y(a+b)]$, if $ya+yb+(-y(a+b)) = p$ over $\mathbb{N}$. Otherwise we apply reduction and obtain $\Phi_3^{-1}(<0,y(a+b),ya>) = [y(a+b),-yb,-ya]$. □

Similarly we obtain:

Remark 2. For $\kappa = [a,b,c,-(a+b+c)]\in K_{4,p}$ and $y\in \mathbb{Z}\setminus\{0\}$ we have:

(i) $y\kappa = [ya,yb,yc,-y(a+b+c)]$, or
(ii) $y\kappa = [ya,y(b+c),-yc,-y(a+b)]$, or
(iii) $y\kappa = [y(a+b),-yb,y(b+c),-y(a+b+c)]$, or
(iv) $y\kappa = [y(a+b),yc,-y(b+c),-ya]$, or
(v) $y\kappa = [y(a+b+c),-yc,-yb,-ya]$, or
(vi) $y\kappa = [y(a+b+c),-yc,-yb,-ya]$. □

We now investigate the sets $K_{3,p}$ and $K_{4,p}$ in detail.

Lemma 6. (i) There is exactly one class $<\kappa>\in<K_{3,p}>$ with $|<\kappa>| = \frac{p-1}{2}$. For this class we have $<\kappa> = <sK_{3,p}>$ and $\kappa = [1,1,-2]$ is a representative.

(ii) If $p\equiv 5 \pmod 6$, then $s^*K_{3,p}$ splits into $\frac{p-5}{6}$ classes $<\kappa>$ with $|<\kappa>| = p-1$ in each case.

(iii) If $p\equiv 1 \pmod 6$, then $s^*K_{3,p}$ splits into $\frac{p-7}{6}$ classes $<\kappa>$ with $|<\kappa>| = p-1$ in each case and one futher class $<\kappa'>$ with $|<\kappa'>| = \frac{p-1}{3}$. In the latter case, $<\kappa'>$ can be represented by $\kappa' = [1,y,y^2]$ with $y^2+y+1 = 0$ in $\mathbb{Z}_p$, up to a reduction.

Proof: We have $sK_{3,p} = \{[a,a,-2a] : a\in\mathbb{Z}_p\setminus\{0\}\}$ and $a\le\frac{p-1}{2}$. On the other hand, if $a, a'\in\{1,2,\dots,\frac{p-1}{2}\}$ and $a\neq a'$, then $[a,a,-2a]\neq[a',a',-2a']$. Hence $|sK_{3,p}| = \frac{p-1}{2}$ and $[1,1,-2]$ is a representative. This proves (i).

In the following we consider the general case $\kappa\in K_{3,p}$. Here we have $|<\kappa>| < p-1$ if and only if there exists a $y\in\mathbb{Z}_p\setminus\{0,1\}$ with $\kappa = y\kappa$.

Let $\kappa = [a,b,-(a+b)] = y[a,b,-(a+b)] \in K_{3,p}$ with $y \in \mathbf{Z}_p \setminus \{0,1\}$.

Remark 1 implies $[a,b,-(a+b)] = [ya,yb,-y(a+b)]$ or $[a,b,-(a+b)] = [y(a+b,-yb,-ya]$.

Now we calculate:

(i,i) if $a = ya$, $b = yb$, $-(a+b) = -y(a+b)$ then $y = 1$ which was excluded,

(i,ii) if $a = yb$, $b = -y(a+b)$, $-(a+b) = ya$ then $y^2 + y + 1 = 0$,

(i,iii) if $a = -y(a+b)$, $b = ya$, $-(a+b) = yb$ then $y^2 + y + 1 = 0$,

(ii,i) if $a = y(a+b)$, $b = -yb$, $-(a+b) = -ya$ then $y = -1$,

(ii,ii) if $a = -yb$, $b = -ya$, $-(a+b) = -y(a+b)$ then $y = -1$,

(ii,iii) if $a = -ya$, $b = y(a+b)$, $-(a+b) = -yb$ then $y = -1$.

In the case $y = -1$, however, κ is symmetric by Lemma 3. For $\kappa \in s^*K_{3,p}$ we have $|\langle\kappa\rangle| < p-1$ iff $\kappa = y\kappa$ with $y^2+y+1=0$ in $\mathbf{Z}_p$, and $y^2+y+1=0$ in $\mathbf{Z}_p$ has a solution in $\mathbf{Z}_p$ iff $p \equiv 1 \pmod 6$. This proves (ii).

Now assume $p \equiv 1 \pmod 6$, and let $\kappa' = [a,b,-(a+b)] \in K_{3,p}$ with $\kappa' = y\kappa'$ and $y^2+y+1=0$, hence $y^3 = 1$ with $y \neq 1$. Here we have $a = -y(a+b)$, $b = -y(a+b)$, $-(a+b) = ya$ or $a = -y(a+b)$, $b = ya$, $-(a+b) = yb$. Therefore, $\kappa' = [yb,b,y^2b] \in \langle[y,a,y^2]\rangle$ or $\kappa' = [a,ya,y^2a] \in \langle[1,y,y^2]\rangle$. Since $-y[y,1,y^2] = [1,y,y^2]$ we have $\langle[y,1,y^2]\rangle = \langle[1,y,y^2]$

It follows from $y^3 = 1$ with $y \neq 1$ that $|\langle\kappa'\rangle| = \frac{p-1}{3}$.

This concludes the proof. □

In particular, Lemma 6 shows that for each $\kappa \in K_{3,p}$ we have $\frac{p-1}{|\langle\kappa\rangle|} \in \{1,2,3\}$. Similarly, for each $\kappa \in K_{4,p}$ we have $\frac{p-1}{|\langle\kappa\rangle|} \in \{1,2,3,4\}$. To be more precise:

<u>Lemma 7.</u> If $p \equiv 1 \pmod{12}$, then

(i) $sK_{4,p}$ consists of exactly one class $\langle\kappa\rangle$ with $|\langle\kappa\rangle| = \frac{p-1}{4}$ (here $\kappa = [1,\frac{2}{y-1},1,\frac{2y}{1-y}]$ with $y^2 = -1$ up to a reduction) and further $\frac{p-5}{4}$ classes $\langle\kappa\rangle$ with $|\langle\kappa\rangle| = \frac{p-1}{2}$, and

(ii) $s^*K_{4,p}$ consists of exactly one class $\langle\kappa\rangle$ with $|\langle\kappa\rangle| = \frac{p-1}{3}$ (here $\kappa = [1,x,\frac{x^2}{1-x},\frac{1}{x-1}]$ with $x^2+x+1=0$ up to a reduction) and further $\frac{(p-1)(p-7)}{4!}$ classes $\langle\kappa\rangle$ with $|\langle\kappa\rangle| = p-1$.

If $p \equiv 5 \pmod{12}$, then

(i) $sK_{4,p}$ consists of exactly one class $\langle\kappa\rangle$ with $|\langle\kappa\rangle| = \frac{p-1}{4}$(here $\kappa = [1,\frac{2}{y-1},1,\frac{2y}{y-1}]$

with $y^2 = -1$ up to a reduction) and further $\frac{p-5}{4}$ classes $<\kappa>$ with $|<\kappa>| = \frac{p-1}{2}$, and

(ii) $s^*K_{4,p}$ consists of exactly one class $<\kappa>$ with $|<\kappa>| = \frac{p-1}{4}$ (here $\kappa = [1,x,\frac{x^2}{1-x},\frac{1}{x-1}]$ with $x^2+x+1 = 0$ up to a reduction) and further $\frac{(p-1)(p-7)}{4!}$ classes $<\kappa>$ with $|<\kappa>| = p-1$.

If $p \equiv 7 \pmod{12}$, then

(i) $sK_{4,p}$ consists of exactly $\frac{p-3}{4}$ classes $<\kappa>$ with $|<\kappa>| = \frac{p-1}{2}$, and

(ii) $s^*K_{4,p}$ consists of exactly one class $<\kappa>$ with $|<\kappa>| = \frac{p-1}{3}$ (here $\kappa=[1,x,\frac{x^2}{1-x},\frac{1}{x-1}]$ with $x^2+x+1 = 0$ up to a reduction) and further $\frac{(p-1)(p-7)}{4!}$ classes $<\kappa>$ with $|<\kappa>| = p-1$.

If $p \equiv 11 \pmod{12}$, then

(i) $sK_{4,p}$ consists of exactly $\frac{p-3}{4}$ classes $<\kappa>$ with $|<\kappa>| = \frac{p-1}{2}$, and

(ii) $s^*K_{4,p}$ consists of exactly $\frac{(p-3)(p-5)}{4!}$ classes $<\kappa>$ with $|<\kappa>| = p-1$.

Proof: We proved Lemma 6 by elementary calculations in $\mathbf{Z}_p$, using Remark 1. In the same way, the proof of Lemma 7 proceeds by calculations in $\mathbf{Z}_p$, using Remark 2. As this is rather straightforward we present only one of the 24 cases as an example.

If $\kappa = [a,b,c,-(a+b+c)] = y\kappa = [ya,y(b+c),-yc,-y(a+b)]$, we obtain (e.g. if $a = -yc$, $b = -y(a+b)$, $c = ya$, $-(a+b+c) = y(b+c)$) $b = -\frac{ay}{1+y}$ and $c = ay$, with $y^2 = -1$. Hence $\kappa \in < [1, \frac{-y}{1+y}, y, \frac{-y}{1+y}] >$. Now for $y^2 = -1$ one has

$$[1, \frac{-y}{1+y}, y, \frac{-y}{1+y}] = \frac{-y}{1+y} [1, \frac{2}{y-1}, 1, \frac{2y}{1-y}].$$

This yields $|<\kappa>| = \frac{p-1}{4}$ for $y^2 = -1$ and $\kappa = y\kappa$. On the other hand, $y^2 = -1$ holds for $y \in \mathbf{Z}_p$ iff $p \equiv 1 \pmod 4$. Since p is a prime number we have $p \equiv 1 \pmod 4$ iff $p \equiv 1, 5 \pmod{12}$.

The other cases can be treated in a similar way. □

As an immediate consequence of Lemma 7 we obtain the following result, where $aS_\lambda(t,k,p)$ denotes an $S_\lambda(t,k,p)$ invariant under the affine group A_p.

Theorem 3. For all prime numbers p there exists an $aS_{12}(2,4,p)$ and an $aS_6(2,4,p)$. For $p \equiv 1 \pmod 6$ there exists an $aS_4(2,4,p)$, and for $p \equiv 1,5 \pmod{12}$ there exists an $aS_3(2,4,p)$.

Proof: Since $A_p = \{x \to ax+b : a,b \in \mathbf{Z}_p,\ a \neq 0\}$ is sharply 2-transitive on $\mathbf{Z}_p$, each $\langle\kappa\rangle$ of $\langle K_{4,p}\rangle$ represents an $aS_\lambda(2,4,p)$ with $\lambda = \frac{|\langle\kappa\rangle| \cdot 4 \cdot 3}{p-1}$ .By Lemma 7, the assertion follows. □

Sufficient Conditions for the Existence of an $aS_{\lambda_0}(3,4,p)$

Firsly, we state two useful remarks.

Remark 3. If $[a,b,-(a+b)] \in K_{3,p}$, then $\langle [a,b,-(a+b)] \rangle$ contains at most six 3-difference-cycles having 1 at least once as a component. These are:

$$[1,\alpha,-(\alpha+1)],\ [1,\tfrac{1}{\alpha},-(\tfrac{1}{\alpha}+1)],\ [1,\tfrac{-\alpha}{\alpha+1},-(\tfrac{-\alpha}{\alpha+1}+1)]\quad,$$
$$[1,-(\alpha+1),\alpha],\ [1,-(\tfrac{1}{\alpha}+1),\tfrac{1}{\alpha}],\ [1,-(\tfrac{-\alpha}{\alpha+1}+1),\ \tfrac{-\alpha}{\alpha+1}],$$

where $\alpha \in \mathbf{Z}_p$ is defined by $\alpha = \frac{b}{a}$.

Proof: It follows from Remark 1 that

$x[a,b,-(a+b)] = [xa,xb,-x(a+b)]$ or $x[a,b,-(a+b)] = [-xa,x(a+b),-xb]$.

For 1 to be a component of $x[a,b,-(a+b)]$, x has to be one of $\pm\frac{1}{a}$, $\pm\frac{1}{b}$, $\pm\frac{1}{a+b}$. This yields, using the two equations of Remark 1, the listed six 3-difference-cycles, which are not necessarily mutually distinct. □

Remark 4. If $[a,b,a,-(2a+b)] \in sK_{4,p}$, then $\langle [a,b,a,-(2a+b)] \rangle$ contains exactly two 4-difference-cycles having the entry 1 in two non-consecutive components. These are $[1,\alpha,1,-(\alpha+2)]$ and $[1,\frac{-\alpha}{\alpha+1},1,\frac{\alpha+2}{\alpha+1}]$, where $\alpha \in \mathbf{Z}_p$ is defined by $\alpha = \frac{b}{a}$.

Proof: Let $\kappa = [a,b,a,-(2a+b)] \in sK_{4,p}$. Considering $y\kappa$ we see that the cases (ii) and (vi) of Remark 2 cannot occur.

Now, looking at case (i) of Remark 2, we must choose $y = \frac{1}{a}$ and get $\frac{1}{a}\kappa = [1,\alpha,1,-(\alpha+2)$. The same result is obtained in case (vi) by putting $y = -\frac{1}{a}$. In case (iii) one gets $\frac{1}{a+b} \cdot \kappa = [1,\frac{-\alpha}{\alpha+1},1,-\frac{\alpha+2}{\alpha+1}]$ for $y = \frac{1}{a+b}$, and we obtain the same result in case (v) by putting $y = \frac{-1}{a+b}$. □

Now we can prove

<u>Lemma 8:</u> $\sum_{\kappa \in sK_{4,p}} \langle\kappa\rangle^3 = 2 \cdot sK_{3,p} + 3 \cdot s^*K_{3,p}$.

<u>Proof:</u> Let $\kappa \in sK_{4,p}$. Then, by Remark 4, $\langle\kappa\rangle$ contains the elements $[1,\alpha,1,-(\alpha+2)]$ and $[1,\frac{-\alpha}{\alpha+1},1,-\frac{\alpha+2}{\alpha+1}]$. Hence $\langle\kappa\rangle^3$ contains $2 \cdot sK_{3,p}$ iff $\alpha = 1$. In this case we have $\langle[1,1,1,-3]\rangle^3 = 2 \cdot sK_{3,p} + \langle[1,2,-3]\rangle$. No element $\kappa \in (sK_{4,p} \setminus \langle[1,1,1,-3]\rangle)$ contains $\tau \in sK_{3,p}$ as a 3-subcycle. On the other hand, each $\tau = [a,b,-(a+b)] \in s^*K_{3,p}$ occurs in exactly 3 symmetric 4-difference-cycles as a 3-subcycle. We have to prove this only for those $\tau \in s^*K_{3,p}$ having 1 as the first component.

Let $\tau = [1,a,-(a+1)]$. Then we have $\tau \in \kappa^3$ with $\kappa \in sK_{4,p}$ iff one of the following cases occurs:

(i) $\kappa = [1,a,1,-(a+2)]$, or

(ii) $\kappa = [1,a-1,1,-(a+1)]$, or

(iii) $\kappa = [1,a,-(2a+1),a]$.

This proves Lemma 8. □

It is now appropriate to define a graph $G_p = (\underset{\cdot}{G}_p, \underset{-}{G}_p)$ for all prime numbers $p > 7$ (here $\underset{\cdot}{G}_p$ denotes the set of vertices of G_p and $\underset{-}{G}_p$ the set of edges of G_p) in the following way:

$\underset{\cdot}{G}_p = \{\langle\alpha\rangle = \{\alpha,-(\alpha+1),\frac{1}{\alpha},-(\frac{1}{\alpha}+1),\frac{-\alpha}{\alpha+1},-(\frac{-\alpha}{\alpha+1}+1)\} : \alpha \in \mathbf{Z}_p \setminus \{0,1,-1,-2,\frac{-1}{2}\}\}$, and

$\{\langle\alpha\rangle,\langle\beta\rangle\} \in \underset{-}{G}_p$ if and only if there exist $a \in \langle\alpha\rangle$ and $b \in \langle\beta\rangle$ with $a = b \pm 1$.

Using these definitions we state a few theorems; examples for these will be given in the last section.

<u>Theorem 4:</u> Let $p \equiv 5 \pmod{12}$. Then for each putative λ, there exists an $aS_\lambda(3,4,p)$ if G_p possesses a 1-factor.

These designs are simple for $\lambda = 2$.

<u>Proof:</u> For $t = 3$, $k = 4$ and $p \equiv 5 \pmod{12}$ we have $\lambda_0 = 2$. Therefore, it is sufficient to prove that the existence of a 1-factor in G_p implies the existence of an $aS_2(3,4,p)$.

By Theorem 2 and Lemma 8, we know that $\sum_{\kappa \in sK_{4,p}} \langle\kappa\rangle^3 \setminus s^*K_{3,p}$ forms an $aS_2(3,4,p)$.

Therefore, we show that the existence of a 1-factor in G_p allows us to choose a subset A from $<sK_{4,p}>$ such that $\sum_{\kappa\in A} <\kappa>^3 = s^*K_{3,p}$. We proceed in the following way: For $<\alpha> \in \underset{\bullet}{G}_p$ we can identify $<\alpha>$ with $<[1,\alpha,-(\alpha+1)]> \in <sK_{3,p}>$. Given $\{<\alpha>,<\beta>\} \in \underline{G}_p$, there are $[1,a,-(a+1)] \in <[1,\alpha,-(\alpha+1)]>$ and $[1,a\pm 1,-(a\pm 1+1)] \in <[1,\beta,-(\beta+1)]>$. This means that $[1,\alpha,-(\alpha+1)]$ and $[1,\beta,-(\beta+1)]$ are contained in $[1,a,1,-(a+2)]$ as 3-subcycles. A 1-factor in G_p corresponds, therefore, to an $A \subseteqq sK_{4,p}$ with $\sum_{\kappa\in A} <\kappa>^3 = s^*K_{3,p}$. □

We have a similar criterion in the case $p \equiv 11 \pmod{12}$ according to the following

Theorem 5. Let $p \equiv 11 \pmod{12}$. Then for each putative λ, an $aS_\lambda(3,4,p)$ exists, if the graph G_p', arising as an induced subgraph of G_p on $\underset{\bullet}{G}_p' = \underset{\bullet}{G}_p \setminus \{<2>\}$, posesses a 1-facto

Proof: For $t = 3$, $k = 4$ and $p \equiv 11 \pmod{12}$ we have $\lambda_o = 4$. Hence we will construct an $aS_4(3,4,p)$. Again by Theorem 2 and Lemma 8, $\sum_{\kappa\in sK_{4,p}} <\kappa>^3 + 2\cdot sK_{3,p} + s^*K_{3,p}$ is an $aS_4(3,4,p)$. Hence we must choose a multiset A from $sK_{4,p}$ satisfying $\sum_{\kappa\in A} <\kappa>^3 = 2\cdot sK_{3,p} + s^*K_{3,p}$. Therefore, we define $A = <[1,1,1,-3]> + B$ with a certain $B \subseteqq sK_{4,p}$. Since $<[1,1,1,-3]>^3 = 2\cdot sK_{3,p} + <[1,2,-3]>$ we must have $\sum_{\kappa\in B} <\kappa>^3 = s^*K_{3,p} \setminus <[1,2,-3]>$. The same consideration as in the proof of Theorem 4 shows that a 1-factor in G_p' corresponds to such a B.

□

The cases $p \equiv 1, 7 \pmod{12}$ are a little bit more complicated.

Theorem 6. Let $p \equiv 7 \pmod{12}$ and $y \in \mathbf{Z}_p$ with $y^2 + y + 1 = 0$. Let $G_p^* = (\underset{\bullet}{G}_p^*, \underline{G}_p^*)$ be the induced subgraph of G_p on $\underset{\bullet}{G}_p^* = \underset{\bullet}{G}_p \setminus \{<y>, <3>, <\frac{1}{y-1}>\}$, and $G_p^{**} = (\underset{\bullet}{G}_p^{**}, \underline{G}_p^{**})$ be the induced subgraph of G_p on $\underset{\bullet}{G}_p^{**} = \underset{\bullet}{G}_p \setminus \{<y>, <2>, <\frac{1}{y-1}>\}$. Then for each putative λ an $aS_\lambda(3,4,p)$ exists, if either G_p^* or G_p^{**} possesses a 1-factor.

Proof: For $t = 3$, $k = 4$ and $p \equiv 7 \pmod{12}$ we have $\lambda_o = 4$. Hence we will construct an $aS_4(3,4,p)$. We have $\sum_{\kappa\in sK_{4,p}\setminus<[1,1,1,-3]>} <\kappa>^3 = 3\cdot s^*K_{3,p} \setminus <[1,2,-3]>$.

On the other hand, $\kappa' = [1, y, \frac{y^2}{1-y}, \frac{1}{y-1}] \in s^*K_{4,p}$ with $|<\kappa'>| = \frac{p-1}{3}$. Hence

$$\sum_{\kappa\in sK_{4,p}\setminus<[1,1,1,-3]>} <\kappa>^3 + <\kappa'>^3 + <[1,1,2,-4]>^3 + \sum_{\kappa\in B} <\kappa>^3$$

forms an $aS_4(3,4,p)$, if the multiset B over $sK_{4,p}$ satisfies

$$\sum_{\kappa \in B} \langle\kappa\rangle^3 = s^*K_{3,p} - \langle[1,y,y^2]\rangle + \langle[1,3,-4]\rangle + \langle[1,\frac{1}{y-1},\frac{-y}{y-1}]\rangle.$$ But such a B is defined by a 1-factor in G_p^*.

Similarly, $\langle[1,1,1,-3]\rangle^3 + \langle\kappa'\rangle^3 + \sum_{\kappa \in sK_{4,p}} \langle\kappa\rangle^3 + \sum_{\kappa \in B^*} \langle\kappa\rangle^3$ is an $aS_4(3,4,p)$, if the multiset B^* over $sK_{4,p}$ is chosen such that it is defined by a 1-factor of G_p^{**}. □

<u>Theorem 7.</u> Let $p \equiv 1 \pmod{12}$ and $y \in \mathbf{Z}_p$ with $y^2+y+1=0$. Let $\bar{G}_p$ be the induced subgraph of G_p on $\bar{\mathfrak{G}}_p = \mathfrak{G}_p \setminus \{\langle y\rangle, \langle y-1\rangle, \langle y-2\rangle, \langle\frac{y}{y-1}\rangle\}$. Then for each putative λ an $aS_\lambda(3,4,p)$ exists, if $\bar{G}_p$ possesses a 1-factor.

<u>Proof:</u> For $t=3$, $k=4$ and $p \equiv 1 \pmod{12}$ we have $\lambda_0 = 2$. Hence we will construct an $aS_2(3,4,p)$. Therefore, we define

$$\kappa' = [1,y,\frac{y^2}{1-y},\frac{1}{y-1}] \quad ; \quad \kappa^* = [1,y-1,1,-(y+1)] \quad ;$$
$$\kappa_1 = [1,y-2,1,-y] \quad ; \quad \kappa_2 = [1,\frac{1}{y-1}, 1, \frac{2y-1}{y-1}] \quad .$$

This gives $|\langle\kappa'\rangle| = \frac{p-1}{3}$, $|\langle\kappa^*\rangle| = |\langle\kappa_1\rangle| = |\langle\kappa_2\rangle| = \frac{p-1}{2}$ and $|\langle[1,y,-(y+1)]\rangle| = \frac{p-1}{3}$.

Now we can calculate, using Lemma 2:

$$\langle\kappa'\rangle^3 = \langle[1,y,-(y+1)]\rangle + \langle[1,y-1,-y]\rangle,$$
$$\langle\kappa^*\rangle^3 = 3\cdot\langle[1,y,-(y+1)]\rangle + \langle[1,y-1,-y]\rangle,$$
$$\langle\kappa_1\rangle^3 = \langle[1,y-1,-y]\rangle + \langle[1,y-2,-(y-1)]\rangle,$$
$$\langle\kappa_2\rangle^3 = \langle[1,y-1,-(y+1)]\rangle + \langle[1,\frac{1}{y-1},\frac{2y-1}{y-1}]\rangle.$$

(The latter follows from $[1,y-1,-(y+1)] \in \langle[1,\frac{1}{y-1},\frac{-y}{y-1}]\rangle$.)

Since

$$\sum_{\kappa \in sK_{4,p}} \langle\kappa\rangle^3 = 2\cdot sK_{3,p} + 3\cdot s^*K_{3,p}$$ the multiset

$$\sum_{\kappa \in sK_{4,p}} \langle\kappa\rangle^3 - \langle\kappa^*\rangle^3 - \langle\kappa_1\rangle^3 - \langle\kappa_2\rangle^3 + 2\cdot\langle\kappa'\rangle^3 - \sum_{\kappa \in B} \langle\kappa\rangle^3$$

forms an $aS_2(3,4,p)$, if a multiset B over $sK_{4,p}$ satisfying

$$\sum_{\kappa\in B} \langle\kappa\rangle^3 = s^*K_{3,p} - \langle[1,y,-(y+1)]\rangle - \langle[1,y-1,-y]\rangle - \langle[1,y-2,-(y-1)]\rangle - \langle[1,\tfrac{y}{y-1},\tfrac{2y-1}{y-1}]\rangle$$

exists. But such a B is represented by a 1-factor in $\bar{G}_p$ if $p \neq 13$.

Many details concerning the structure of the graphs G_p can be found in [5].

The case p < 100

Now we apply Theorems 4,5,6 and 7 and show that, for "small" primes, the necessary conditions for the existence of an $aS_\lambda(3,4,p)$ are also sufficient. Thus we prove the following

Theorem 8. For all primes p with $11 \le p \le 97$ there exists an $aS_\lambda(3,4,p)$ if and only if λ is a putative parameter and $p \neq 13$.

Proof: In the case p = 13 we have

$$\langle K_{3,p}\rangle = \langle[1,1,11]\rangle + \langle[1,3,9]\rangle + \langle[1,2,10]\rangle,$$
$$\langle sK_{4,p}\rangle = \langle[1,1,1,10]\rangle + \langle[1,4,1,7]\rangle + \langle[1,2,1,9]\rangle \text{ and}$$
$$\langle s^*K_{4,p}\rangle = \langle[1,2,6,4]\rangle + \langle[1,1,2,9]\rangle + \langle[1,1,4,7]\rangle + \langle[1,1,3,8]\rangle.$$

After having made the following calculations

$$\langle[1,1,1,10]\rangle^3 = 2\cdot\langle[1,1,11]\rangle + \langle[1,2,10]\rangle$$
$$\langle[1,4,1,7]\rangle^3 = \langle[1,2,10]\rangle,$$
$$\langle[1,2,1,9]\rangle^3 = 3\cdot\langle[1,3,9]\rangle + \langle[1,2,10]\rangle,$$
$$\langle[1,2,6,4]\rangle^3 = \langle[1,3,9]\rangle + \langle[1,2,10]\rangle,$$
$$\langle[1,1,2,9]\rangle^3 = 4\cdot\langle[1,1,11]\rangle + 3\cdot\langle[1,3,9]\rangle + \langle[1,2,10]\rangle,$$
$$\langle[1,1,4,7]\rangle^3 = 2\cdot\langle[1,1,11]\rangle + 3\cdot\langle[1,2,10]\rangle, \text{ and}$$
$$\langle[1,1,3,8]\rangle^3 = 2\cdot\langle[1,1,11]\rangle + 3\cdot\langle[1,3,9]\rangle + 2\cdot\langle[1,2,10]\rangle,$$

one sees immediately, using Theorem 2, that there does not exist an $aS_2(3,4,13)$.
A further case must be treated separatly, too: the graphs $\bar{G}_{61}$ posses <16> as an isolated vertex. Therefore, there is no 1-factor in $\bar{G}_{61}$. Nevertheless one can construct an $aS_2(3,4,61)$ in the following way.

Since $\langle K_{3,61}\rangle = \langle[1,1,59]\rangle + \langle[1,13,47]\rangle + \langle[1,16,44]\rangle + \langle[1,21,39]\rangle + X$
with $X = \{\langle[1,i,-(i+1)]\rangle : i = 2,3,\dots,8\}$,
there is a multiset L over $\langle K_{4,61}\rangle$ which represents an $aS_2(3,4,61)$. This is

$$L = \langle[1,10,1,49]\rangle + \langle[1,6,1,53]\rangle + \langle[1,7,1,52]\rangle + \langle[1,8,1,51]\rangle + 2\cdot\langle[1,15,1,44]\rangle + \langle[1,20,1,39]\rangle + 2\cdot\langle[1,4,9,47]\rangle + \langle[1,2,3,55]\rangle.$$

Indeed, this multiset L defines an $aS_2(3,4,61)$ as one can see by considering Theorem 2 and the following equations

$$\langle[1,10,1,49]\rangle^3 = \langle[1,5,55]\rangle,$$
$$\langle[1,6,1,53]\rangle^3 = \langle[1,6,54]\rangle + \langle[1,7,53]\rangle\ ,$$
$$\langle[1,7,1,52]\rangle^3 = \langle[1,7,53]\rangle + \langle[1,8,52]\rangle\ ,$$
$$\langle[1,8,1,51]\rangle^3 = \langle[1,6,54]\rangle + \langle[1,8,52]\rangle\ ,$$
$$2\cdot\langle[1,15,1,44]\rangle^3 = 2\cdot\langle[1,3,57]\rangle + 2\cdot\langle[1,16,44]\rangle\ ,$$
$$\langle[1,20,1,39]\rangle^3 = \langle[1,2,58]\rangle + \langle[1,21,39]\rangle\ ,$$
$$2\cdot\langle[1,4,9,47]\rangle^3 = 2\cdot\langle[1,13,47]\rangle + 2\cdot\langle[1,4,56]\rangle\ , \text{ and}$$
$$\langle[1,2,3,55]\rangle^3 = 2\cdot\langle[1,1,59]\rangle + \langle[1,2,58]\rangle + \langle[1,5,55]\rangle + \langle[1,21,39]\rangle.$$

In all other cases the graph corresponding to the prime p has a 1-factor. This can easily be seen by applying one of the Theorems 4,5,6 or 7 to the figures below.

$G'_{11} = G^*_{19} = \emptyset$; G_{17} =

G'_{23} =

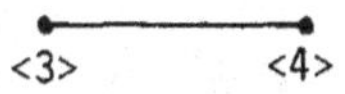

; G_{29} =

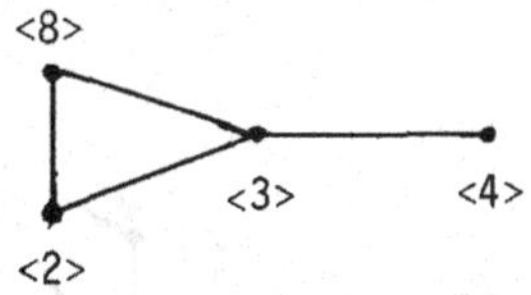

G^*_{31} =

; $\bar{G}_{37}$ =

G_{41} =

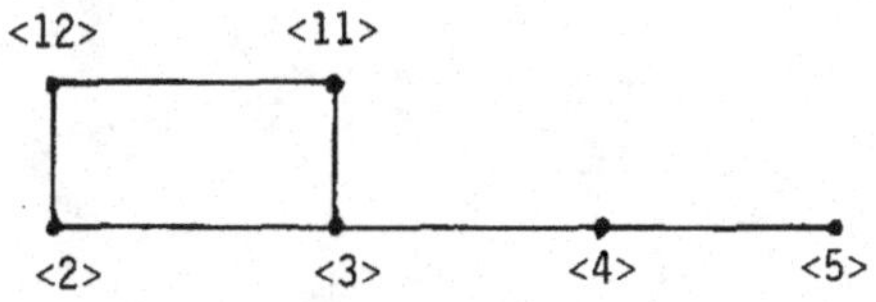

G^*_{43} =

G'_{47} =

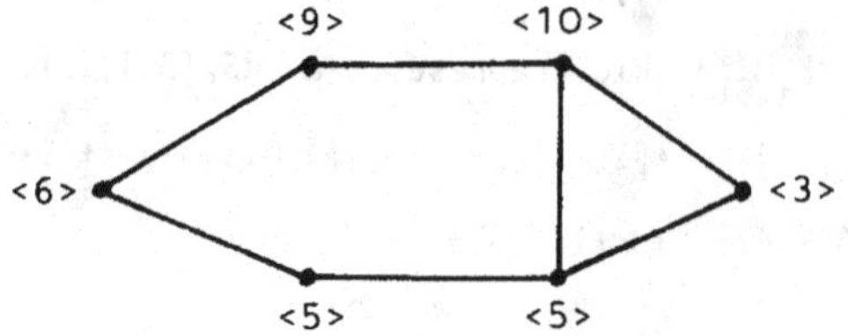

G_{53} =

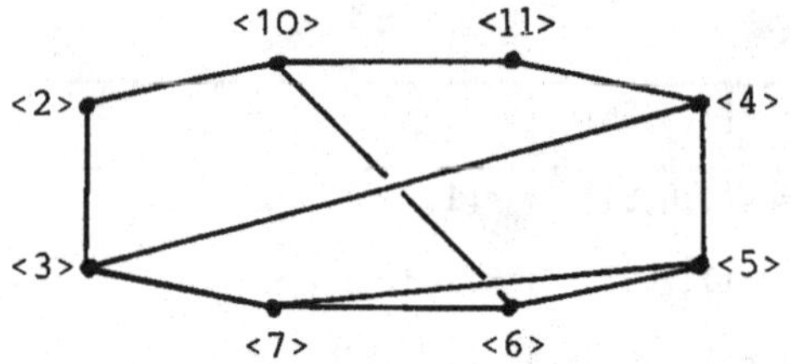

G'_{59} =

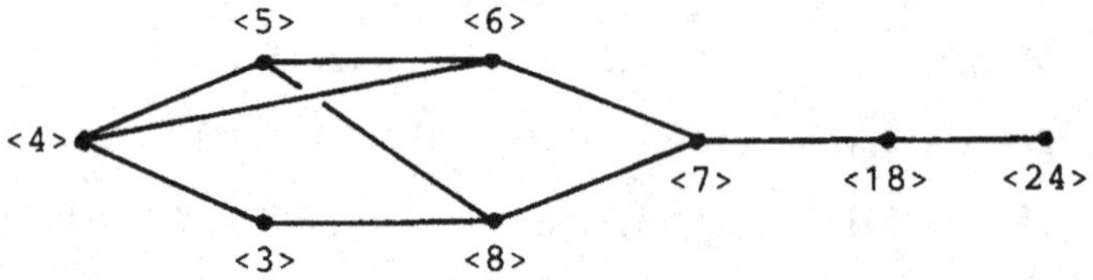

G^*_{67} =

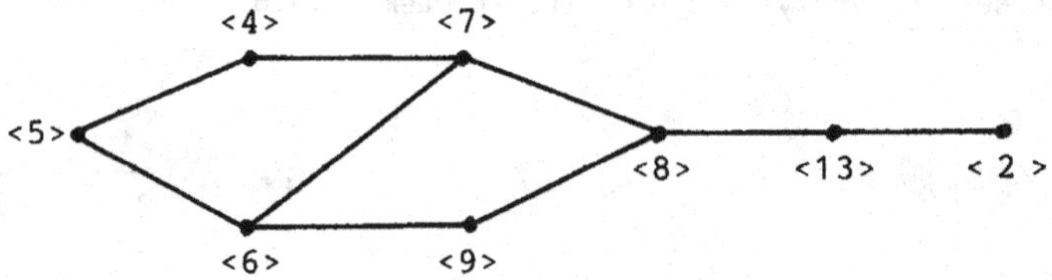

G'_{71} =

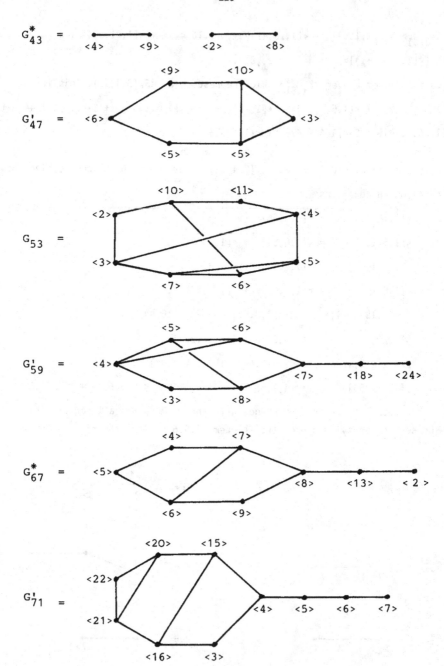

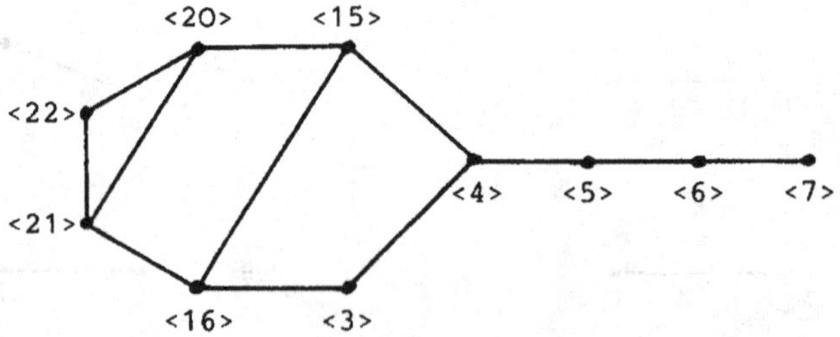

$\bar{G}_{73}$ =

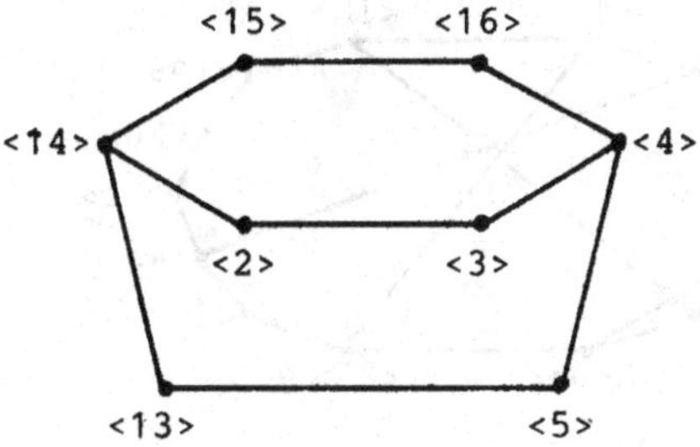

G^{**}_{79} =

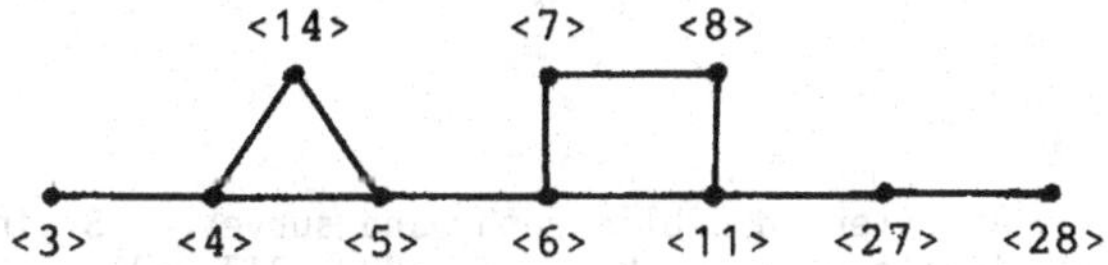

G'_{83} =

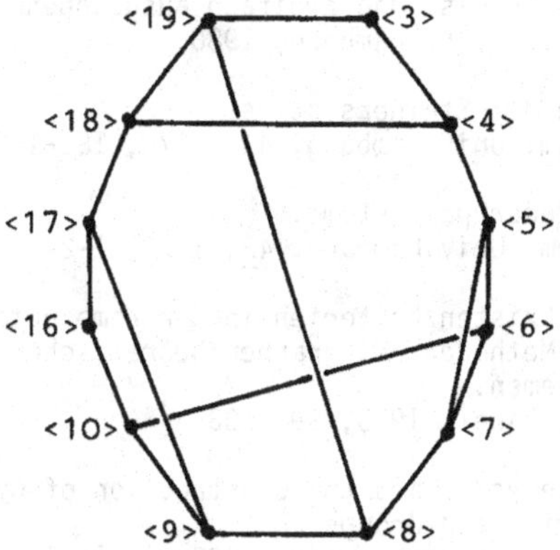

G_{89} =

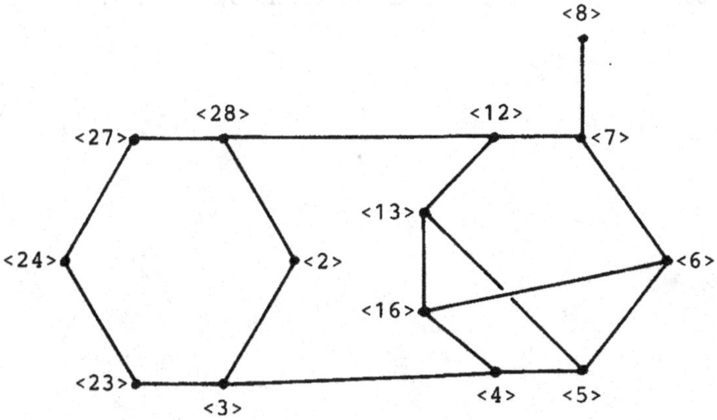

G'_{97} =

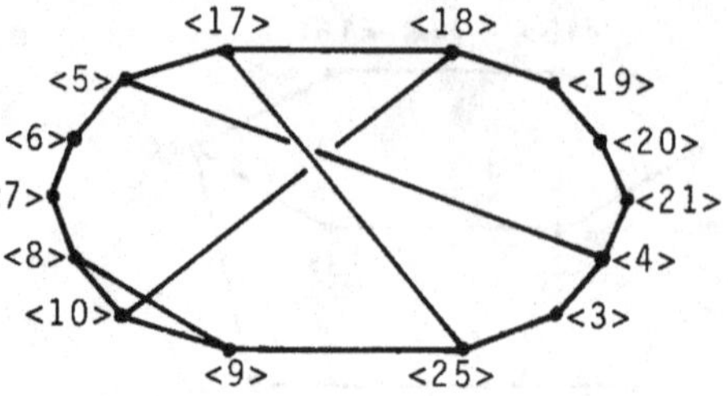

References

1. J.Doyen, A.Rosa — An extended bibliography and survey of Steiner systems. Annals of Discr.Math., 7, 1980, 317-349.

2. M.Kleemann — k-Differenzenkreise und zweifach ausgewogene Pläne. Diplomarbeit, Univ. Hamburg, 1980.

3. E.Köhler — Zur Theorie der Steinersysteme. Abh.Math.Sem. Univ.Hamburg, 43, 1975, 181-185.

4. E.Köhler — Zyklische Quadrupelsysteme. Abh.Math.Sem. Univ.Hamburg, 48, 1979, 1-24.

5. E.Köhler — Numerische Existenzkriterien in der Kombinatorik. Numerische Methoden bei graphentheoretischen und kombinator schen Problemen. Birkhäuser, Basel, 1975, 99 - 108.

6. E.Köhler — k-Differencecycles and the Construction of Cyclic t-Designs in: Geometries and Groups. Lecture Notes in Mathematics, 893, 1981, 195 - 203, Berlin - Heidelberg - New York.

7. C.Lindner, A. Rosa — Steiner Quadrupel Systems. Discrete Math., 22, 1978, 147 - 181.

Arcs and Ovals in Steiner Triple Systems

Hanfried Lenz
Freie Universität , Königin-Luise-Str. 24-26, D-1000 Berlin 33
and
Herbert Zeitler
Universität Bayreuth, Universitätsstr. 30, D-8580 Bayreuth
F.R. Germany

Several constructions of Steiner triple systems (STS) with ovals are given. For every $v \equiv 3$ or 7 mod 12 there are STS's with hyperovals, for every $v \equiv 1$ or 3 mod 6 there are STS's with ovals, and for infinitely many $v \equiv 1$ or 3 mod 6 there are STS's without ovals. The ovals may be classified by their complementary sets, the so-called counterovals. Several questions remain open.

1. Introduction

Up to now arcs and ovals were mainly investigated in projective planes. In classical projective planes of odd order the famous theorem of B. Segre holds that each oval is a conic [13], [7]. Of course these concepts make sense in linear spaces resp. in partial linear spaces. A partial linear space is a finite incidence structure $(V,\mathfrak{B})$ with point set V and line set $\mathfrak{B}$ with at most one line through any two points and at least two points on every line. It becomes a linear space if every unjoined point pair is considered as a new line. We write v for $|V|$ and b for $|\mathfrak{B}|$.

Examples of partial linear spaces are the so-called group divisible designs (GDD), where the points are partitioned into classes such that two points are joined iff they are in distinct classes; in particular the transversal designs (TD) with $k > 2$ classes such that every line intersects every point class. It is well known that then each class has exactly g points and there are g^2 lines. Such a TD is called a TD[k;g]. The existence of a TD[k;g] is equivalent to the existence of $k - 2$ mutually orthogonal Latin squares. In the sequel we assume that in a partial linear space at least one line has more than two points.

Definitions: An arc in a partial linear space is a point set which intersects no line in more than two points. Obviously every arc is a subset of a maximal arc. For any point set B a line L is called a

$$\left.\begin{array}{l}\underline{\text{subline}}\\ \underline{\text{secant}}\\ \underline{\text{tangent}}\\ \underline{\text{passant}}\end{array}\right\} \text{ of B if } |L \cap B| = \left\{\begin{array}{l}|L|\\ 2\\ 1\\ 0.\end{array}\right.$$

An arc is called a hyperoval if it has no tangents, and an oval if there are tangents but at most one through any point of it. Let r_p be the number of all lines through a given point p. If B is an arc and $p \in B$, then the number of tangents of B through p is $r_p - |B| + 1$. If H is a hyperoval and $x \notin H$, then there are exactly $|H|/2$ secants through x, and the number of tangents in a point $p \in H$ is $0 = r_p - |H| + 1$, i.e. $r_p = |H| - 1$.

An oval B in a linear space can be extended to a hyperoval only if each point $x \in B$ is on exactly one tangent and all these tangents have a point in common.
If r_p is independent of p (e.g. in Steiner systems S(2,k;v) with exactly k points on every line, or in GDD's GD[k,g;v] with exactly k points on every line and g points in every class, in particular in case $v = kg$, i.e. in transversal designs TD[k;g]), then each point of an oval is on exactly one tangent. The number $t_B(x)$ of tangents through a point $x \notin B$ is odd iff $r = r_p = |B|$ is odd, and even otherwise. In a Steiner system S(2,k;v) it is well known (e.g. Hall [5]) that

$$(1.1) \qquad r = \frac{v-1}{k-1}, \quad b = \frac{vr}{k} = \frac{v(v-1)}{k(k-1)} .$$

There is a huge literature on Steiner systems S(t,k;v), see the book [9] edited by Lindner and Rosa, in particular the bibliography by Doyen and Rosa (in this book [9]) with more than 700 titles. In case $k = 3$ we get a Steiner triple system (STS) with $r = (v-1)/2$, $b = v(v-1)/6$. Let STS be the set of $v \in \mathbb{N}$ for which an STS(v) exists. It is well known that $STS = 6\mathbb{N}_0 + \{1,3\}$.

In this paper we shall show that for each $v \equiv 3$ or 7 mod 12 there are STS's with hyperovals, for each $v \in STS$ there are STS's with ovals, and for for almost all $v \in STS$ there are STS's without ovals. The proof of the last two assertionswas considerably improved by several remarks of W. Piotrowski [12].

2. Hyperovals in Steiner Triple Systems

Theorem 2.1: An STS(v) has a hyperoval iff it has a sub-STS(r) with $r = (v-1)/2$ points. H is a hyperoval iff $V \setminus H$ is a sub-STS(r).

Proof: I. Let U be a subspace of order (= number of points) r, and $p \notin U$ a point. Each line $L \ni p$ intersects U, since otherwise there would be at least $r+1$ points $x \notin U \cup \{p\}$ on the r lines through p, hence $v > 2r+1$, a contradiction. Hence $V \setminus U$ is a hyperoval.

II. Let H be a hyperoval and $U := V \setminus H$. As H has no tangents, any line through two points of U is contained in U, i.e. U is a subspace of order $v - |H| = r$. □

Theorem 2.2: An STS(v) with a hyperoval exists iff $v \equiv 3$ or 7 mod 12.

Proof: The necessity of the condition $v \equiv 3$ or 7 mod 12 follows from the fact that r must be odd. The sufficiency can be shown by several classical constructions. In the sequel we shall present three of them.

Construction 2.1: Let K_{r+1} be the complete graph with $r+1$ vertices (=points). It is well known that K_{r+1} can be factorized. E.g. put the vertices into the centre and the corners of a regular r-gon. Then each parallel class consists of one side S of the r-gon, the $(r-3)/2$ diagonals parallel to S, and the radius from the centre to the remaining corner. For the number of possible factorizations see Lindner-Mendelsohn-Rosa [10]. Now add a new point p_i to each parallel class $\mathcal{F}_i$ $(i=1,\dots,r)$ such that p_i forms lines together with the egdes of $\mathcal{F}_i$. Moreover form an STS(r) on the new points, say U. Thus we have constructed an STS(2r + 1) with a subspace of order r, i.e. with a hyperoval. □

Construction 2.2 (doubling construction): Given the incidence matrix M of an STS(r), replace the three 1's in each column by the auxiliary matrices

$$(2.1) \qquad A = \begin{pmatrix}1&1&0&0\\0&0&1&1\end{pmatrix},\ B = \begin{pmatrix}1&0&1&0\\0&1&0&1\end{pmatrix},\ C = \begin{pmatrix}1&0&0&1\\0&1&1&0\end{pmatrix} \text{ or else } \begin{pmatrix}0&1&1&0\\1&0&0&1\end{pmatrix},$$

and zeroes by matrices $\begin{pmatrix}0&0&0&0\\0&0&0&0\end{pmatrix}$. Thus we get the incidence matrix of a GD[3,2;2r] which is completed to an STS(2r + 1) by a new point ∞ which forms new lines together with the r point classes. □

If the original STS(r) has a hyperoval, say in the first (r+1)/2 rows of the incidence matrix (then $r \equiv 3$ or $7 \bmod 12$), then the $r+1$ first rows of the large new matrix form a hyperoval in the constructed STS$(2r+1)$. In this case we have a lot of free choice in our construction: for each of the $r(r-1)/6$ columns of M there is free choice between $\begin{pmatrix}1&0&0&1\\0&1&1&0\end{pmatrix}$ and $\begin{pmatrix}0&1&1&0\\1&0&0&1\end{pmatrix}$ for C.

Remark: Construction 2.2 works in case $v \equiv 1$ or $9 \bmod 12$ too. The first, third, ..., $(2r-1)^{th}$ row of the large matrix form a sub-STS, say U, isomorphic to the original STS(r), together with the first, fifth, ..., $(4r-3)^{th}$ column, if for C always $\begin{pmatrix}1&0&0&1\\0&1&1&0\end{pmatrix}$ is chosen.

This doubling construction extends each line of the original STS(r) to a subspace U' of order 7, containing ∞. The intersection H' of U' with the hyperoval $H := V \setminus U$ is a quadrangle, i.e. a hyperoval in U'. Deleting a point $p \in H'$ yields an oval $B := H \setminus \{p\}$ whose intersection with U' is a triangle, i.e. an oval in U'.

Theorem 2.3: An STS(v) with $v \equiv 3$ or $19 \bmod 24$ has at most one hyperoval.

Proof: By a well known lemma of Doyen [3] any two subspaces of order r have an intersection of order $(r-1)/2$, but $(r-1)/2$ is not the order of an STS. Hence there are no two subspaces of order r, i.e. no two hyperovals. □

Corollary: For each $v \equiv 3$ or $19 \bmod 24$ there are STS(v)'s with exactly one hyperoval. For $v \equiv 7$ or $15 \bmod 24$ there are STS(v)'s with more than one hyperoval as the doubling construction shows.

Lemma 2.1: Let $v = 3u - 2w$, and let $(V,\mathcal{B})$ be an STS(v) with two subspaces U,U' of order u such that $|U \cap U'| = w < u$. Then the complementary set U" of $(U \cup U') \setminus (U \cap U')$ is a third subspace of order u.

Proof: If $p \in U \setminus U'$ then each line L(p,x) through p and a point $x \in U' \setminus U$ contains a third point $z \notin U \cup U'$. As there are only $v - |U \cup U'| = u - w$ such points, every line through a point $z \notin U \cup U'$ intersects both subspaces U,U' or none of them. The assertion easily follows. □

Theorem 2.4: If an STS has two hyperovals then it has at least three hyperovals.

This follows from theorem 2.1 and lemma 2.1 with $u = r$, $w = (r-1)/2$. □
Next we consider a third classical construction which yields STS's with hyperovals and which will be needed in the sequel several times.

Construction 2.3 (tripling construction): Let an STS(u) with a sub-STS(w) be given ($w < u$, hence $w < u/2$). By the theorem of Doyen and Wilson ([4], see also [14]) this is possible if $u, w \in$ STS and $w < u/2$. But we will also include the case $w = 0$ where no subspace is considered.

First we construct a transversal design TD[3;u-w], using a Latin square Q of order $g = u - w$. This is well known and works as follows.

Every Latin square Q defines a quasigroup operation $\circ$, say on the set $\{1,\dots,g\}$ such that the number $x \circ y$ appears in the xth row and y^{th} column of the square Q. The point set of the desired TD[3;g] is

$$\{1,\dots,g\}\times\{1,2,3\} = \{x_i : x \in \{1,\dots,g\} \text{ and } i \in \{1,2,3\}\}.$$

Lines of the TD are the triples $\{x_1, y_2, (x \circ y)_3\}$. On the other hand it is easy to reconstruct the Latin square from a given TD[3;g].

Now one forms an STS(u) with a subspace W of order w, which is possible by the Doyen-Wilson-theorem. First assume $w \geq 3$. Let

$$(2.2) \qquad \begin{array}{|c|c|}\hline L & O \\ \hline M & N \\ \hline\end{array}$$

be the incidence matrix of this STS(u), and N the incidence matrix of the subspace W. Then the submatrix M has at most one 1 in each column. Let

$$(2.3) \qquad \begin{array}{|c|}\hline F \\ \hline G \\ \hline H \\ \hline\end{array}$$

be the incidence matrix of a TD[3;g], where the $g \times g$ -submatrices F,G,H satisfy the equations (I is the g-rowed unit matrix and J is an all-one matrix)

$$FF^T = GG^T = HH^T = g \cdot I,$$

$$FG^T = GF^T = GH^T = HG^T = HF^T = FH^T = J.$$

Then the matrix

$$(2.4) \qquad X = \begin{array}{|c|c|c|c|c|}\hline F & L & O & O & O \\ \hline G & O & L & O & O \\ \hline H & O & O & L & O \\ \hline O & M & M & M & N \\ \hline\end{array}$$

is an incidence matrix of an STS(3u - 2w) with three sub-STS(u)'s pairwise intersecting in a common sub-STS(w).

In the exceptional cases $w = 0$ and $w = 1$ let L be the incidence matrix of an STS(u) resp. of a GD[3,2;u-1]. Then

(2.5)

F	L	O	O
G	O	L	O
H	O	O	L

is the incidence matrix of the desired STS(3u) with three sub-STS(u)'s resp. of a GD[3,2;3u-3]. In the latter case the desired STS(3u - 2) is obtained by introduction of a new point. □

Now we apply this construction to the problem of finding STS(v)'s with hyperovals. Assume $w \in$ STS, $u = 2w + 1$, $v = 2u + 1 = 4w + 3 = 3u - 2w$. Then construction 2.3 yields an STS(v) with three subspaces of order $u = r = (v-1)/2$, hence with (at least) three hyperovals. Of course $v \equiv 7$ or 15 mod 24.

In these cases the tripling construction yields many distinct STS's with hyperovals. The number of distinct such STS(v)'s obviously exceeds the number L(g) of Latin squares on 1,...,g, and it is well known that

(2.6) $\quad L(g) > g^{\alpha g^2}$,

where $\alpha > 0$ is a positive number which can be found in Wilson's paper [15]. Note that the value of α has been improved by Egorychev's proof of van der Waerden's conjecture on the permanent, see Knuth [8].

3. Steiner Triple Systems without Ovals

Definitions: For each $v \in$ STS let $\mathcal{S}(v)$ be the set of STS(v), and for each $\mathfrak{D} \in \mathcal{S}(v)$ denote by

$\alpha(\mathfrak{D})$ the minimal size $|B|$ of a maximal arc B,

$\beta(\mathfrak{D})$ the maximal size of a (maximal) arc, moreover

(3.1) $\quad \alpha(v) := \min\{\alpha(\mathfrak{D}) : \mathfrak{D} \in \mathcal{S}(v)\}$,

(3.2) $\quad \alpha'(v) := \max\{\alpha(\mathfrak{D}) : \mathfrak{D} \in \mathcal{S}(v)\}$,

(3.3) $\quad \beta'(v) := \min\{\beta(\mathfrak{D}) : \mathfrak{D} \in \mathcal{S}(v)\}$,

(3.4) $\quad \beta(v) := \max\{\beta(\mathfrak{D}) : \mathfrak{D} \in \mathcal{S}(v)\}$.

Examples: a) By theorem 2.2

(3.5) $\quad \beta(v) = \frac{v+1}{2}$ for $v \equiv 3$ or 7 mod 12.

b) In this section we shall prove that $\beta'(v) < \frac{v-1}{2}$ for infinitely many $v \in STS$, and in the next section that

(3.6) $\beta(v) = \frac{v-1}{2}$ for all $v \equiv 1$ or $9 \bmod 12$.

c) Obviously $\alpha(7) = \alpha'(7) = \beta'(7) = \beta(7) = \alpha(9) = \alpha'(9) = \beta'(9) = \beta(9) = 4$. In [16] it was shown that $\beta'(13) = \beta(13) = 6$.

d) (3.7) $\alpha(3^n) \leq 2^n$ for $n \in \mathbb{N}$.

<u>Proof</u>: In affine n-space $AG_n(3)$ over GF(3) the 2^n points $(x_1,\ldots,x_n)$ with $x_i \neq 0$ for $i=1,\ldots,n$ form a maximal arc. Note that three points in $AG_n(3)$ are on a line iff their sum is the zero vector. □

e) For $v \in STS$

(3.8) $\alpha(v)^2 + \alpha(v) \geq 2v$.

<u>Proof</u>: An arc B with α points has $\binom{\alpha}{2}$ secants. In case $v > \binom{\alpha}{2} + \alpha$ there must exist a point p which is on no secant of B, hence $B \cup \{p\}$ is an arc and B is not maximal. This implies the assertion. □

f) This example is important for the sequel [12].

(3.9) $\beta'(27) \leq 9$.

<u>Proof</u>: Let $\mathfrak{D} = AG_3(3)$. Then we shall show that

(3.10) $\beta(\mathfrak{D}) = 9$.

In order to show this, let B be an arc through two points x,y. There are four planes containing x,y. Each of them contains at most two points of $B \setminus \{x,y\}$, hence $\beta(\mathfrak{D}) \leq 10$. Assume $|B| = 10$. Among three parallel planes each must contain at most four and at least two points of B, and one of them at most three points, among them x,y (w.l.o.g). Now the same reasoning as above yields $\beta(\mathfrak{D}) \leq 9$. The existence of an arc B with 9 points is shown by the example [write xyz for (x,y,z)]

$B = \{001, 002, 010, 101, 102, 110, 210, 221, 222\}$. □

g) (3.11) $\beta(AG_{m+n}(3)) \geq \beta(AG_m(3)) \cdot \beta(AG_n(3))$, in particular

(3.12) $2\beta(AG_n(3)) \leq \beta(AG_{n+1}(3)) \leq 3\beta(AG_n(3))$.

<u>Proof [12]</u>: If A resp. B are arcs in $AG_m(3)$ resp. $AG_n(3)$, then $A \times B$ is an arc in $AG_m(3) \times AG_n(3) \cong AG_{m+n}(3)$. □

h) Our knowledge of $\beta(AG_n(3))$ for $n > 3$ is unsatisfactory, e.g.

(3.13) $\quad 18 \leq \beta(AG_4(3)) \leq 24.$

Proof: The first inequality follows from (3.10) and (3.12). Now let B be an arc in $AG_4(3)$.

Case 1: Every plane contains at most 3 points of B. Then the number of planes containing 3 points of B is $\binom{|B|}{3}$ and does not exceed 1170, the number of all planes. Hence $|B| \leq 20$.

Case 2: There is a plane E with $|B \cap E| = 4$. There are exactly four hyperplanes containing E. Each of them contains at most five points of $B \setminus E$. Hence $|B| \leq 4 + 4 \cdot 5 = 24$. □

Lemma 3.1: If $v \equiv 0 \bmod 27$, then there is an STS(v) which is the disjoint union of v/27 sub-STS(27)'s.

Remark: If an STS(v) is the disjoint union of sub-STS(27)'s, then each of these sub-STS's may be replaced by an STS(27) isomorphic to $AG_3(3)$ (in many ways) and we have got an STS(v) which is the disjoint union of affine 3-spaces $AG_3(3)$.

Proof of lemma 3.1: Let the incidence matrix of a TD[3;9] be given by (2.3) with $g = 9$, and let D be the 9×12-incidence-matrix of an STS(9).

Set $u := v/9$. It is well-known that there is an STS(u) which is the disjoint union of u/3 lines, e.g. a Kirkman system of schoolgirls. In the incidence matrix M of such an STS(u) replace the three 1's of each column by the auxiliary matrices F,G,H, and the zeroes by 9×81-zero-matrices. The result is the incidence matrix of a GD[3,9;v] which may be completed to the desired STS(v), e.g. as follows. Write a u-rowed unit matrix I to the right of the original incidence matrix M, replace each 1 in this matrix by D and each 0 by a 9×12-zero-matrix. □

Corollary: If $v \equiv 0 \bmod 27$, then

(3.14) $\quad \beta'(v) \leq \frac{v}{3}.$

Proof: By lemma 3.1 there is an STS(v) which is the disjoint union of sub-$AG_3(3)$'s. Each of them contains at most 9 points of an arc, which implies the assertion. □

Conjecture: For almost all $v \in STS$ there are STS(v)'s without ovals, i.e. $\beta'(v) < (v-1)/2$ for almost all $v \in \mathbb{N}$.

By (3.14), this is true if 27 divides v. It is also true for $v = 18n + 9$, $n > 0$. This follows from

Lemma 3.2: If v is divisible by 9, then there is an STS(v) which is the disjoint union of v/9 sub-STS(9)'s.

The proof is analogous to that of Lemma 3.1, only more simple.

Corollary: If 9 divides v, then $\beta'(v) \leq 4v/9$.

Lemma 3.3: Let $u \in STS$ and $w \in \{0,1,3\}$. Then

(3.15) $\beta'(3u - 2w) \leq 3\beta'(u)$.

Proof: The tripling construction 2.3 yields an STS(3u-2w) which is the union of three STS(u)'s. Since the intersection of an arc with a subspace U is an arc in U, the assertion follows. □

Corollary 1: If $u \in STS \setminus \{1,3,7\}$, then $\beta'(3u) < (3u-1)/2$.

Proof: Doyen [3] has shown that for each $u \in STS$ there is an STS(u) which is generated by each of its triangles. In case $u > 7$ such an STS cannot have a hyper oval, by theorem 2.1. Hence $\beta'(u) \leq (u-1)/2$, and (3.15) implies the assertion. □

Corollary 2: If $u \in STS$ and $\beta'(u) < (u-1)/2$, then $\beta'(v) < (v-1)/2$ for $v = 3u-2$ and $v = 3u-6$.

Hence it is easily seen that $\beta'(v) < (v-1)/2$ for $v = 54n + 3,7,9,21,25,27,45$. But these examples do not suffice to prove the above conjecture. Moreover, some additional information on the function $\beta'(v)$ would be very desirable.

4. Recursive Constructions of Steiner Triple Systems with Ovals

Theorem 4.1: For each $v \in STS$ there is an STS(v) with an oval.

The proof will need a few lemmas. Note that in case $v \equiv 3$ or $7 \bmod 12$ the theorem follows from theorem 2.2.

Definitions: Let P be the set of all $v \in STS$ for which there exists an STS(v) with an oval. The assertion is $P = STS$. By an oval in STS(1) we understand the empty set.

An $OSTS_u(v)$ is, by definition, an STS(v) with a subspace $U, |U| = u$, and with an oval B such that $U \cap B$ is an oval in U, i.e.

$|U \cap B| = (u-1)/2$. Let P_u be the set of $v \in STS$ for which an $OSTS_u(v)$ exists.

Examples: a) Obviously $P_1 = P$.

b) (4.1) $P_3 = P \setminus \{1\}$.

Proof: Let B be an oval in an STS(v) and U a tangent of B. Then $U \cap B$ is an oval in U. □

c) The remark after construction 2.2 shows that

(4.2) $12n+3,\ 12n+7 \in P_7$ for all $n \in \mathbb{N}$.

d) $13 \in P$ follows from [16].

Lemma 4.1: If $u \in P_w$ then $P_u \subseteq P_w$.

Proof: Assume $u \in P_w$ and $v \in P_u$. There is an $OSTS_u(v)$, say $\mathfrak{D} = (V, \mathfrak{B})$, with an oval B and a subspace U, $|U| = u$, such that $B' := B \cap U$ is an oval in U. Because of $u \in P_w$ we can replace the lines in U by other lines such that U has a subspace W with $|W| = w$, B' remains an oval in U, and $B' \cap W = B \cap W$ is an oval in W. This was to be shown. □

Lemma 4.2: If $u \equiv 3$ or $7 \bmod 12$, then $3u \in P_u$.

Proof: We use construction 2.3 with $w = 0$ and a special Latin square Q or order $g = u$, such that the first $(u-1)/2 =: m$ rows and columns of Q form a Latin subsquare of order m, say with entries $1,\dots,m$. This is possible, see Dénes-Keedwell [2]. At the right of and below this subsquare only the entries $m+1,\dots,2m+1$ appear. As an arc in the TD[3;u] of construction 2.3 we define the set of points

$$(4.3) \qquad \begin{array}{ll} x_1 & (x = 1,\dots,m), \\ y_2 & (y = m+1,\dots,2m+1), \\ z_3 & (z = 1,\dots,m). \end{array}$$

Then never $z = x \circ y$, hence B is an arc in the TD[3;u]. Now the construction of the STS's on the point classes can be done in such a way that the points in the three rows of (4.3) form arcs in the respective STS(u)'s, i.e. ovals in the first and third one and a hyperoval in the second one. Thus B becomes an oval in the constructed STS(3u), and the intersection of B with the first sub-STS(u) is an oval in this subspace, q.e.d. □

Example: $21 \in P_7$, moreover

(4.4) $36n+9,\ 36n+21 \in P_7$ for all $n \in \mathbb{N}$.

Proof: For $u = 12n+3$ resp. $12n+7$ lemma 4.2 implies $36n+9$, $36n+21 \in P_u$. By (4.2) and lemma 4.1 the assertion follows. □

Lemma 4.3: If $u \in P_w$ and $u > w > 0$, then

(4.5) $3u - 2w \in P_u \cap P_w$.

Proof: Again construction 2.3 is applied. Set

$$g := u - w = 2m.$$

We use a Latin square Q of order 2m consisting of four Latin subsquares of order m:

(4.6) $Q = \begin{array}{|c|c|} \hline C & D \\ \hline D & C \\ \hline \end{array}$.

W.l.o.g. C has the entries 1,2,...,m, and D has the entries m+1,...,2m. The rows with numbers $m+1$ to $2m$, $3m+1$ to $4m$, and $5m+1$ to $6m$ of the incidence matrix of the TD[3;g] of construction 2.3 form an arc A in this TD, since for $x,y,z > m$ always $z \neq x \circ y$.

The completion of the TD to an STS(3u - 2w) by construction 2.3 can be achieved as follows. By the hypothesis $u \in P_w$ the submatrices L, M,N in (2.2) can be chosen such that N is the incidence matrix of the subspace W, and the (u-1)/2 rows with numbers $m+1,\dots,2m,\dots,2m+\frac{w-1}{2}$ form an oval in the STS(u). Then the rows $2m+1,\dots,2m+\frac{w-1}{2}$ form an oval in W (which is empty in case $w = 1$). Now the matrix (2.4) yields the desired STS(3u - 2w), with an oval B in the rows $m+1$ to $2m$, $3m+1$ to $4m$, and $5m+1$ to $6m+\frac{w-1}{2}$. Hence $v := 3u - 2w \in P_u \cap P_w$. □

Examples:

u	9		13		21			25	
w	1	3	1	3	1	3	7	1	3
3u - 2w	25	21	37	33	61	57	49	73	69

If $12n+9$ resp. $12n+13 \in P$, then $36n+25$ resp. $36n+37 \in P$ (Lemma 4.3 with $w = 1$) and $36n+21$ resp. $36n+33 \in P\,(w = 3)$.
If $12n+9 \in P_7$, then $36n+13 \in P_7$ $(w = 7)$.

These results are not quite sufficient for the proof of theorem 4.1. We need one more lemma.

<u>Lemma 4.4</u>: If $u \in P$ and $u \geq 7$, then $3u - 6 \in P_7$.

Proof: In case $u \equiv 3$ or $7 \bmod 12$ the assertion is true by (4.2). Now assume $u \equiv 1$ or $9 \bmod 12$. By hypothesis and by (4.1) there is an STS(u) with an oval A and a line W such that $|W \cap A| = 1$ (i.e. W is a tangent of A). The incidence matrix of this STS(u) is given by (2.2) where

$$N = \begin{pmatrix} 1 \\ 1 \\ 1 \end{pmatrix} .$$

We may assume, w.l.o.g., that the oval A belongs to the rows no. $m+1, \ldots, 2m, 2m+1$ with $m = \frac{u-3}{2}$. Consider the lines through the point p in the last, i.e. $(2m+3)^{rd}$ row. Since m is odd there are at most $\frac{m-1}{2}$ such lines $G \neq W$ with both points of $G \setminus \{p\}$ in the oval A and at most $\frac{m-1}{2}$ such lines $G \neq W$ with both points of $G \setminus \{p\}$ outside $A \cup W$. The total number of lines $G \neq W$ is $r - 1 = m$, hence there is a line $\{p,a,b\}$ with $a \notin A \cup W$ and $b \in A \setminus W$. We may assume, w.l.o.g., that $\{a,b,p\}$ belongs to the rows no. 1, $m+1$, and $2m+3$ of the matrix (2.2).

Now proceed as in the proof of lemma 4.3, construct the TD[3;2m] with incidence matrix (2.3), and complete it to the STS(3u - 6) with incidence matrix (2.4).

Since the first and $(m+1)^{th}$ rows and columns of the Latin square (4.6) form a Latin subsquare of order 2, it is easily seen that the 7 rows no. 1, $m+1$, $2m+1$, $3m+1$, $4m+1$, $5m+1$, $6m+3$ form a sub-STS(7) whose intersection with the oval B is a triangle, i.e. an oval in the STS(7), corresponding to the rows no. $m+1$, $3m+1$, $5m+1$. This was to be shown. □

<u>Corollary</u>: If $12n + 13 \in P$, then $36n + 33 \in P_7$ $(n \in \mathbb{N} \cup \{0\})$, in particular $33, 69, 105 \in P_7$.

<u>Proof of Theorem 4.1</u>: Assume the theorem to be false. Then there is a smallest $v \equiv 1$ or $9 \bmod 12$ with $v \notin P$. By (4.4) and the examples to Lemma 4.3 $v \geq 85$, and

$$v \equiv 1,\ 13,\ 25, \text{ or } 33 \bmod 36 .$$

The cases $v = 36n + 25$, 33, or 37 are covered by the examples after Lemma 4.3. Hence only the case $v = 36n + 13$ remains open. By (4.4) and by the corollary to Lemma 4.4

(4.7) $12n + 9 \in P_7$ for all $n \in \mathbb{N}$.

Again by the above examples (Lemma 4.3 with $w = 7$) $36n + 13 \in P$, a contradiction proving the theorem. □

It may be worthwile to seek more information on the sets P_w, $w \geq 7$.

5. Counterovals

The above existence proof does not give any information on the question whether distinct STS(v)'s with given ovals are isomorphic, and of characterizing the isomorphy classes of STS(v)'s with ovals. In this generality the problem appears hopeless.

In [16] it was shown that for $v = 13$ there are exactly three isomorphismen classes, and that the complementary structure of an oval in an STS(13) always has one of the two structures

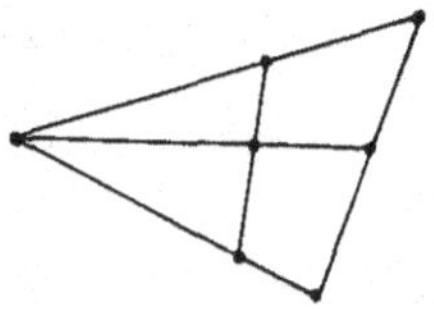

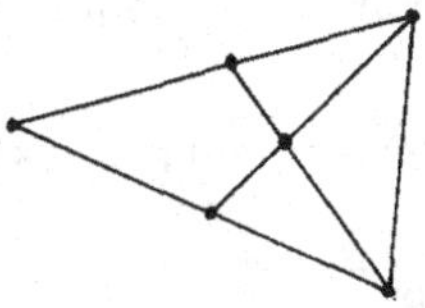

with 7 points, 5 lines, and 6 secants. We generalize this observation as follows.

Definitions: A partial Steiner triple system (PSTS) is a finite incidence structure with at most one line through any two points and with exactly three points on every line. A PSTS with v points is a PSTS(v), and a PSTS(v) with s secants (i.e. unjoined pointpairs) is a PSTS(v,s). We shall prove the following lemma.

Lemma 5.1: The complementary set of an oval in an STS(v) is a PSTS(r + 1,r). If an STS(v) contains a PSTS(r + 1,r), say A, as a substructure, then the complementary set $B := V \setminus A$ is an oval.

Hence we define a counteroval as a PSTS(r + 1,r), regardless of the question whether it can be embedded into an STS(2r + 1). By r always $\frac{v-1}{2}$ is meant.

Examples of counterovals: a) $r = 4$, $v = 9$: A counteroval consists of 5 points on two intersecting lines.

b) We shall show that the counterovals are exactly the PSTS(v)'s with exactly $\frac{r^2-r}{6}$ lines.

c) If $v \equiv 3$ or 7 mod 12 and if an STS(v) with a subspace STS(r) is given, then this STS(r) together with an arbitrary additional point is a counteroval.

The first part of lemma 5.1 is easily proved. Let B be an oval and $A := V \setminus B$. Then $|A| = r+1$. The number x of unjoined point pairs in A is the number of tangents of B, i.e. $x = r$.

For the second part we need some notation. Let $(V,\mathfrak{B})$ be an STS(v), and $A, B \subset V$ with $B = V \setminus A$. Denote by

$l_A(x)$ the number of sublines of A through a point $x \in V$,

$s_A(x)$ the number of secants of A through a point $x \in V$,

$t_A(x)$ the number of tangents of A through a point $x \in V$,

$p_A(x)$ the number of passants of A through a point $x \in V$,

and analogously for B instead of A. Obviously

$$(5.1) \qquad l_A(x) = p_B(x), \quad s_A(x) = t_B(x) \quad \text{etc.}$$

Moreover denote by

$l_A = p_B$ the number of all sublines $L \subset A$

$s_A = t_B$ the number of all secants of A,

$t_A = s_B$ the number of all tangents of A,

$p_A = l_B$ the number of all passants of A. Obviously

$$(5.2) \qquad 3l_B = \sum_{x \in B} l_B(x),$$

$$(5.3) \qquad 2s_B = \sum_{x \in B} s_B(x),$$

$$(5.4) \qquad t_B = \sum_{x \in B} t_B(x)$$

$$(5.5) \qquad 2l_B(x) + s_B(x) = |B| - 1 \qquad \text{for } x \in B,$$

$$(5.6) \qquad l_B(x) + s_B(x) + t_B(x) = r \qquad \text{for } x \in B,$$

$$(5.7) \qquad l_B + s_B + t_B + p_B = b = \frac{v(v-1)}{6}.$$

Hence $3p_A = 3l_B = \sum_{x\in B} (r - s_B(x) - t_B(x)) = \sum_{x\in B} (r - |B| + 1 + 2l_B(x) - t_B(x))$

$$= |B|(r - |B| + 1) + 6l_B - t_B,$$

(5.8) $\quad 3l_B = t_B - |B|(r - |B| + 1),$ and similarly

(5.9) $\quad 3p_B = 3l_A = t_A - |A|(r - |A| + 1) = s_B + (v - |B|)(r - |B|).$

Using (5.7) and the equality $vr = 3b$, a short calculation yields the equations

(5.10) $\quad t_B + s_B = 3(l_B + p_B) + |B|(r - |B| + 1) - (v - |B|)(r - |B|),$

(5.11) $\quad l_B + p_B = \frac{r(r-1)}{6} + \frac{1}{2}(|B| - r)(|B| - r - 1).$

We note some consequences of (5.11): If $|B| = r + 1$ and $l_B = \frac{r(r-1)}{6}$, then $p_B = l_A = 0$, i.e. $A = V \setminus B$ is an oval. If $|B| = r + 1$ and $s_B = r$, then by (5.5), (5.2), and (5.3)

(5.12) $\quad 6l_B + 2s_B = |B|(|B| - 1) = r^2 + r,$

(5.13) $\quad l_B = \frac{r(r-1)}{6},$

hence A is an oval. This proves the second part of Lemma 5.1. □

Note that the equation

$$3l_B + s_B = \binom{|B|}{2}$$

holds for every partial Steiner system B regardless of the question of embeddability into an STS$(2r + 1)$. In case $|B| = r + 1$ this means

(5.14) $\quad s_B = r \iff l_B = \frac{r(r-1)}{6}.$

Hence counterovals PSTS$(r + 1, r)$ can only exist for $r \equiv 0, 1 \bmod 3$.

Moreover (5.11) implies

(5.15) $\quad l_A = p_B \leq \frac{r(r-1)}{6} + \frac{1}{2}(|B| - r)(|B| - r - 1)$

$$= \frac{r(r-1)}{6} + \frac{1}{2}(|A| - r)(|A| - r - 1),$$

with equality iff B is an arc.

Now the question arises which counterovals can be embedded into STS$(2r + 1)$'s. We are quite unable to give an answer. A few examples will follow in section 6.

The existence of PSTS(r + 1,r)'s for any $r \equiv 0$ or 1 follows from the literature, e.g. Hanani [6], last chapter, using the obvious fact that a PSTS(r + 1,r) is obtained from a PSTS(r + 1,s) with $s < r$ by deleting some lines such that $\frac{r(r-1)}{6}$ lines remain, see (5.14). An easy and elementary existence proof for counterovals works as follows.

From an STS(v) delete at most two points together with the lines containing them, and perhaps some more lines. This works in case $r \equiv 0$, 1 or 4 mod 6. In case $r \equiv 1$ or 3 mod 6 a counteroval is obtained by adding an isolated point to an STS(r).

Of course the oval constructions of sections 2 and 4 also yield counterovals. Since apparently there is an abundance of counterovals we note a few other constructions.

Some more constructions of counterovals

a) A given counteroval A can easily be transformed into another one. If {a,b} and {a,c} are secants of A and if L = {b,c,d} is a line then replace L by {a,b,c}. But if b and c are not on a line, replace an arbitrary line of A by {a,b,c}.

b) The complete graph K_{2m} with 2m vertices can be extended to a PSTS(2m + s) by adding s new points $u_1,\dots,u_s$ such that u_i forms a line with the edges of the parallel class $\mathcal{P}_i$ (i=1,...,s), of course $s \leq 2m - 1$. Then more lines on the s new points may be formed, yielding a PSTS(s).

Examples: α) 2m = 6, s = 5, and the PSTS(5) has two lines. Thus we obtain a PSTS(11) with 17 lines. By deleting any two of them we get a counteroval.

β) Construction 2.1.

c) The difference method may be used to construct counterovals too. Set $V := \mathbb{Z}_{16} \cup \{\infty\}$. Then the 40 lines

$$\{\infty,0,8\},\{0,2,5\}\ ,\ \text{and}\ \{0,1,7\}\ \text{mod}\ 16$$

form a PSTS(17,16).

Similarly $V := \mathbb{Z}_{10} \cup \{\infty\}$ with the 15 lines

$$\{\infty,0,5\}\ \text{and}\ \{0,1,4\}\ \text{mod}\ 10$$

forms a PSTS(11,10),

and $V := \mathbb{Z}_{12}$ with the 22 lines

$$\{\infty,0,6\},\ \{0,4,8\},\ \text{and}\ \{0,2,5\} \bmod 12$$

forms a PSTS(13,12).

d) The last few examples can be generalized. Each base line $\{0,a,a+b\}$ in $\mathbb{Z}_m$ corresponds to a difference triple (a,b,c) with $0 < a,b,c \le \frac{v-1}{2}$ and $c = a+b$ or $c = m-a-b$, and each difference triple of this kind gives rise to a base line $\{0,a,a+b\}$. Hence it is convenient to describe STS(v)'s and related structures with a cyclic automorphism group by their difference triples instead of their base lines. Of course this is well-known.

Examples: α) $m = 12n+10$, $V = \mathbb{Z}_m \cup \{\infty\}$. The base line $\{\infty,0,6n+5\}$ and the $2n+1$ difference triples

$$\begin{array}{ll} (1,5n+3,5n+4), & (2,3n+1,3n+3), \\ (3,5n+2,5n+5), & (4,\ 3n\ ,3n+4), \\ \dots\dots\dots\dots & \dots\dots\dots\dots \\ (2n+1,4n+3,6n+4), & (2n,2n+2,4n+2) \end{array}$$

generate $(2n+1)\cdot(12n+10)+6n+5 = \frac{(12n+10)(12n+9)}{6}$ lines, i.e. a counteroval.

β) $m = 12n+4$, $V = \mathbb{Z}_m \cup \{\infty\}$. The base line $\{\infty,0,6n+2\}$ and the 2n difference triples

$$\begin{array}{ll} (1,5n+1,5n+2) & (2,3n,3n+2) \\ (3,\ 5n\ ,5n+3) & (4,3n-1,3n+3) \\ \dots\dots\dots\dots & \dots\dots\dots\dots \\ (2n-1,4n+2,6n+1) & (2n,2n+1,4n+1) \end{array}$$

generate $2n(12n+4)+6n+2 = \frac{(12n+4)(12n+3)}{6}$ lines, i.e. a counteroval.

γ) $m = 12n+6$. The base lines $\{\infty,0,6n+3\}$ and $\{0,4n+2,8n+4\}$, and the 2n difference triples

$$\begin{array}{ll} (1,5n+2,5n+3) & (2,3n,3n+2) \\ (3,5n+1,5n+4) & (4,3n-1,3n+3) \\ \dots\dots\dots\dots & \dots\dots\dots\dots \\ (2n-1,4n+3,6n+2) & (2n,2n+1,4n+1) \end{array}$$

generate $2n(12n+6)+(6n+3)+(4n+2) = \frac{(12n+6)(12n+5)}{6}$ lines, i.e. a PSTS(12n+7,12n+6).

δ) $m = 12n+12$. The base lines $\{\infty, 0, 6n+6\}$ and $\{0, 4n+4, 8n+8\}$ and the $n+1$ difference triples

$$\begin{array}{ll} (1,3n+2,3n+3) & (2,5n+4,5n+6) \\ (3,3n+1,3n+4) & (4,5n+3,5n+7) \\ \dots\dots\dots\dots & \dots\dots\dots\dots \\ (2n+1,2n+2,4n+3) & (2n,4n+5,6n+5) \end{array}$$

generate a $PSTS(12n+13, 12n+12)$.

For each $r \in STS \setminus \{9\}$, R. Peltesohn [11] found an $STS(r)$ with $\mathbb{Z}_r$ as an automorphism group by the above difference method.
Adding an isolated point yields a $PSTS(r+1, r)$ with $\mathbb{Z}_r$ as automorphism group.

6. Construction of Steiner Triple Systems from given Counterovals

I. Given a counteroval $(A, \mathcal{A})$, one can try to extend it to a Steiner triple system $(V, \mathcal{B})$ with $A \subset V$, $\mathcal{A} \subset \mathcal{B}$ as follows. Let $A = \{0, 1, \dots, r\}$ (w.l.o.g.). It has exactly r secants $\{x,y\}$, which we denote by $\frac{x}{y}$ instead (or $\frac{y}{x}$). These r secants are the tangents of the desired $STS(2r+1)$, and they can be identified with the points of the oval $B = V \setminus A$, since each oval point $q \in B$ uniquely determines its tangent. Hence V consists of A and the r unjoined pairs $\frac{x}{y} \in \binom{A}{2}$, and the tangents of the oval B are triples $\{x, y, \frac{x}{y}\}$ for which $\frac{x}{y}$ is a secant of A. The problem is to find the $\binom{r}{2}$ secants of B. These secants are triples $\{\frac{x}{y}, \frac{z}{u}, p\}$ with $p \in A \setminus \{x,y,z,u\}$. One can proceed as follows by trial and error:
Let B be the set of secants $\frac{x}{y} \in \binom{A}{2}$. For each pair $\{\frac{x}{y}, \frac{z}{u}\} \in \binom{B}{2}$ find a point $p(x,y,z,u) = p(y,x,z,u) = p(z,u,x,y) = \dots\dots$ such that always

$$\frac{z}{u} \neq \frac{z'}{u'} \Rightarrow p(x,y,z,u) \neq p(x,y,z',u'). \tag{6.1}$$

Whether this choice is successful we do not know in general. If it succeeds then it yields $\binom{r}{2}$ distinct lines $\{\frac{x}{y}, \frac{z}{u}, p(x,y,z,u)\}$ such that no two of them have a point in common. Together with the given $\frac{r(r-1)}{6}$ lines in the counteroval A and the r tangents of B (i.e. secants of A) the total number of lines becomes

$$\frac{r(r-1)}{6} + r + \binom{r}{2} = \frac{(2r+1)2r}{6} \quad ,$$

hence each point pair $\{x,y\} \in \binom{V}{2}$ is on exactly one line and the desired STS(2r + 1) with oval B is constructed.

Since we cannot give a general answer to the question whether the search for the p(x,y,z,u)'s works let us present two examples.

Example a) For r = 7 let the counteroval A be given by the figure

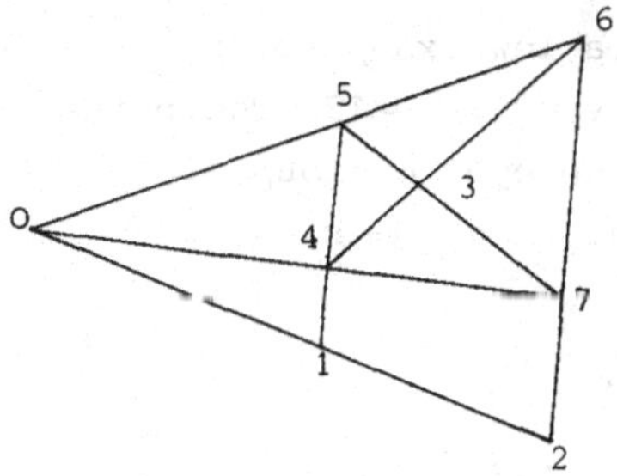

Then $B = \{{0 \atop 3}, {1 \atop 3}, {1 \atop 6}, {1 \atop 7}, {2 \atop 3}, {2 \atop 4}, {2 \atop 5}\}$.

The points p(x,y,z,u) are in one row of the following table where we write ${xz \atop yu}$ instead of $\{{x \atop y}, {z \atop u}\}$. Distinct rows contain distinct solutions, all of them are found by trial and error.

01	01	01	02	02	02	11	11	12	12	12	11	12	12	12	12	12	12	22	22	22
33	36	37	33	34	35	36	37	33	34	35	67	63	64	65	73	74	75	34	35	45
2	4	5	1	6	7	5	4	6	7	0	2	7	0	3	0	3	6	5	4	1
4	2	5	1	6	7	5	2	6	7	0	4	7	0	3	0	3	6	5	4	1
7	5	2	4	1	6	2	5	6	0	4	4	7	3	0	0	6	3	5	1	7
5	2	4	1	6	7	4	2	0	7	6	5	7	0	3	6	3	0	5	4	1
4	2	5	1	7	6	5	2	0	6	7	4	6	0	3	7	3	0	4	5	1

Note that each point $x \in A$ occurs exactly $l_A(x)$ times as a point p(x,y,z,u) in any solution. Indeed $s_B(x) = t_A(x)$, and subtraction of (5.6) from (5.5) [with A instead of B] yields

$$l_A(x) - t_A(x) = |A| - 1 - r = 0.$$

Example b) For r = 0 let the counteroval A consist of the 11 points $\infty,0,1,2,3,4,0',1',2',3',4'$ and the 15 lines $\{\infty,0,0'\}$, $\{0,1',4'\}$, $\{0,2',3'\}$ mod 5 [i.e. $\{\infty,0,0'\},\{\infty,1,1'\},\{\infty,2,2'\},\ldots,\{0,1',4'\}$, $\{1,2',0'\},\ldots$].

The oval B consists of the 10 points ${x \atop y}$ $(x,y \in \{0,1,2,3,4\})$. The oval secants (found by trial and error) are the following 45 triples.

$\{\infty, 0_1, 0_3\}$ $\{0_4, 1_4, 0'\}$ $\{2_4, 3_4, 0'\}$

$\{0_3, 1_2, 0'\}$ $\{0_2, 1_3, 0'\}$ $\{0_1, 2_3, 0'\}$ mod 5.

$\{0_2, 0_4, 3\}$ $\{1_4, 2_4, 3\}$ $\{0_1, 1_2, 3\}$

II. It is also possible to construct STS(v)'s with ovals by the difference method, although we can only give a few examples, no general result. We begin with an example for $v = 25$, $r = 12$. The point set V consists of two copies $\mathbb{Z}_{12}$ and $\mathbb{Z}'_{12}$ of the cyclic group of order 12, and one more point ∞, with the usual rules $\infty + x = \infty + x' = \infty$ for $x \in \mathbb{Z}_{12}$, $x' \in \mathbb{Z}'_{12}$.

The points of $\mathbb{Z}_{12}$ with the 22 lines

$\{\infty,0,6\}$, $\{0,4,8\}$, $\{0,2,5\}$ mod 12

form a PSTS(13,12), i.e. a counteroval A. The points of $\mathbb{Z}'_{12}$ form the oval $B = V \setminus A$. The 12 tangents of B (i.e. secants of A) are the triples $\{0,1,1'\}$ mod 12. The 66 secants of B are the triples $\{\infty,0',6'\}$, $\{0',1',2\}$, $\{0',2',8\}$, $\{0',3',7\}$, $\{0',4',9\}$, $\{0',5',3\}$ mod 12. That we have indeed constructed an STS(25) with the counteroval A and with the automorphism group $\mathbb{Z}_{12}$ is checked by Bose's method of pure and mixed differences [1].

<u>Three similar examples:</u>

a) $V = \mathbb{Z}_{16} \cup \mathbb{Z}'_{16} \cup \{\infty\}$.

<u>Counteroval A:</u> $\{\infty,0,8\}$, $\{0,4,5\}$, $\{0,3,9\}$ mod 16.

<u>Tangents of $B = V \setminus A$:</u> $\{0,2,0'\}$ mod 16

<u>Secants of B:</u> $\{\infty,0',8'\}$, $\{0',1',12\}$, $\{0',2',3\}$, $\{0',3',7\}$, $\{0',4',10\}$, $\{0',5',13\}$, $\{0',6',15\}$, $\{0',7',5\}$ mod 16

b) $V = \mathbb{Z}_{24} \cup \mathbb{Z}'_{24} \cup \infty$.

<u>Counteroval A:</u> $\{\infty,0,12\}$, $\{0,8,16\}$, $\{0,1,6\}$, $\{0,2,11\}$, $\{0,3,7\}$ mod 24, of course on the point set $\mathbb{Z}_{24} \cup \{\infty\}$.

<u>Tangents of B:</u> $\{0,10,12\}$ mod 24

<u>Secants of B:</u> $\{\infty,0',12'\}$, $\{0',1',2\}$, $\{0',2',8\}$, $\{0',3',3\}$, $\{0',4',14\}$, $\{0',5',4\}$, $\{0',6',21\}$, $\{0',7',20\}$, $\{0',8',19\}$, $\{0',9',18\}$, $\{0',10',17\}$, $\{0',11',16\}$ mod 24.

c) $V = \mathbb{Z}_{28} \cup \mathbb{Z}'_{28} \cup \{\infty\}$.

Counteroval A: $\{\infty,0,14\}$, $\{0,4,10\}$, $\{0,3,8\}$, $\{0,7,16\}$, $\{0,2,13\}$ mod 28.

Tangents of B: $\{0,1,0'\}$ mod 28.

Secants of B: $\{\infty,0',14\}$, $\{0',1',27\}$, $\{0',2',14\}$, $\{0',3',6\}$, $\{0',4',8\}$, $\{0',5',2\}$, $\{0',6',16\}$, $\{0',7',24\}$, $\{0',8',23\}$, $\{0',9',22\}$, $\{0',10',21\}$, $\{0',11',20\}$, $\{0',12',19\}$, $\{0',13',18\}$ mod 28.

7. Problems and Open Questions

Section 2: Find STS's with hyperovals and large automorphism groups.

Section 3: Determination resp. estimation of the numbers $\alpha(v)$, $\beta'(v)$, in particular

$$\inf_{v\in STS} \frac{\log \alpha(v)}{\log v} \quad \liminf_{v\to\infty} \frac{\beta'(v)}{v} \quad \text{and} \quad \limsup_{v\to\infty} \frac{\beta'(v)}{v}.$$

Let $n(v)$ be the number of non-isomorphic STS(v)'s and $n_o(v)$ the number of non-isomorphic STS(v)'s with an oval. Is

$$\lim_{v\to\infty} n_o(v)/n(v) = 0?$$

The analogous question for hyperovals.

Are there STS(v)'s without ovals for $v = 15, 19, 21, 25$?

Section 4: Are there STS(v)'s with ovals and with point-transitive automorphism group ? Of course some examples are known such as the projective spaces over GF(2). Are there STS(v)'s (for given v) with a lot of ovals? Find precise answers to this question. Get more information on the sets P_w, $w \geq 7$.

Section 5: How many non-isomorphic counterovals PSTS(r+1,r) are there for given $r \equiv 0 \bmod 3$? Are there counterovals with large automorphism groups?

Section 6: Which counterovals PSTS(r+1,r) can be embedded into STS(2r + 1)'s? All of them? Find more STS(v)'s with ovals by direct constructions such as the difference method.

References

1. R.C. Bose — On the construction of balanced incomplete bloc designs. Ann. Eugenics 9 (1939), 353-399.

2. J. Dénes, A.D. Keedwell — Latin squares and their applications. Academic Press, New York 1976.

3. J. Doyen — Sur la structure de certains systèmes triples d Steiner. Math. Z. 111 (1969), 289-300.

4. J. Doyen, R.M. Wilson — Embedding of Steiner triple systems. Discrete Math. 5 (1973), 229-239.

5. M. Hall — Combinatorial Theory. 2nd ed., Blaisdell, Waltham, Mass. 1975.

6. H. Hanani — Balanced incomplete block designs and related designs. Discrete Math. 11 (1975), 255-369.

7. J.W.P. Hirschfeld — Projective Geometries over Finite Fields. Oxford University Press 1979.

8. D.E. Knuth — A permanent inequality. Amer. Math. Monthly 1981, 731-740.

9. C.C. Lindner, A. Rosa (eds.) — Topics on Steiner systems. Annals Discrete Math. 7 (1980), 317-349.

10. C.C. Lindner, E. Mendelsohn, A. Rosa — On the number of 1-factorizations of the complete graph. J. Comb. Th. (B) 20 (1976), 265-282.

11. R. Peltesohn — Eine Lösung der beiden Heffterschen Differenzen probleme. Composito Math. 6 (1939), 251-167.

12. W. Piotrowski — Oral communication.

13. B. Segre — Lectures on modern geometry. Cremonese 1960.

14. G. Stern, H. Lenz — Steiner systems with given subspaces; another proof of the Doyen-Wilson-theorem. Boll. Un. Mat. Ital. (5) 17-A (1980), 109-114.

15. R.M.Wilson — Nonisomorphic Steiner triple systems. Math. Z. 135 (1974), 303-313.

16. H.Zeitler — Ovals in STS(13). Math. Semesterber., to appear

Note added in proof. M.J. de Resmini has considered arcs (and more general substructures) in Steiner systems S(2,l;v) in her paper "On k-sets of type (m,n) in a Steiner system S(2,l;v)" (In: Finite Geometrie and Designs, LMS Lecture Notes 49 (1981), 1o4-113). In particular, her paper includes our Theorem 2.2, which in fact (as Prof. de Resmini mentions) goes back to Kirkman.

ON DEDEKIND NUMBERS

by

Heinz Lüneburg

For $n \in \mathbb{N}$ we denote by $\pi(n)$ the set of all prime divisors of n. If $X \subseteq \pi(n)$, then $n(X) = n/\prod_{p\in X} p$. With these conventions we define the Dedekind numbers D(n,q), where n and q are integers ≥ 1, by

$$D(n,q) = (1/n) \sum_{X\subseteq\pi(n)} (-1)^{|X|} q^{n(X)}.$$

If q is a power of a prime, then D(n,q) is known to be the numer of monic irreducible polynomials of degree n over GF(q), the Galois field with q elements (see e. g. [7, Satz 6.5, p. 33]). As Dedekind was the first to prove this (Dedekind [3]), we call these numbers Dedekind numbers.

On the other hand, if F is the free group on q generators and if $F^{(1)} = F$, $F^{(2)}$, ... is the lower central series of F, then $F^{(n)}/F^{(n+1)}$ is a free abelian group and D(n,q) is its rank, as was proved by Witt in [10]. Witt found this coincidence "merkwürdig", but this noteworthy coincidence is not accidental, as we are going to show. Moreover, we shall also show that the explanation of this coincidence yields at the same time an algorithm producing all irreducible polynomials of degree n over GF(q), given $GF(q^n)$, i. e., given one irreducible polynomial of degree n.

Let A be a non-empty alphabet and denote by A^+ the set of non-empty words over A and by A_n^+ the set of all words of length n. If C_n is the cyclic group of order n, then we let C_n operate on A_n^+ by the rule

$$a_1 \ldots a_n \to a_2 \ldots a_n a_1.$$

We have the following well-known theorem (see e. g. Cohn [2, p. 296]) the author of which I don't know.

THEOREM. *If* $|A| = q$, *then the number of orbits of length* n *of* C_n *on* A_n^+ *is equal to* D(n,q).

Proof. Let T be an orbit of C_n on A_n^+. Then $n = |C_n| = |T||(C_n)_a|$, where $(C_n)_a$ denotes the stabilizer of $a \in T$. Furthermore, if $|T| = t$, then $a_i = a_{i+t}$ for all i, where the indices have to be reduced modulo n. This yields that a is equal to u^s, where $u = a_1 \ldots a_t$ and $s = n/t$. It follows immediately that there is a bijection of the set of all the orbits of length t of C_n on A_n^+ onto the set of all orbits of length t of C_t on A_t^+ for all divisors t of n. Therefore, if $\alpha(t,q)$ is the number of orbits of length t of C_t on A_t^+, we have

$$q^n = \sum_{t|n} t\alpha(t,q).$$

Möbius inversion yields

$$\alpha(n,q) = (1/n) \sum_{t|n} \mu(t) q^{n/t},$$

where μ denotes the Möbius function. Hence $\alpha(n,q) = D(n,q)$, q. e. d.

COROLLARY. $D(n,q)$ *is the number of irreducible polynomials of degree* n *over* $GF(q)$.

Proof. There exists a normal basis of $GF(q^n)$ over $GF(q)$ (see e. g. Jacobson [4, vol III, p. 61]). Let $b_1, \ldots, b_n$ be such a basis indexed in such a way that $b_i^q = b_{i-1}$ for $i = 2, \ldots, n$ and $b_1^q = b_n$. If $x \in GF(q^n)$, then $x = \sum_{i=1}^n x_i b_i$ with $x_i \in GF(q)$. Moreover $x^q = \sum_{i=1}^n x_i b_i^q = \sum_{i=1}^n x_i b_{i-1}$ with $b_0 = b_n$, i. e., $x^q = \sum_{i=1}^n x_{i+1} b_i$ with $x_{n+1} = x_1$. Hence, if we set $A = GF(q)$, the Galois group $Aut(GF(q^n):GF(q))$ induces the operation of C_n on A_n^+ described above. Therefore, the orbits of length n of C_n on A_n^+ are in a one-to-one correspondence with the orbits of length n of $Aut(GF(q^n):GF(q))$. As each of these orbits is the set of zeros of an irreducible polynomial of degree n over $GF(q)$ and as $GF(q^n)$ is the splitting field of each irreducible polynomial of degree n over $GF(q)$, we see that $D(n,q)$ is also the number of irreducible polynomials of degree n over $GF(q)$, q. e. d.

This proof gives an algorithm to compute all the irreducible polynomials of degree n over $GF(q)$ which can roughly be described as follows:

1) Determine an irreducible polynomial of degree n over $GF(q)$.

2) Construct a normal basis $b_1, \ldots, b_n$ of $GF(q^n)$ over $GF(q)$ using f.

3) Determine a representative for each orbit of length n of C_n on A_n^+ where $A = GF(q)$.

4) Compute the minimal polynomial of $x = \sum_{i=1}^n x_i b_i$ for all the representatives $(x_1, \ldots, x_n)$ determined under 3).

Algorithms to achieve 1) and 4) are to be found in Berlekamp [1]. Jacobson's proof of the Normal-Basis-Theorem for Galois fields yields a good algorithm for 2). Here we shall say more only about 3).

Let $q \in \mathbb{N}$ and $A = \{0, 1, \ldots, q\}$. (The number q has now another meaning. The old q is one larger than the new one.) Order A_n^+ lexicographically, i. e., if $a, b \in A_n^+$, then $a < b$, if and only if there exists an $i \in \{1, \ldots, n\}$ such that $a_j = b_j$ for $j = 1, \ldots, i-1$ and $a_i < b_i$. If T is an orbit of length n of C_n, then pick the largest element of T as a representative of T. Given $a \in A_n^+$, then it is easily checked whether a be larger than all its cyclic conjugates. If a is larger than all its cyclic conjugates, then a representative has been found. In order to find all the representatives, one need not check all the $a \in A_n^+$, for, if a is larger than all its cyclic conjugates, then $a_1 \geq a_i$ for $i = 2, \ldots, n - 1$ and $a_1 > a_n$, as is easily seen. Therefore, given $a_1 \in \{1, \ldots, q\}$, one has to check only those $a_1 \ldots a_n$ with $a_2 \ldots a_n \in \{0, \ldots, a_1\}^{n-2} \times \{0, \ldots, a_1 - 1\}$. Hence the number of words to be checked is $\sum_{i=1}^{q} i(i+1)^{n-2}$, still a lot. The words in $\{0, \ldots, a_1\}^{n-2} \times \{0, \ldots, a_1 - 1\}$ can be generated by a Gray-code. If one uses the one described by Joichi, White and Williamson in [5], then the last word is $a_2 \ldots a_n = a_1 \ldots a_1(a_1 - 1)$ if a_1 is even, and $= a_1 0 \ldots 0$ if a_1 is odd. This follows easily from [9, Satz 1 and Satz 5]. (Here we operate on Gray-codes from right to left instead of from left to right as we did in [9].) Therefore $a_1 \ldots a_n$ is a representative in either case. This yields that the following procedure will generate recursively all the representatives in A_n^+. This procedure is written in PASCAL. The calling program has to provide the type vector = array[1..t] of integer where t is a constant $\geq$ n. The variable a will assume the representatives, whereas s is needed for the generation of the Gray-code as described in [5] or [9]. Moreover, type menge = set of 1..t. The variable x is also used for the Gray-code algorithm. The function reg tests *in situ* whether a is a representative or not. Everything else explains itself provided one knows enough about Gray-codes.

```
          procedure regwort(var a, s: vector; var x: menge;
          var q1, p: integer; q, n: integer;
          var anfang, anfgray, ekm: boolean);

          function reg(var a: vector): boolean;
    var i, k, ipk: integer;
        rg: boolean;
    begin k := 0;
          repeat i := 1; k := k + 1; ipk := i + k;
                 while (a[ipk] =  a[i]) and (i < n) do
                 begin i := i + 1; ipk := ipk + 1;
                       if ipk > n then ipk := ipk - n
                 end;
                 rg := a[ipk] < a[i]
          until (not rg) or (k = n - 1);
          reg := rg
    end; (* reg *)

          procedure neuanf;
    var i: integer;
    begin q1 := q1 + 1; a[1] := q1;
          for i := 2 to n do
          begin a[i] := 0; s[i] := 1 end;
```

```
        x := [1];
        for i := 2 to n - 1 do x := x + [i];
        if q1 > 1 then x := x + [n];
        p := n;
        while not (p in x) do p := p - 1;
        anfgray := p = 1;
    end;

        procedure gray;
    var i: integer;
    begin a[p] := a[p] + s[p];
        if (n > p) and (q1 > 1) then x := x + [n];
        i := n - 1;
        while i > p do
        begin x := x + [i];
            i := i - 1
        end;
        if (a[p] = 0) or (a[p] = q1) then
        begin x := x - [p]; s[p] := -s[p] end;
        if (p = n) and (a[n] = q1 - 1) then
        begin x := x - [n]; s[n] := -s[n] end;
        p := n;
        while not (p in x) do p := p - 1;
        anfgray := p = 1;
        ekm := (p > 1) or (q > q1)
    end;

    begin if anfang then
        begin anfang := false;
            anfgray := true;
            q1 := 0
        end;
        if anfgray then neuanf
        else repeat gray until reg(a)
    end; (* regwort *)
```

There is another way to produce all the representatives or regular words as they are called. Let $(A,\leq)$ be a linearly ordered alphabet. We extend the ordering $\leq$ to A^+ in the following way. Let $a = a_1 \ldots a_s$ and $b_1 \ldots b_t$ be words in A^+. Then $a < b$, if and only if one of the following conditions is satisfied:

α) There exists $c \in A^+$ such that $a = bc$.

β) There exists $r \leq s, t$ such that $a_i = b_i$ for $i = 1, \ldots, r - 1$ and $a_r < b_r$. This ordering is really strange: beggar occurs earlier in the dictionary than beg. On the other hand, the restriction of this ordering to A_n^+ yields the lexicographic ordering.

Next we consider G(A) the free groupoid on A. We denote the elements of G(A) by (u). If all parentheses are removed in (u), we obtain a word in A^+ which will be denoted by u, i. e., if $(u) = ((((a_1)(a_2))(a_3))(a_4))$, then $u = a_1a_2a_3a_4$. We define standard products in G(A) recursively on the length of the words as follows:

1) Products of length 1, i. e., elements $(a) \in G(A)$ with $a \in A$ are standard products.

2) Let $(a) = ((b)(c))$ be a product of length n. Then (a) is a standard product, if and only if either $b \in A$ and (c) is a standard product with $b > c$ or (b) and (c) are standard products, $(b) = ((u)(v))$ with $b > c \geq v$.

Now, if (a) is a standard product, then a is a regular word, and conversely, if b is a regular word, then there exists exactly one standard product (a) such that $a = b$ (see e. g. Cohn [2, Lemma 6.1, p. 291]). It is an interesting exercise in dynamic programming to write a program which produces all the standard products and hence all the regular words of length $\leq n$ on an alphabet with q letters. Such a program involves less computations than the former one, but it has the disadvantage that it requires a lot of storage: One has to have at hand all the regular words of length $\leq n - 1$ in order to compute the regular words of length n.

Let R be a commutative ring with 1 and denote be R[A] the free groupoid algebra and by LR[A] the free Lie-algebra on A over R. Then there exists an epimorphism of R[A] onto LR[A]. Let J be its kernel. Then Širsov has proved (see e. g. Cohn [2, Theorem 6.2, p. 292]) that $\{(a) + J \mid (a)$ is a standard product in $G(A)\}$ is a free basis of the R-Modul $R[A]/J = M$. Hence $\{(a) + J \mid (a)$ is a standard product of length $n\}$ is a free basis of the n-th homogeneous component M_n of M. Moreover, if $R = \mathbb{Z}$, then M_n is isomorphic to $F^{(n)}/F^{(n+1)}$, as was proved by Witt [10]. Therefore, the rank of $F^{(n)}/F^{(n+1)}$ is equal to the number of regular words which is D(n,q) if F is generated by q elements. These considerations show that there is a common source for the two theorems on the number of irreducible polynomials of degree n over GF(q) and the rank of $F^{(n)}/F^{(n+1)}$.

Finally a word about the computation of D(n,q). As

$$D(n,q) = (1/n) \sum_{X \subseteq \pi(n)} (-1)^{|X|} q^{n(X)},$$

one has to determine $\pi(n)$ wich does not offer any problem, as n is small for all D(n,q) within our reach. Given $\pi(n)$, one has to produce all the subsets of it. This can be achieved by the binary reflected Gray-code (see e. g. [9]). Set $\varepsilon_X = (-1)^{|X|}$. Then $\varepsilon_\emptyset = 1$. Let X' be the successor of X. Then there exists a prime p such that

$X' = X - \{p\}$ or $X' = X \cup \{p\}$. In either case $|X'| \equiv |X| + 1 \bmod 2$, whence $\varepsilon_{X'} = -\varepsilon_X$. If $X' = X - \{p\}$, then $n(X') = n(X)p$, i. e., $q^{n(X')} = (q^{n(X)})^p$. If $X' = X \cup \{p\}$, then $n(X') = n(X)/p$, i. e., $q^{n(X')} = (q^{n(X)})^{1/p}$. The well known multiplication algorithm of the Russian peasant is a good algorithm to compute $(q^{n(X)})^p$. The question is whether there is an algorithm to compute $(q^{n(X)})^{1/p}$ as easily.

PROPOSITION. *Let* $a, p \in \mathbb{N}$ *with* $p \geq 2$. *Then* $[((p - 1)a + [d/a^{p-1}])/p] \geq [d^{1/p}]$ *for all* $d \in \mathbb{N}$.

Proof. Assume that $[((p - 1)a + [d/a^{p-1}])/p] < [d^{1/p}]$. Then $((p - 1)a + d/a^{p-1})/p \leq [d^{1/p}] + (p - 1)/p < [d^{1/p}]$. It follows $[d/a^{p-1}] < p[d^{1/p}] - (p - 1)a$. Moreover $d/a^{p-1} \leq [d/a^{p-1}] + (a^{p-1} - 1)/a^{p-1} < [d/a^{p-1}] + 1$. Hence $d/a^{p-1} < p[d^{1/p}] - (p - 1)a \leq pd^{1/p} - (p - 1)a$. This yields $((p - 1)a + d/a^{p-1})/p < d^{1/p}$. On the other hand $d^{1/p} \leq ((p - 1)a + d/a^{p-1})/p$ by the inequality between the geometric and arithmetic means (see e. g. [8, Satz 9.1, p 68]). This contradiction proves the proposition.

PROPOSITION. *Let* $a, d, p \in \mathbb{N}$. *If* $a > [d^{1/p}]$, *then* $[((p - 1)a + [d/a^{p-1}])/p] < a$.

Proof. Assume $[((p - 1)a + [d/a^{p-1}])/p] \geq a$. Then $((p - 1)a + d/a^{p-1})/p \geq a$. It follows $d/a^{p-1} \geq a$ and hence $d^{1/p} \geq a > [d^{1/p}]$. This contradicts the fact that $d^{1/p} - [d^{1/p}] < 1$.

Using these two propositions, we get the following result.

THEOREM. *If* $d, a, p \in \mathbb{N}$ *and if* $a \geq d^{1/p}$, *then*

```
repeat w := a;
       a := ((p - 1)*w + d div w^(p-1)) div p
until a ≥ w;
```

At the exit of the repeat-loop we have $w = [d^{1/p}]$.

If one starts with an a between $d^{1/p}$ and $((3p - 2)/(2p - 2))d^{1/p}$, then the number of times the statement in the repeat-loop is executed is bounded by

$$2 + \log_2((1/p)\log_2 d - \log_2((p - 1)/p)).$$

If one starts with an a between $d^{1/p}$ and $2d^{1/p}$, then this number is bounded by

$$2 + p\ln 2 + \log_2((1/p)\log_2 d + 1/((p - 1)\ln 2)).$$

This can be proved with the methods described in [6].

BIBLIOGRAPHY

[1] E. R. Berlekamp, Algebraic Coding Theory. New York etc. 1968.
[2] P. M. Cohn, Universal Algebra. 1st ed., New York etc. 1964.
[3] R. Dedekind, Abriß einer Theorie der höheren Congruenzen in Bezug auf einen reellen Primzahl-Modulus. J. Reine angew. Mathematik 54, 1-26 (1857).
[4] N. Jacobson, Lectures in Abstract Algebra. Reprint, New York etc. 1964.
[5] J. T. Joichi, D. E. White & S. G. Williamson, Combinatorial Gray-Codes. SIAM J. Comp. 9, 130-141 (1980).
[6] A. Klostermair & H. Lüneburg, Ein Algorithmus zur Berechnung der größten Ganzen aus Wurzel d. Erscheint in den Math. Semesterberichten.
[7] H. Lüneburg, Galoisfelder, Kreisteilungskörper und Schieberegisterfolgen. Mannheim 1979.
[8] H. Lüneburg, Vorlesungen über Analysis. Mannheim 1981.
[9] H. Lüneburg, Gray-Codes. Erscheint in Abh. Math. Seminar Univ. Hamburg.
[10] E. Witt, Treue Darstellung Liescher Ringe. J. reine angew. Math. 177, 152-160, (1937).

Address of the author:

FB Mathematik der Universität
Paffenbergstraße 95

D-6750 Kaiserslautern

REGULAR SETS AND QUASI-SYMMETRIC 2-DESIGNS

A. Neumaier

Institut für Angewandte Mathematik
Universität Freiburg

West-Germany

Abstract. The paper presents a classification of quasi-symmetric 2-designs, and sufficient parameter information to generate a list of all feasible "exceptional" parameter sets for such designs with at most 40 points. The main tool is the concept of a regular set in a strongly regular graph.

Acknowledgment. I am indebted to the Dutch Organization for Pure Scientific Research, Z.W.O., who made this work possible by supporting a four month stay at the University of Eindhoven (Netherlands), and to F. Bussemaker who did most of the numerical calculations.

1. Regular sets in strongly regular graphs

Throughout the paper, all graphs are finite, undirected, without loops or multiple edges. A graph Γ is strongly regular (see e.g. [9], [11], [16]) if (i) every vertex is adjacent with exactly k other vertices, and (ii) the number of vertices adjacent with two distinct vertices x and y is λ or μ, depending on whether x and y are adjacent or not. Related to a graph is its adjacency matrix $M = (m_{xy})$, indexed by the vertices, with $m_{xy} = 1$ if $\overline{xy}$ is an edge, $m_{xy} = 0$ otherwise. If I,J denote the identity and the all-one matrix (of suitable size) then a graph is strongly regular iff its adjacency matrix satisfies

$$MJ = kJ, \quad M^2 = (\lambda-\mu)M+(k-\mu)I+\mu J. \tag{1}$$

The adjacency matrix of a connected strongly regular graph has just three distinct eigenvalues k (valency), r (≥ 0), s (≤ -1); the eigenvalue k is simple and has the all-one vector $\mathbf{j}$ as an associated eigenvector. In terms of r, s, and μ, the other parameters of a strongly regular graph can be expressed by

$$v = (k-r)(k-s)/\mu, \quad k = \mu - rs, \quad \lambda = \mu + r + s, \tag{2}$$

where v denotes the total number of vertices. The multiplicity of the eigenvalue r is given by

$$f = \frac{k(k-s)(-s-1)}{\mu(r-s)} . \tag{3}$$

Now let Γ be a strongly regular graph with parameters (2). A nonempty set B of vertices of Γ is a <u>regular set</u> with <u>valency</u> d and <u>nexus</u> e if the number of vertices of B adjacent with a point $x \in \Gamma$ is d ($< n$) or e (> 0), depending on whether $x \in B$ or not. We call a regular set <u>positive</u> if $d \geq e$, and <u>negative</u> if $d < e$. It is easy to see that the complement of a regular set is also regular, with same sign, valency d and nexus e', where

$$d' = k - e, \quad e' = k - d. \tag{4}$$

Also, a subset B of Γ is regular iff the subgraphs induced on B and its complement are both regular. In the terminology of Delsarte [6], a regular set is a <u>1-design in</u> Γ, and the pair $(B, \Gamma \setminus B)$ is a <u>regular bipartition</u> of Γ.

Denote by M_1 the adjacency matrix of the graph induced on a regular set B of Γ. Then the adjacency matrix of Γ can be written as

$$M = \begin{pmatrix} M_1 & N \\ N^T & M_2 \end{pmatrix} ,$$

and the properties of a regular set imply

$$M_1\mathbf{j} = d\mathbf{j} , \qquad M_2\mathbf{j} = (k-e)\mathbf{j} ,$$

$$N\mathbf{j} = (k-d)\mathbf{j} , \quad N^T\mathbf{j} = e\mathbf{j} .$$

These relations imply that the vector

$$\begin{pmatrix} (k-d)\mathbf{j} \\ -e\,\mathbf{j} \end{pmatrix}$$

is an eigenvector of M for the eigenvalue $d-e < k$. Hence $d-e \in \{r,s\}$, and we have

<u>Proposition 1</u>

The parameters of a regular set B satisfy the relation

$$e = d-r \quad \text{if B is positive,}$$

$$e = d-s \quad \text{if B is negative.} \quad \square$$

In particular, if a strongly regular graph contains a regular set then the eigenvalues are integers.

<u>Proposition 2</u>

The number of vertices of a regular set B of valency d is

$$K = (k-s)(d-r)/\mu \quad \text{if B is positive,}$$

$$K = (k-r)(d-s)/\mu \quad \text{if B is negative.}$$

<u>Proof</u>. We count in two ways the number of edges $\overline{xy}$ with $x \notin B$, $y \in B$ and get $(v-K)e = K(k-d)$, whence

$$K = ve/(k-d+e). \tag{5}$$

Now use Proposition 1 and equation (2) and simplify. □

<u>Examples</u>. 1. If Γ is a disjoint union of cliques, a positive regular set is a union of classes $(e=0,\ d=k)$, and a negative regular set is a set with e points from every class $(d=e-1)$.

2. If Γ is a complete multipartite graph, a positive regular set is a set with i points from every class $(d=e=K-i)$, and a negative regular set is a union of classes $(e=K,\ d=K-m)$.

3. In the Petersen graph, the 12 pentagons are positive regular sets

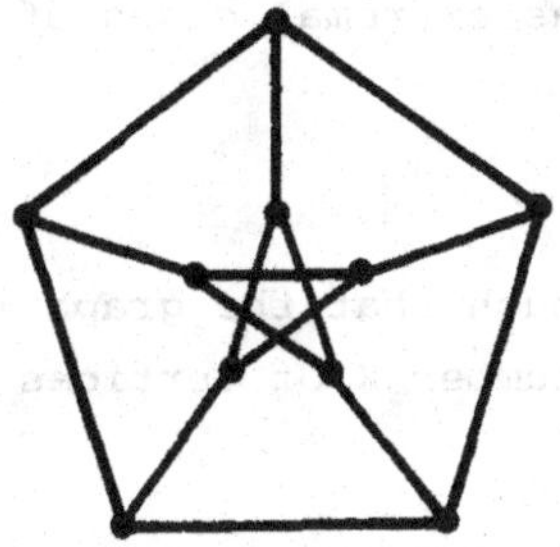

with $K = 5$, $d = 2$, $e = 1$, and the 5 cocliques of size 4 are negative regular sets with $K = 4$, $d = 0$, $e = 2$.

4. In the lattice graph $L_2(n)$, the union of e parallel lines form

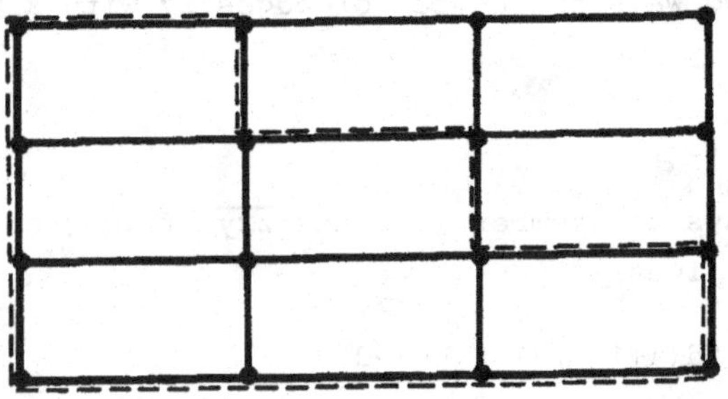

positive regular sets with $k = en$, $d = n+2-e$, and the union of t disjoint transversals form examples of negative regular sets with $K = tn$, $d = 2t$, $e = 2t+2$. For $t = 2$, the polygon indicated in the figure is one of several possibilities.

5. If B is a positive (negative) regular set of size K, valency d, and nexus e in Γ, then, in the complementary graph $\overline{\Gamma}$, B is a negative (positive) regular set with valency $\overline{d}$ and nexus $\overline{e}$ given by

$$\overline{d} = K-1-d, \quad \overline{e} = K-e.$$

This explains the similarity in the first two examples.

6. Many examples of regular cliques ($d = K-1$) are given in Neumaier [13]. Regular cliques are always positive. Complementarily, regular cocliques ($d = 0$) are always negative.

Regular sets can be viewed as extremal cases of induced regular subgraphs:

Proposition 3

Let B be a set of vertices such that the graph induced on B is regular of valency d. Then the number K of vertices of B satisfies the inequality

$$(k-s)(d-r)/\mu \leq K \leq (k-r)(d-s)/\mu. \tag{6}$$

The lower (upper) bound is attained iff B is a positive (negative) regular set.

Proof. For $x \notin B$, denote by e_x the number of vertices of B adjacent with x. Counting in two ways the number of edges $\overline{xy}$ with $x \notin B$, $y \in B$ gives

$$\sum e_x = K(k-d), \tag{7}$$

and counting in two ways the number of paths $\overline{zxy}$ of length 2 with $x \notin B$, $y,z \in B$, $y \neq z$ gives

$$\sum e_x(e_x-1) = Kd(\lambda+1-d) + K(K-1-d)\mu. \tag{8}$$

Here the sum is over all $x \notin B$. Using (2), (7), and (8) gives

$$\sum e_x^2 = K^2\mu - K(d-r)(d-s). \tag{9}$$

From (7), the averave value of e_x is

$$e := \frac{K(k-d)}{v-K}, \tag{10}$$

and we compute

$$\begin{aligned}\sum (e_x-e)^2 &= K^2\mu - K(d-r)(d-s) - 2eK(k-d) + e^2(v-K)\\ &= K^2\mu - K(d-r)(d-s) - K^2(k-d)^2/(v-K)\\ &= -K(\mu K-(k-r)(d-s))(\mu K-(k-s)(d-r))/(\mu(v-K)),\end{aligned} \tag{11}$$

where we simplifies with (2). Now the sum of squares is nonnegative, whence μK must lie between $(k-r)(d-s)$ and $(k-s)(d-r)$. But $(k-r)(d-s) - (k-s)(d-r) = (r-s)(k-d) > 0$, whence (6) holds. If equality holds in (6) then $\sum(e_x-e)^2 = 0$, whence $e_x = e$ for all $x \notin B$. Therefore,

B is a regular set, and from (2) and (10) we find $e = d-r$ (resp. $e = d-s$) if the lower (resp. upper) bound is attained. □

Note that this proof contains a matrix-free proof of Proposition 1 and 2.

2. Semiregular partially balanced designs

A partially balanced design (with two associate classes) is a pair $(\Gamma, \mathcal{B})$ consisting of a connected strongly regular graph Γ (whose v vertices are now called points) and a collection $\mathcal{B}$ of subsets of Γ (called blocks) such that (i) every block contains K points ($2 \leq K \leq v-1$), (ii) every point is in R (> 0) blocks, and (iii) two distinct points x,y are in q or p common blocks ($p \neq q$) depending on whether x,y are adjacent or not. For other, equivalent definitions see e.g. [2], [15]. Associated with a partially balanced design is its incidence matrix $A = (a_{xB})$ indexed by points and blocks, with $a_{xB} = 1$ if $x \in B$, $a_{xB} = 0$ otherwise. The v×n-matrix $N = AA^T$ has three nonnegative eigenvalues, among them the simple eigenvalue $\lambda = RK$. A partially balanced design is called semiregular (in [2]: special) if $\det(N) = 0$, i.e. if $\lambda = 0$ is an eigenvalue of N. The results of Neumaier [12; Section 3] imply that every $1\frac{1}{2}$-design with two connection number p and q (< R) is a semiregular partially balanced design; the converse follows easily from the following result of Bridges and Shrikhande [2]:

Proposition 4
A partially balanced design is semiregular iff there are numbers d and e such that every block is a regular set with valency d and nexus e. □

Proposition 5
If Γ is a rank 3 graph then the orbit of every regular set is a semiregular partially balanced design.

Proof. The automorphism group of Γ is transitive on vertices, edges, and nonedges. This implies (ii) and (iii) in the definition of a partially balanced design. Obviously, automorphic images of a regular set are regular sets with the same parameters; this provides (i) and semiregularity. □

Proposition 6

The parameters of a semiregular partially balanced design can be written in terms of d and e as

$$b = \frac{R(k-r)}{d-r}, \quad t = \frac{R(k-d)}{k(-s-1)}, \; p = R+st, \quad q = R+(s+1)t \tag{12}$$

if all blocks are positive, and as

$$b = \frac{R(k-s)}{d-s}, \quad t = \frac{R(k-d)}{k(r+1)}, \; p = R-rt, \quad q = R-(r+1)t \tag{13}$$

if all blocks are negative. In particular,

$$\begin{aligned} &p < q \quad \text{iff the blocks are positive,} \\ &p > q \quad \text{iff the blocks are negative.} \end{aligned} \tag{14}$$

Proof. For fixed $z \in \Gamma$, we count in two ways the number of pairs (x,B) with $x \in B$, resp. with $x,z \in B$, x adjacent with z, resp. with $x,z \in B$, x not adjacent with z, and obtain

$$\begin{aligned} &Kb = Rv, \\ &kq = Rd, \\ &(v-1-k)p = R(K-1-d). \end{aligned} \tag{15}$$

Now assume that the blocks are positive. Then $e = d-r$, $K = (k-s)(d-r)/\mu$ by Proposition 1 and 2, and with (2) we find $K-1-d = -(r+1)(k+ds)/\mu$, $v/K = (k-r)/(d-r)$. From (2), we also find $v-1-k = k(r+1)(-s-1)/\mu$, whence by (15),

$$b = \frac{R(k-r)}{d-r}, \quad q-R = \frac{Rd}{k} - R = \frac{-R(k-d)}{k} = (s+1)t,$$

$$p-R = \frac{-R(k+ds)}{k(-s-1)} - R = \frac{Rs(k-d)}{k(-s-1)} = st.$$

This implies (12). Since $d < k$, $s \leq -1$, and $R > 0$ we have $t > 0$, hence $p < q$. The case of negative blocks follows by interchanging the eigen values r and s, and replacing t with $-t$. □

Remarks. 1. Since $t = \pm(p-q)$, the number t in (12) resp. (13) must be a positive integer.

2. If $(\Gamma,\mathcal{B})$ is a partially balanced design and Γ is not complete multipartite then $\overline{\Gamma}$ is connected, whence $(\overline{\Gamma},\mathcal{B})$ is a partially balanced design with p and q interchanged. Hence for the proper choice of Γ we will have $p < q$, and all blocks are positive.

Proposition 7

In a semiregular partially balanced design with positive blocks, the number b of blocks satisfies

$$b \geq f+1, \tag{16}$$

where f is given by (3). Equality holds iff any two blocks intersect in the same number of points.

Proof. This is a special case of a theorem for $1\frac{1}{2}$-designs given in Neumaier [12]. □

In the terminology of statisticians, $b = f+1$ characterizes the linked designs. If we dualize a linked design we obtain a 2-design with only two intersection numbers p and q, i.e. a quasi-symmetric 2-design. This is the topic of the next section.

3. Quasi-symmetric 2-designs

A 2-(v^*,k^*,λ^*)-design consists of a set P of v^* points and a collection $\mathcal{B}$ of b^* blocks such that each block consists of k^* points and every pair of points is in λ^* blocks. Then every point is in a constant number r^* of points, and the relations

$$b^*k^* = r^*v^* \ , \quad r^*(k^*-1) = \lambda^*(v^*-1) \tag{17}$$

hold (see e.g. Raghavarao [15]). A 2-design is called quasi-symmetric if any two blocks have either p or q common points, $p < q$, and if both possibility occur. Goethals and Seidel [7] showed that the graph Γ whose vertices are the blocks, adjacent if they have q common points (the block graph) is strongly regular. We denote its parameters as in Section 1.

Proposition 8

For each point $x \in P$, the set $S(x) := \{B \in \mathcal{B} \mid x \in B\}$ is a positive regular set of the block graph with valency d and nexus e given by

$$d=((k^*-1)(\lambda^*-1)-(r^*-1)(p-1))/(q-p), \quad e=(k^*\lambda^*-r^*p)/(q-p). \tag{18}$$

Proof. Fix $x \in P$. For each block B, denote by e_B the number of blocks through x adjacent with B. We count in two ways the number $s(x,B)$ of pairs (y,C) such that $x,y \in C$, $y \in B$, $y \neq x$, $C \neq B$. If $B \in S(x)$ then $x \in B$ and $e_B(q-1)+(r^*-1-e_B)(p-1) = s(x,B) = (k^*-1)(\lambda^*-1)$ whence $e_B = (k^*-1)(\lambda^*-1)-(r^*-1)(p-1)/(q-p)$. If $B \notin S(x)$ then $x \notin B$ and $e_Bq+(r^*-e_B)p = s(x,B) = k^*\lambda^*$ whence $e_B = (k^*\lambda^*-r^*p)/(q-p)$. Hence each set $S(x)$ is a regular set with valency and nexus given by (18). By Proposition 4, the dual of a quasi-symmetric 2-design is a semiregular partially balanced design. Hence, since $p < q$ by definition, $S(x)$ is positive by Proposition 6. □

Proposition 9

The parameters of a quasi-symmetric 2-designs can be expressed in terms of the parameters of the block graph as follows:

$$v^* = f+1, \quad k^* = (f+1)e/(k-r), \quad p = k^*+st, \quad q = k^*+(s+1)t, \tag{19}$$

$$b^* = v \quad , \quad r^* = ve/(k-r) \quad , \quad \lambda^* = r^*-(r-s)t, \quad d = e+r \quad , \tag{20}$$

with a positive integer

$$t = \frac{k^*(k-r-e)}{k(-s-1)} \quad . \tag{21}$$

Proof. The results of the last section apply with

$$v = b^*, \quad b = v^*, \quad K = r^*, \quad R = k^*. \tag{22}$$

By Proposition 7, $v^* = b = f+1$ since the dual of a quasi-symmetric 2-design satisfies the equality condition. If we solve the first equation of (12) for R and substitute (22) we find $k^* = ve/(k-r) = (f+1)e/(k-r)$ and obtain (19).

From Proposition 1 we have $d = e+r$. From (17) we find $r^* = b^*k^*/v^* = vk^*/(f+1) = ve/(k-r)$ and $(v^*-1)(r^*-\lambda^*) = r^*(v^*-k^*) = ve(f+1)(k-e-r)/(k-r)^2 = tvk(-s-1)/(k-r) = tk(k-s)(-s-1)/\mu = t(r-s)f = t(r-s)(v^*-1)$, using (2) and (3), whence $r^*-\lambda^* = t(r-s)$, $\lambda^* = r^*-t(r-s)$. Therefore (20) holds.

Substitution of $R = k^*$ and $d = e+r$ into the second equation of (12) gives (21). Finally, $t = q-p$ is a positive integer. □

For further reference we note the formula

$$v^* = f+1 = -\frac{(k-r)(\mu+s(k-s))}{(r-s)\mu} \tag{23}$$

which follows from (3) by a simple calculation.

Proposition 10

For a quasi-symmetric 2-design with connected block graph,

$$b^* \leq \frac{1}{2}v^*(v^*-1).$$

Proof. For connected strongly regular graphs $s < -1$ whence $q < k^*$, so a result of Cameron and van Lint [5; Prop. 3.4] applies. □

Proposition 11

The complement of a quasi-symmetric 2-design is again a quasi-symmetric 2-design; the corresponding block graphs are isomorphic.

Proof. The new blocks are the complements of the old blocks. Two adjacent old blocks have complements intersecting $\bar{p} = v^*-2k^*+p$ points two nonadjacent old blocks have complements intersecting in $\bar{q} = v^*-2k^*+q$ points. Since $\bar{p} < \bar{q}$ the two block graphs are isomorphic.

We now consider some particular classes of quasi-symmetric 2-designs.

Class 1. Multiples of symmetric 2-designs. In a symmetric 2-design, every block contains k^* points and any two blocks intersect in $\lambda < k^*$ points. The design consisting of $m > 1$ copies of the blocks has intersection numbers $p = \lambda$ and $q = k^*$, hence is quasi-symmetric; the block graph is a disjoint union of cliques.

Class 2. Strongly resolvable 2-designs. A 2-design with v^* points and b^* blocks is strongly resolvable if the blocks can be partitioned into (the minimal number of) b^*-v^*+1 classes such that every point occurs in the same number of blocks of each class. By a theorem of Hughes and Piper [10], strongly resolvable 2-designs are quasi-symmetric, and the block graph is a complete multipartite graph.

Class 3. Steiner systems with $v^* > k^{*2}$. A Steiner system $S(2,k^*,v^*)$ is the same as a $2-(v^*,k^*,\lambda^*)$-design with $\lambda^* = 1$. Since two points are on a unique block, two blocks intersect in 0 or 1 point. Hence Steiner systems are quasi-symmetric. Their block graphs are the Steiner graphs, cf. [11]. The excluded Steiner systems with $v^* \leq k^{*2}$ are affine planes ($v^* = k^{*2}$) belonging to class 2, projective planes ($v^* = k^{*2}-k^*+1$) with only one intersection number, and the designs with only one block ($v^* = k^*$) with no intersection number.

Class 4. Residuals of biplanes. By results of Hall and Connor [8; Lemma 4.1, Thm. 3.2], every 2-design with parameters $v^* = \binom{n}{2}$, $k^* = n-1$, $\lambda^* = 2$, $r^* = n+1$, $b^* = \binom{n+1}{2}$ is quasi-symmetric with intersection numbers $p = 1$, $q = 2$, and is the residual design of a unique biplane (= symmetric 2-designs with $\lambda = 2$). The block graph is the complement of a triangular graph $T(n+1)$. The known biplanes (see Cameron [4]) realize the cases $n = 3,4,5,6,7,10,12,14$, sometimes with several non-isomorphic solutions. The Bruck-Ryser-condition for biplanes excludes infinitely many values of n, starting with $n = 8,9,11,13,\ldots$.

Theorem Q

(i) A quasi-symmetric 2-design with disconnected block graph is of class 1.

(ii) A quasi-symmetric 2-design with complete multipartite block graph is of class 2.

(iii) A quasi-symmetric 2-design with $p = o$, $q = 1$ is of class 3.

(iv) A quasi-symmetric 2-design with $p = 1$, $q = 2$ is of class 4, or a $2-(5,3,3)$-design.

Proof. (i) A disconnected strongly regular graph is a disjoint union of ≥ 2 cliques of the same size m. By Example 1 of Section 2, positive regular sets have $d = k$. By Proposition 8 and equations (14), (20) we hence have $kq = Rd = k^*k$, or $q = k^*$. Therefore, adjacent blocks contain the same points, and the blocks of the design form copies of another 2-design $\mathcal{B}'$. Since two nonadjacent blocks intersect in the same number p of points, $\mathcal{B}'$ must be a symmetric 2-design.

(ii) This is part of Theorem 5.3 of Beker and Haemers [1].

(iii) $p = 0$, $q = 1$ implies that two blocks have at most one common point. But two distinct points are in $\lambda^* \geq 1$ blocks whence $\lambda^* = 1$.

(iv) If $p = 1$, $q = 2$ then (19) implies that $t = 1$, $k^* = 1-s$, and using (22),

$$e = \frac{(k-r)(1-s)}{f+1} = \frac{(s-1)(r-s)\mu}{\mu+s(k-s)} . \tag{24}$$

Now (20), (2) and (24) imply that $e+sr^* = e+sve/(k-r) = e+s(k-s)e/\mu = (\mu+s(k-s))e/\mu = (s-1)(r-s)$ whence $e \equiv -r \bmod s$. Hence for a suitable integer i,

$$e = -si-r, \quad d = -si. \tag{25}$$

Equation (21) implies $1 = (1-s)(k-d)/k(-s-1)$ whence $k(-s-1) = (1-s)k-(1-s)d$, $2k = d(1-s)$, and by (25),

$$k = \frac{1}{2}\, is(s-1). \tag{26}$$

If we insert (25) and (26) into (24), observe that $\mu = k+rs$ (by (2)), and simplify, we find the relation

$$(2r+s(i-1))^2 = (i+1)(2i-s^2(i-1)). \tag{27}$$

Now (26) implies that $i > 0$.
If $i = 1$ then by (27), (25), (26), (23), and (2) we find

$$r=1,\ s=-d,\ e=d-1,\ k=\binom{d+1}{2},\ \mu=\binom{d}{2},\ f+1=\binom{d+2}{2},\ v=\binom{d+3}{2}$$

whence by (19) and (20),

$$v^* = \binom{d+2}{2}, \quad k^* = d+1, \ \lambda^* = 2.$$

Therefore, the design is of class 4.
If $i > 1$ then (27) implies $0 \leq 2i-s^2(i-1)$ whence $(s^2-2)(i-1) \leq 2$. This is only possible if $i = 2$, $s = -2$. In this case we obtain as before

$$r=1,\ s=-2,\ e=3,\ k=6,\ \mu=4,\ f+1=5,\ v=10,$$

$$v^*=5,\ k^*=3,\ \lambda^*=3,$$

which is the second alternative in the statement. □

Note that there is a unique 2-(5,3,3)-design, consisting of 5 points and the 10 possible point triples. Its complement is of class (iii).

4. Exceptional quasi-symmetric 2-designs with few points

We call a quasi-symmetric 2-design $\mathcal{B}$ exceptional if neither $\mathcal{B}$ nor its complement is in class 1, 2, 3, or 4. There are fairly many feasible exceptional parameter sets with a small number of points. By Proposition 11 it is sufficient to consider designs with $k^* \leq \frac{1}{2}v^*$, and a list of all possibilities with $2k^* \leq v^* \leq 40$ was compiled as follows. Using the necessary conditions given in Neumaier [11], [14], we calculated the possible parameter sets for strongly regular graphs with f $(= v^*-1) \leq 39$ which were connected and not complete multipartite. For each "graph" obtained we checked whether there are one or more values of e such that the parameters resulting from Proposition 9 are integral, and $2k^* \leq v^*$. Then the designs belonging to class 3 and class 4 were deleted. To rule out some of the remaining 36 "designs" two further existence tests were applied; they can be considered as analogues of the Krein condition [16] and the improved absolute bound [14] for strongly regular graphs.

Proposition 12

The parameters of a quasi-symmetric 2-designs satisfy the inequality

$$B(B-A) \leq AC, \tag{28}$$

where

$$A = (v^*-1)(v^*-2), \quad B = r^*(k^*-1)(k^*-2), \tag{29}$$

$$C = r^*d(q-1)(q-2)+r^*(r^*-1-d)(p-1)(p-2). \tag{29}$$

Equality holds in (28) iff any three distinct points are in a constant number of blocks.

Proof. For distinct points x,y,z, denote by λ_{xyz} the number of blocks containing x,y, and z. Now fix a point x, and take the following sums over all pairs (y,z) with $x \neq y \neq z \neq x$. By counting suitable configurations in two ways we find $\sum 1 = A$, $\sum \lambda_{xyz} = B$, $\sum \lambda_{xyz}(\lambda_{xyz}-1) = C$, given by (29). Hence the average value of λ_{xyz} is $\bar{\lambda} = B/A$, and $0 \leq \sum(\lambda_{xyz}-\bar{\lambda})^2 = (C+B)-2\bar{\lambda}B+\bar{\lambda}^2A = C+B-B^2/A = (AC-B(B-A))/A$, from which the assertion follows. □

Proposition 13

If for a quasi-symmetric 2-design

$$b^* = \frac{1}{2}v^*(v^*-1) \tag{30}$$

then (28) holds with equality.

Proof. By a result of Cameron and van Lint [5; Prop. 3.6], equation (30) implies that the design is a 4-design. In particular, the equality condition of Proposition 12 is satisfied. □

Proposition 12 is quite powerful, and eliminates 12 of the 36 cases. As an example, for the parameter sets

$$v^* = 27,\quad k^* = 7,\quad \lambda^* = 21,\quad r^* = 91,\quad b^* = 351,$$
$$p = 1,\quad q = 3,\quad d = 60,\quad e = 28,$$

equation (30) holds but (28) is satisfied with strict inequality. Unfortunately, all parameter sets with $b^* < \frac{1}{2}v^*(v^*-1)$ pass Proposition 12. But one of them,

$$v^* = 19,\quad k^* = 7,\quad \lambda^* = 7,\quad r^* = 21,\quad b^* = 57,$$
$$p = 1,\quad q = 3,\quad d = 18,\quad e = 14$$

is impossible since no strongly regular graph with corresponding parameters

$$v = 57,\quad k = 42,\quad \lambda = 31,\quad \mu = 30,\quad r = 4,\quad s = -3$$

No.	Ex?	v^*	k^*	λ^*	p	q	v	k	λ	μ	d	e
1	?	19	9	16	3	5	76	45	28	24	25	18
2	?	20	10	18	4	6	76	35	18	14	21	14
3	?	20	8	14	2	4	95	54	33	27	27	18
4	?	21	9	12	3	5	70	27	12	9	15	9
5	?	21	8	14	2	4	105	52	29	22	26	16
6	yes	21	6	4	0	2	56	45	36	36	15	12
7	yes	21	7	12	1	3	120	77	52	44	33	22
8	?	22	8	12	2	4	99	42	21	15	21	12
9	yes	22	6	5	0	2	77	60	47	45	20	15
10	yes	22	7	16	1	3	176	105	68	54	45	28
11	yes	23	7	21	1	3	253	140	87	65	60	35
12	?	24	8	7	2	4	69	20	7	5	10	5
13	?	28	7	16	1	3	288	105	52	30	45	20
14	yes	28	12	11	4	6	63	32	16	16	16	12
15	?	29	7	12	1	3	232	77	36	20	33	14
16	yes	31	7	7	1	3	155	42	17	9	18	7
17	?	33	15	35	6	9	176	45	18	9	27	15
18	?	33	9	6	1	3	88	60	41	40	20	15
19	?	35	7	3	1	3	85	14	3	2	6	2
20	?	35	14	13	5	8	85	14	3	2	8	4
21	yes	36	16	12	6	8	63	30	13	15	15	12
22	?	37	9	8	1	3	148	84	50	44	28	18
23	?	39	12	22	3	6	247	54	21	9	27	12

Table 1. Quasi-symmetric 2-(v^*,k^*,λ^*)-designs with intersection numbers p,q and block graph parameters v,k,λ,μ; subgraphs induced by a point have valency d and nexus e in the block graph.
The list covers all designs with $2k^* \leq v^* \leq 40$ not characterized by Theorem Ω.

exists (see Wilbrink and Brouwer [17]). There remained 23 parameter sets, listed in Table 1. The entry 'yes' under the heading 'Ex ?' indicates that a quasi-symmetric 2-design with the stated parameters is known.

The designs No. 6, 7, 9, 10, 11 are well-known classical designs, related to the binary Golay code (see Goethals and Seidel [7]). Examples 9 and 11 must be the unique Steiner systems S(3, 6, 22) and S(4, 7, 23) constructed by Witt [18]; indeed for No. 9, 10, and 11, relation (28) is satisfied with equality, whence we have 3-designs, and a counting argument similar to that of Proposition 12 shows that No. 11 must be a 4-design.

Designs No. 14 and 21 were constructed by Peter Cameron (personal communication) from the symplectic group Sp(6,2), and design No. 16 was realized by Andries Brouwer (personal communication) as the set of all planes in the projective space PG(4,2); in fact, these are the first members of 3 infinite families of quasi-symmetric designs.

For No. 4, 17, and 23, no designs are known, but the block graphs of Steiner triple systems with 21, 33, and 39 points, respectively, have the parameters needed for the block graphs of No. 4, 17, and 23. Perhaps this can be used for a construction.

It is hoped that Table 1 will challenge some readers to construct a few more quasi-symmetric 2-designs, or to devise new existence tests which eliminate some of the undecided cases.

Finally, we mention one more interesting feasible parameter set:

$$v^* = 56, \quad k^* = 16, \quad \lambda^* = 6, \quad r^* = 32, \quad b^* = 77,$$

$$p = 4, \quad q = 6, \quad d = 6, \quad e = 4.$$

The block graph has parameters

$$v^* = 77, \quad k^* = 16, \quad \lambda^* = 0, \quad \mu^* = 4.$$

These are the complementary parameters of the block graph of S(3, 6, 22), which might be a good start for a construction.

References

1. H. Beker and W. Haemers, 2-designs having an intersection number k-n, J. Combin. Theory (Ser. A) 28 (1980), 64-82.

2. W.G. Bridges and M.S. Shrikhande, Special partially balanced incomplete block designs and associated graphs, Discrete Math. 9 (1974), 1-18.

3. P.J. Cameron, Extending symmetric designs, J. Combin. Theory 14 (1973), 215-220.

4. P.J. Cameron, Biplanes, Math. Z. 131 (1973), 85-101.

5. P.J. Cameron and J.H. van Lint, Graphs, Codes and Designs. London Math.Soc. Lecture Note Series 43, Cambridge Univ. Press 1980.

6. P. Delsarte, An algebraic approach to the association schemes of coding theory, Phil. Res. Rep. Suppl. 10 (1973), 1-97.

7. J.M. Goethals and J.J. Seidel, Strongly regular graphs derived from combinatorial designs, Can. J. Math. 22 (1970), 597-614.

8. M. Hall, jr., and W.S. Connor, An embedding theorem for balanced incomplete block designs, Can. J. Math. 6 (1953), 35-41.

9. X. Hubaut, Strongly regular graphs. Discrete Math. 13 (1975), 357-381.

10. D.R. Hughes and F.C. Piper, On resolutions and Bose's theorem, Geom. dedicata 5 (1976), 129-133.

11. A. Neumaier, Strongly regular graphs with smallest eigenvalue -m, Arch. Math. 33 (1979), 392-400.

12. A. Neumaier, $t\frac{1}{2}$ - designs, J. Combin. Theory (Ser. A) 28 (1980), 226-248.

13. A. Neumaier, Regular cliques in graphs and special $1\frac{1}{2}$ - designs. In: Finite Geometries and Designs, London Math. Soc. Lecture Note Series 49, Cambridge Univ. Press 1981, pp. 244-259.

14. A. Neumaier, New inequalities for the parameters of an association scheme. In: Combinatorics and Graph Theory, Lecture Notes in Mathematics 885, Springer Verlag, 1981, pp. 365-367.

15. D. Raghavarao, Constructions and Combinatorial Problems in Design of Experiments. Wiley, New York 1971.

16. J.J. Seidel, Strongly regular graphs, an introduction. In: Surveys in Combinatorics, London Math. Soc. Lecture Note Series 38, Cambridge Univ. Press 1979, pp. 157-180.

17. H.A. Wilbrink and A.E. Brouwer, A (57,14,1) strongly regular graph does not exist. Math. Centrum Report ZW 121/78, Amsterdam 1978.

18. E. Witt, Über Steinersche Systeme, Abh. Math. Sem. Hamburg Univ. 12 (1938), 265-275.

Anschrift des Verfassers:

Dr. A. Neumaier
Institut für Angewandte Mathematik
der Universität Freiburg i.Br.
Hermann-Herder-Straße 10

D-7800 Freiburg i.Br.

Asymptotic 0-1 Laws in Combinatorics

W.Oberschelp
RWTH Aachen, Templergraben 64
5100 Aachen, Federal Republic of Germany

Abstract. The paper considers a special chapter of the theory of asymptotic methods in enumeration. While the general theory has been covered by an excellent exposition of Bender [1], we mainly consider relative frequencies for relational systems of a special kind within a general class of configurations. We give a survey of results and try to emphasize the intuitive ideas behind the formal results.

Contents

1. Introduction. Let $\{c_n\}_{n\in\mathbb{N}}$ be a sequence of natural numbers, where c_n is interprete as the cardinality of a set $C(n)$ of configurations with parameter n. The general problem is to determine the behaviour of c_n for $n \to \infty$.
The special problem is to compare $C(n)$ with a set $B(n)$ of basic configurations which are counted by a sequence $\{b_n\}_{n\in\mathbb{N}}$, and to determine c_n with the help of b_n. In particular, we are interested in the quotient $q_n = c_n/b_n$. If the limit of q_n exists for $n \to \infty$ and is equal to 0 or 1, we say that $C(n)$ fulfills an asymptotic 0-1 law in the basic class $B(n)$, and in the 1-case we write for short $c_n \sim b_n$ as usual.
In some applications, b_n is easily computed. Then an asymptotic 0-1 law yields information on the growth of c_n. It determines the order of growth in the case of a 1-law and gives at least an upper bound in the case of a 0-law. Thus, we are not ambitious enough to calculate c_n via b_n up to an additive error; we only expect first approximations for c_n with respect to quotient behaviour. On the other hand, it is our aim to find general results for a wide range of basic configurations and - given a class $B(n)$ - for as many types of configurations $C(n)$ in $B(n)$ as possible.
In general, we assume that $B(n)$ is determined by some characteristic property or basic condition $\mathcal{L}$, and that $C(n)$ is defined by what we call a special condition $\mathcal{L}$. The results will depend on the growth of b_n which can be measured by properties of generating functions. By

$$B(z) = \sum_{n=0}^{\infty} b_n z^n \quad \text{and} \quad b(z) = \sum_{n=0}^{\infty} \frac{b_n}{n!} z^n$$

now we cannot even fit corollaries of their result into the general proof method which we want to describe here. In the following sections we shall investigate, whether (and if yes how) these examples of asymptotic 0-1 laws fit into a more general framework.

2. Parametric conditions and Blass-Fagin Properties.

It has been proved that the asymptotic equality $B_n \sim \frac{1}{n!} b_n$ is correct for a large class of basic properties; here b_n and B_n mean the number of labelled and unlabelled basic configurations, respectively. Example 2 can be generalized to the so-called basic parametric conditions.

In order to keep things simple, we only consider configurations over one k-ary relation R with $k \geq 2$. As an example of a parametric condition for a basic property in the language of one binary relation we take

$$\mathcal{L} \equiv \forall x_1 (\neg R x_1 x_1) \wedge \forall x_1 \, x_2 (x_1 \neq x_2 \rightarrow (R x_1 x_2 \rightarrow R x_2 x_1)).$$

This condition means that R is an ordinary graph.

Another example is the following one for tournaments.

$$\mathcal{L} \equiv \forall x_1 (\neg R x_1 x_1) \wedge \forall x_1 \forall x_2 (x_1 \neq x_2 \rightarrow (R x_1 x_2 \leftrightarrow \neg R x_2 x_1)).$$

In general, by a parametric condition (cf.Oberschelp[17],p.298) we understand a conjunction of universal formulae

$$\begin{aligned} \mathcal{L} \equiv\ & \forall x_1 M_1(x_1) \\ & \wedge \forall x_1 \forall x_2 (\neq (x_1, x_2) \rightarrow M_2(x_1, x_2)) \\ \wedge \ldots & \wedge \forall x_1 \forall x_2 \ldots \forall x_r (\neq (x_1, x_2, \ldots, x_r) \rightarrow M_r(x_1, x_2, \ldots, x_r)) \\ \wedge \ldots & \wedge \forall x_1 \forall x_2 \ldots \forall x_k (\neq (x_1, x_2, \ldots, x_k) \rightarrow M_k(x_1, x_2, \ldots, x_k)). \end{aligned}$$

Here each M_r is a purely propositional formula in atomic expressions $R x_{i_1} x_{i_2} \ldots x_{i_k}$ such that for the sets of variables we have

$$\{x_{i_1}, \ldots, x_{i_k}\} = \{x_1, \ldots, x_r\}.$$

Furthermore, $\neq (x_1, \ldots, x_r)$ is an abbreviation for the formula expressing that all the variables $x_1, \ldots, x_r$ have different values. The properties of being a direct graph, tournament, m-graph, plex etc. can be expressed by parametric conditions. The idea behind this concept is the following. A parametric property defines a class of relations which can be determinded by the independent choice of values (parameters) in fixed regions of the adjacency array. Thus, for instance, a directed graph, possibly with loops, (i.e. a binary relation) is determined by fixing one of the values 1 or 0 in each position (x_1) of the diagonal (which means, that x is a loop or not) and one

we understand the ordinary and the exponential generating functions for the sequence $\{b_n\}_{n\in\mathbb{N}}$. It is well-known that - if the radius of convergence of the indicated power series of the complex variable z is greater than zero - the function theoretic behavior of B(z) and b(z) yields information on the growth of b_n.
We start with three examples.

Example 1:
Let $\mathcal{B}(n)$ be the set of all binary relations over a finite domain N, say N = {1,...,n}. Define $\mathcal{C}(n)$ to be the class of those binary relations which fulfill the special condition

$$\mathcal{L} \equiv \forall x \exists y\ Rxy.$$

Formally and systematically the condition $\mathcal{L}$ for the basic property is the empty condition in the language of one binary relation symbol R. Obviously, we have

$$b_n = 2^{n^2}, \text{ and it is easy to see that } c_n = (2^n - 1)^n,$$

since in every row of the $n \times n$ adjacency matrix of R there are $2^n - 1$ choices of zeros and ones - the only forbidden row is the row consisting of zeros only. Furthermore, it is easy to see that for the special condition

$$\bar{\mathcal{L}} \equiv \exists x \forall y\ Rxy$$

the number of configurations is $\bar{c}(n) = 2^{n^2} - (2^n - 1)^n$.
For instance, for n = 10 we have

$$b_n = 1.26765 \cdot 10^{30}$$
$$c_n = 1.25533 \cdot 10^{30}$$
$$\bar{c}_n = 0.01233 \cdot 10^{30}.$$

In our terminology, $\mathcal{C}(n)$ fulfills a 1-law and $\bar{\mathcal{C}}(n)$ fulfills a 0-law in the basic set of binary relations.

Example 2:
We consider the so-called asymptotic counting problem for unlabelled binary relations, graphs, tournaments and other simple types of basic configurations. In other words, we are counting isomorphism types.

For instance, let $\mathcal{B}(n)$ be the class of isomorphism types of binary relations, and let $\mathcal{C}(n)$ be the class of those isomorphism types which are invariant only under the trivia vertex permutation ε. Those structures are sometimes called rigid. It is well-known that here again we have an asymptotic 1-law (cf., for instance, Oberschelp [16], Wright [19] and Harary-Palmer [10]).

The problem of enumerating unlabelled structures deserves a special comment. The solution is usually given by Polya's counting theory which interpretes $B(n)$ as the set of orbits of a permutation group Γ acting on the class of all labelled basic configurations. We will denote the unlabelled numbers by capital letters.
The well-known Frobenius-Burnside-Lemma, which is the heart of Polya's theory, counts the number of those orbits according to the formula

$$B_n = \frac{1}{|\Gamma|} \sum_{g \in \Gamma} f_1(g) ,$$

where $f_1(g)$ is the number of unlabelled basic configurations which remain fixed under g. In this example, we cannot evaluate B_n directly. In order to prove the asymptotic 1-law we show as a central proposition that

$$(*) \qquad \lim_{n\to\infty} \frac{\frac{1}{|\Gamma|} f_1(\varepsilon)}{B_n} = 1.$$

This means that the first term (corresponding to the trivial element ε of Γ) of the Frobenius-Burnside formula for B_n determines as a main term the asymptotic growth of B_n. Here $b_n = f_1(\varepsilon)$ is of course the number of labelled basic structures (in this example $b_n = 2^{n^2}$), since each basic structure is fixed under the trivial permutation, and $|\Gamma|$ is $n!$.
It is easy to prove the asymptotic 1-law $C_n \sim B_n$ using (*). We note that

$$n!C_n + \frac{n!}{2}(B_n - C_n) \geq b_n,$$

since applying the group Γ to one of the $B_n - C_n$ isomorphism types, which are invariant under some nontrivial element of Γ, yields at most $\frac{n!}{2}$ differently labelled basic configurations. This is equivalent to

$$C_n \geq 2\frac{b_n}{n!} - B_n \qquad \text{or} \qquad \frac{C_n}{B_n} \geq 2\frac{b_n}{n!B_n} - 1.$$

For $n\to\infty$ the right side tends to 1 by (*), and, since $C_n \leq B_n$, we have $C_n \sim B_n$.

As an example for (*) we note for the case of binary relations that

$B_8 = 4.582971 \cdot 10^{14}$, while for $\frac{1}{|\Gamma|} b_n$ we have the value

$$\frac{2^{64}}{8!} = 4.575085 \cdot 10^{14}.$$

Example 3:
As a further example for the technique of getting information on the growth of b_n via the order of growth of c_n we mention the famous theorem of Kleitman and Rothschild [11] on the asymptotic behaviour of the number b_n of partial orders. They count partia orders of a special kind by c_n and show that almost all partial orders are of this special kind. There is a lot of ingenious ad hoc argumentation in their proof; up to

of 4 values in each upper right position (x_1, x_2). Each value codes one of the 4 possibilities

$$Rx_1x_2 \wedge Rx_2x_1$$

or $$Rx_1x_2 \wedge \neg Rx_2x_1$$

or $$\neg Rx_1x_2 \wedge Rx_2x_1$$

or $$\neg Rx_1x_2 \wedge \neg Rx_2x_1 \; .$$

The figure shows the two parameter regions for binary relations.

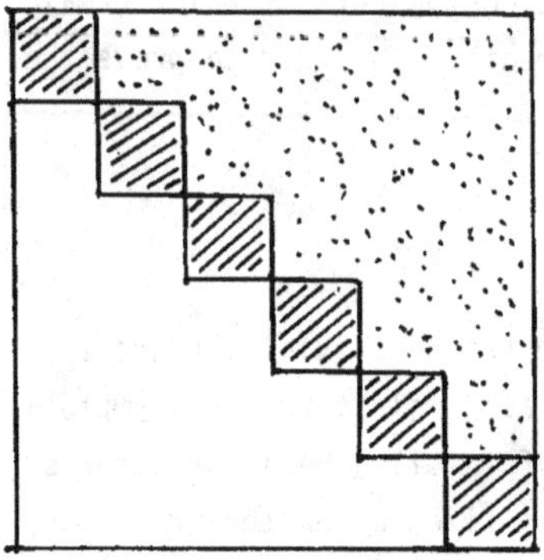

Figure 1

Parameter regions

If we consider ordinary graphs or tournaments, the parameter regions remain the same. This time, however, we only have one choice in the diagonal and two choices in the right upper half. Though the two choices in the second parameter region have different meanings in the case of graphs and tournaments, we have essentially the same combinatorics for the number of ordinary graphs and of tournaments.

A simple counterexample, which is not parametric, is a formulation of transitivity, which is needed to define partial orders for instance:

$$\forall x_1 \forall x_2 \forall x_3 ((x_1 \neq x_2 \wedge x_1 \neq x_3 \wedge x_2 \neq x_3) \rightarrow (Rx_1x_2 \wedge Rx_2x_3 \rightarrow Rx_1x_3)) \; .$$

Here the atoms such as Rx_1x_2 do not contain all the variables of the prefix. In general, no quantifier sequence which is longer than the place number of the relation R can fulfill the condition for parametric properties.

<u>Theorem 1</u>: If $\mathcal{L}$ is parametric, then $B_n \sim \frac{1}{n!} b_n$

<u>Proof:</u> See Oberschelp [17]. □

According to this theorem, relation (*) of section 1 holds. We have, therefore, again the result that almost all parametric relations are rigid. Moreover, we can calculate

b_n explicitly from a certain normalization of condition $\mathcal{B}$, as explained in the proof.

As a generalization of Example 1 we consider the following situation.

Definition: Let $\mathcal{B}$ be a basic property. If for each condition $\mathcal{L}$, written in the language of first order logic with identity, there is a 0-1 law, i.e.

$$\lim_{n\to\infty} \frac{c_n}{b_n} \quad \text{exists and is 0 or 1 ,}$$

then $\mathcal{B}$ is called a Blass-Fagin property, or BF for short.
Here we understand by c_n the number of models with n vertices for the condition $\mathcal{L} \wedge \mathcal{B}$, i.e. we restrict ourselves to basic configurations. Blass-Harary [2] and Fagin [9] have shown among other results the following

Theorem 2: The empty condition, the graph condition, etc., considered as basic property, are BF.

In the light of this result, the 0-1 theorem in Example 1 is a corollary since the conditions $\mathcal{L}$ and $\mathcal{B}$ used there are of first order. Thus BF-properties are a good source for getting asymptotic approximations. If $\lim q_n = 1$ (this can even be decided effectively), then we have approximated c_n and b_n by each other.
As a generalization of the results of Blass-Fagin we have

Theorem 3: All parametric conditions, considered as basic properties, are BF.
The proof is sketched in Oberschelp [18].

□

It is plausible that not every condition $\mathcal{B}$, written in the language of first order logic with identity, can be BF, since the spectrum of those values n which are cardinalities of models for $\mathcal{B}$ can be very irregular. As an example - admittedly usually not written in the language of one single relation - choose $\mathcal{B}$ as the conjunction of the axioms for the theory of fields.

As an aid in the search for more BF-properties beyond the class of parametric conditions let us now note some properties of parametric conditions $\mathcal{B}$.

(i) In all non-trivial cases the numbers b_n of (labelled) n-vertex models for $\mathcal{B}$ are strictly increasing. Moreover, the functions b_n are growing very fast such that the convergence radius of the generating series

$$B(z) = \Sigma\, b_n z^n$$

and even of the exponential generating series

$$b(z) = \Sigma\, \frac{b_n}{n!}\, z^n$$

is zero. In fact, it follows from the explicit formulae for b_n mentioned above that b_n grows at least as 2^{cn^2} to infinity for some $c > 0$.

(ii) The growth of b_n is monotonically regular in the sence that

$$n \cdot \frac{b_{n+1}}{b_n} \text{ tends to zero.}$$

Thus there are no essential breaks in growth rapidity.

(iii) Models for $\mathcal{L}$ are always closed under induced substructures. This means that we can remove vertices of models with their adjoining "edges" without violating property $\mathcal{L}$. This fact follows immediately from the definition type for parametric relations via universal quantifiers alone.

(iv) Configurations with a parametric property fulfill a condition of "internal richness". What we understand by this notion is explained best by analyzing Blass's proof of the graph property being BF. In order to show that almost all models of $\mathcal{L}$ also have property $\mathcal{L}$, we try to give a constructive proof of $\mathcal{L}$ from $\mathcal{L}$. We try to succeed by elimination of quantifier changes of the form

$$\forall x_1 \ldots \forall x_n \exists y_1 \ldots \exists y_m C(x_1,\ldots,x_n,y_1,\ldots,y_m)$$

within the condition $\mathcal{L}$, beginning with the inner parts. If we could always do this, we would obtain after a finite number of such steps a quantifier-free kernel which could be decided in the usual way. Now we can show that in almost all models of a parametric condition $\mathcal{L}$ any statement of the form given above is true. This means in the special case of graphs that, for any selection of n vertices $x_1,\ldots,x_n$ there are m vertices $y_1,\ldots,y_m$ in the graph $\mathcal{G}$ which have the <u>interconnection pattern</u> (among $x_1,\ldots,x_n$ and $y_1,\ldots,y_m$) which is expressed by the kernel $C(x_1,\ldots,x_n,y_1,\ldots,y_m)$. It can be shown that richness conditions are highly probable if $\mathcal{G}$ is big enough. Interestingly enough, it is difficult to give explicit examples for graphs which fulfill richness conditions. The so-called Paley graphs are essentially the only known models with such a behaviour, which is true for big "random" graphs. As an example, we consider the Paley graph with 17 vertices and $\frac{1}{2}\binom{17}{2} = 68$ edges. Here two vertices are joined if their difference is a quadratic residue mod 17. The graph in Figure 2 fulfills the richness conditions

$$\forall x_1 \forall x_2 \forall x_3 \exists y_1((x_1 \neq x_2 \wedge x_1 \neq x_3 \wedge x_2 \neq x_3) \rightarrow (Rx_1y_1 \wedge Rx_2y_1 \wedge \neg Rx_3y_1))$$

and

$$\forall x_1 \forall x_2 \forall x_3 \exists y_1((x_1 \neq x_2 \wedge x_1 \neq x_3 \wedge x_2 \neq x_3) \rightarrow (Rx_1y_1 \wedge \neg Rx_2y_1 \wedge \neg Rx_3y_1)).$$

This was shown by Exoo [8].

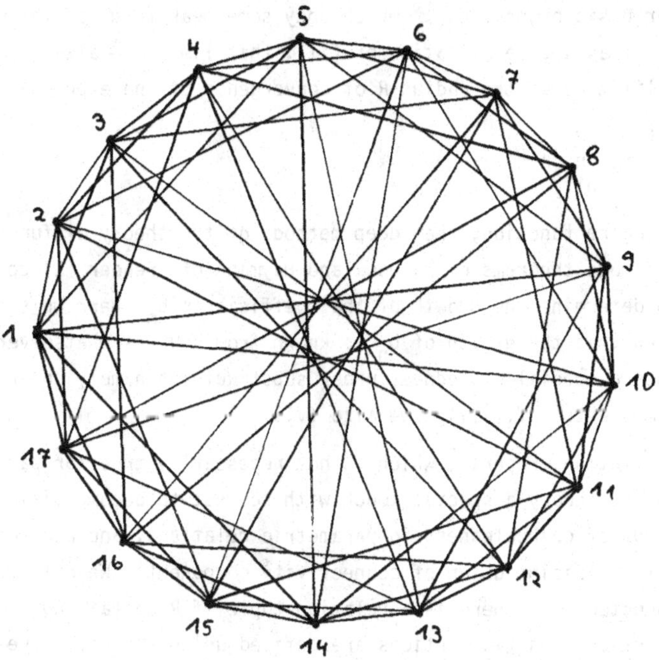

Figure 2:

Paley graph of order 17

It is only fair to remark that also negations of richness conditions would suffice to prove an asymptotic 0-1 law, since we could try to refute $\mathcal{L}$ dually and would then end up with an asymptotic 0-law. Thus we can summarize: the validity of either a richness or a poorness condition is the key to success in proving that a condition $\mathcal{L}$ is BF.

3. <u>Compton's theory of slowly growing numbers</u>. We have extracted the properties (i) to (iv) in the last section, since it is our feeling that properties of this kind could lead to BF properties. On the other hand they characterize what we would like to call an elementary data structure. Such structures should be available in big numbers (cf. property (i) and (ii)), they should be highly resistant against structure manipulations (like property (iii)) and they should - if big enough - also be rich enough to contain almost always all types of interconnection patterns between vertices (as explained in (iv)). Thus we should like to maintain the working thesis that elementary data structures are exactly those configurations, which fulfill a BF condition. It is in the spirit of this program that we try to find a complete characterization of BF properties under assumptions such as conditions (i) to (iv). In particular, we would like to discover BF-conditions $\mathcal{L}$ beyond the parametric relations.

In a remarkable paper by Compton [5] an explanation of this kind is given. Compton considers those first order basic properties $\mathcal{L}$ which obey some weak forms of the above conditions. He assumes, as a generalization of (ii), that $\lim n\frac{b_{n+1}}{b_n}$ always exists. In this case, the limit gives the radius R of convergence of the exponential generating function

$$b(z) = \Sigma \frac{b_n}{n!} z^n.$$

If $R > 0$, or if b(z) is an entire function, then deep methods of the theory of function (like Tauber-theorems and famous theorems of Darboux and Hayman (cf. Bender [1] pp.498 and pp.506)) can be used to determine the growth of the coefficients b_n. Here we simplify the situation by assuming that the growth of b_n is known from sources whatsoever. Furthermore, Compton assumes a special closedness under substructures namely under removal of connectivity components (cf.(iii), see also (v)).

Finally, he assumes a manipulation property, which is not necessarily true for parametric relations but gives the theory a special touch with respect to decomposition: (v) There is a natural notion of connectedness in parametric relations, and each class defined by $\mathcal{L}$ is closed unter disjoint union of connectivity components: We call vertices x and y <u>directly connected</u> if there is a relation tuple of R containing x and y. Then arbitrary connections between two vertices are defined using the reflexive and transitive closure of the direct connection relation; finally, the equivalence classes arising from connectivity are called the <u>components</u> of a parametric configuration.

The results of Compton characterize a basic property $\mathcal{L}$ as BF in terms of a radius of convergence of b(z). In the proofs there appear analogues of richness-poorness conditions in the sense of condition (iv). They are needed in proof-refutation attempts for $\mathcal{L}$ and turn out to be almost always true. But in contrast to the situation in parametric relations, the theory of Compton only works for slowly growing coefficients It is, therefore, not surprising that the instances of condition (iv) which occur in the proofs appear to be poorness conditions. They state that for a certain type $\mathcal{K}$ of finite connected substructures with condition $\mathcal{L}$, there are not exactly j components in the structure, for which we want to prove or disprove condition $\mathcal{L}$. There is a lot of technical model theoretic argumentation in the proofs which are behind this idea.

The main result of Compton can thus be summarized as follows.

<u>Theorem 4</u>: Let the assumptionsbe as above. If the radius R of convergence of b(z) satisfies $R > 0$, then

$$\mathcal{L} \text{ is BF if and only if } R = \infty.$$

Before giving examples, we comment on the appearance of the exponential generating function in this proposition. At first glance the trick of dividing by the factorials seems to give just the right measure to bring the growth of b_n and the BF-property into proper coincidence. But inherent in the proofs is also always the well-known exponential theorem, which counts under manipulations - right in the technique of exponential generating functions - the numbers b_n of models of a given cardinality via the numbers d_n of connected models as follows (cf.Harary-Palmer [8], p.8 for graphs and Compton [5], p. 19 for the general case):

Theorem 5: If $b(z) = \Sigma \frac{b_n}{n!} z^n$ and $d(z) = \Sigma \frac{d_n}{n!} z^n$, then

$$b(z) = \Sigma \frac{(d(z))^m}{m!} = e^{d(z)} .$$

A corresponding technique is not available for ordinary generation functions.

Theorem 4 covers the case of labelled enumeration. We mention that for unlabelled enumeration there is an analogous theory and an analogous concept of BF-property, which uses ordinary generating functions. Since the numbers B_n are positive integers from their combinatorial meaning, an ordinary generating function

$$B(z) = \Sigma B_n z^n$$

in this field can never have a radius of convergence S greater than 1. Again we get a characterization using the shortest possible growing order with respect to convergence radius.

Theorem 6: Let the assumptions be as above.
If the radius S of convergence of B(z) satisfies $S > 0$, then

$$\mathcal{L} \text{ is unlabelled-BF if and only if } S = 1. \quad \square$$

The application of Compton's theorems yields essentially negative results: If the number of basic configurations is growing not too fast (i.e. $R > 0$ or $S > 0$), only the case of slow growth (i.e. $R = \infty$ or $S = 1$) yields the BF-property. In the labelled case, for instance, the range of applicability of Theorem 4 begins somewhat beneath the region $b_n = 2^{n\log n\, O(1)}$. Therefore, there is no application to the case of partial orders, where b_n is about $2^{\frac{n^2}{4}(1+O(1))}$ (this follows trivially from the Kleitman-Rothschild result mentioned in Example 3). The question, whether or not the basic property of being a partial order is BF, cannot be answered by the general theory so far. Nevertheless, Compton has announced (private communication) that he has proved the BF property of partial orders directly, using the results of Kleitmann-Rothschild and the methods of Blass-Fagin.

Neither can the BF-question for various types of trees and forests be answered positively by Compton's theorems. Either the growth of b_n is irregular such that assumption (ii) fails (cf. for instance,[5], example 1.9), or the growth is regular but the numbers b_n are growing too fast. The latter is the case for rooted trees, where R is positive and finite. Note that the asymptotics for tree enumerations are well-known since the fundamental work of Polya and Otter (cf. [1] sections 7.2 and 7.4). Thus, by Theorem 4, we have definitely no BF-property in these cases.

One of the few new BF-properties is the case of equivalence relations (partitions of a finite set). Here the b_n - known usually as the Bell numbers - are growing just fast enough to guarantee that $R = \infty$ (as is well-known from work of Moser and Wyman [15]). Here there is a clear indication of poorness in $\mathcal{B}(n)$, since we have only one type of connectivity components for every cardinality; from this only few models arise in the general non-connected case by the exponential theorem (Theorem 5).

For the unlabelled case, we have here the numbers B_n (usually named p_n) of partitions of the number n. It is also well-known that in this case $S = 1$, i.e. the power series $p(z) = \Sigma p_n z^n$ has radius of convergence 1, since

$$p(z) = \frac{1}{\prod_n (1-z^n)} \quad .$$

In fact, by Ramanujan's work it follows that $p_n \sim \frac{1}{4n\sqrt{3}} e^{\left(\pi\sqrt{\frac{2}{3}}\sqrt{n}\right)}$.

Thus, by Theorem 6, we have an asymptotic 0-1 law for every first oder property of partitions . This result is weak, but general. Of course, the deep properties of partition theory are not first order. One applicable condition would be to postulate a fixed number of components; to postulate an even number of components would, however, not be an application.

4. Systems with a priori structure and Lynch's theory.

There is another possible extension of the situation in section 1, where we asymptotically counted (binary) relations which fulfill an additional special first order condition $\mathcal{L}$. Now we assume that the vertex domain $N = \{1,...,n\}$ has, a priori, a certain structure which is described by the diagram of some relation S. We allow $\mathcal{L}$ to be written with the additional use of this relation S.

Example 4: Let S be the (trivial) relation which is true for all x,y. Then any condition $\mathcal{L}$ using S can equivalently be written without using S. Since the basic set of binary relations is BF, we have an asymptotic 0-1 law for each condition $\mathcal{L}$.

Example 5: Let S be the cyclic successor, i.e., Sxy means $y = x + 1 \mod n$. Consider a condition such as

$$\mathcal{L} \equiv \forall x \, \exists y (Sxy \wedge Rxy).$$

General problem: Under which assumptions on S does a 0-1 law exist for every condition $\mathcal{L}$ in the first order language of R and S?

For the special example $\mathcal{L}$, the relative frequence of the number of relations which fulfill $\mathcal{L}$ in the class of all relations is $q_n = \frac{1}{2^n}$ which tends to zero. Thus we have an asymptotic 0-law for this condition $\mathcal{L}$.

Example 6: We take S to be the natural $\leq$-relation on N with smallest element 1 and largest element n. An easy computation shows that the relative frequency for the same $\mathcal{L}$ has the value $q_n = \Pi(1 - \frac{1}{2^i})$. This converges to a limit 0.288787... different from 0 and 1.

In this case we cannot have an asymptotic 0-1 law for each first oder condition $\mathcal{L}$.

Lynch [13] has given an explanation for these different situations with respect to 0-1 laws. More specific, he gave a sufficient condition for the validity of a 0-1 law in the situation described above (cf.[13], Corollary 5.10). Lynch defines a notion of k-extendibility of the structure S. This condition is rather technical, but there seems to be a clear intuitive background. The successor is a poor structure which could be realized by few interchanges in different ways. On the other hand, for linear orders the so-called Ehrenfeucht game cannot be won (by the second of two players). This game means intuitively that the second of two players triesto answer to vertex-choices of the first player in such a way that two isomorphic structures of k vertices (where k is the number of quantifiers in the prenex normal form of $\mathcal{L}$) have been created in the end. The rule is that the first player can always decide which of the two structures in progress he wants to complete, while the second player has to work with the other structure at this step. Roughly summarized: The validity of an asymptotic 0-1 law for all $\mathcal{L}$ appears as a consequence of the fact that here is a large stock of choices with respect to the problem of finding many isomorphic substructures.

5. Further results on asymptotic 0-1 laws: Random graphs.

The results on asymptotic 0-1 laws reported so far keep the range of $\mathcal{L}$ fixed within the first order conditions and try to be as general as possible for basic properties $\mathcal{B}$. In this section we specialize on graphs.

Let us first keep the class of all graphs as the basic set of configurations. There are special conditions $\mathcal{L}$ which cannot be formulated in first order language, but for which there is an asymptotic 0-1 law; for instance, connectedness $\mathfrak{Z}$, hamiltonicity $\mathfrak{H}$ (cf. Moon [14]) and rigidity $\mathfrak{R}$ (cf. Harary-Palmer [10] and the remarks about Example 2)

While the result can be proved with respect to $\mathfrak{Z}$ as an implicational consequence from the analogous first order result for connectedness with diameter two (cf. Blass-Harary [2], Corollary 13), it can be proved that the properties $\mathfrak{H}$ and $\mathfrak{R}$ are not implied by any first order condition $\mathfrak{L}$ with an asymptotic 1-law. Therefore, no direct applicatio of the Blass-Fagin theory is possible (cf.Blass-Harary [2], chapter 3). Thus there are positive asymptotic results which have been proved by special methods only.

Secondly, we turn to the most important field, where general first order arguments have failed so far. This is the theory of random graphs, which was introduced by Erdös and Renyi [6]. It is not our aim to give an exposition of this beautiful theory. For a summary we refer to Bollobás ([3], pp. 144). We only want to give some indications how the results of this theory fit into the framework for asymptotic 0-1 laws which we hav developed.

Usually, in random graph theory, the basic configurations for a given vertex cardinali ty n are the graphs with $M = M(n)$ edges. Therefore, the edge function $M(n)$ characterizes the class $\mathcal{B}$ of basic configurations. The class of all graphs with n vertices and $M(n)$ edges is denoted by $\mathcal{G}(n,M(n))$.

Some edge functions are of special interest in this theory. In particular, we define $M_\alpha(n) = \lfloor \alpha n \log n \rfloor$ for a fixed number α.

Other edge functions such as $\lfloor \alpha n \rfloor$ or $\lfloor \alpha \binom{n}{2} \rfloor$ are also common in the theory of random graphs.

For several of those basic structures there are again asymptotic 0-1 laws with respect to single special conditions.

We consider the following example:

$$\mathfrak{Z} \text{ has a 0-law in the basic set } G(n,M_{\frac{1}{2}-\varepsilon}(n))$$

$$\mathfrak{Z} \text{ has a 1-law in the basic set } G(n,M_{\frac{1}{2}+\varepsilon}(n)).$$

This is a corollary from early observations of Erdös and Renyi who located the exact threshold where the probability of being connected jumps from 0 to 1. It can also be proved directly by using richness arguments (cf. Bollobás [3],p.139 and exercise 13, p.143).

The same proposition with condition $\mathfrak{H}$ instead of $\mathfrak{Z}$ can be deduced as a corollary from recent results of Korshumov (cf. Bollobás, [3] pp. 141).

The introduction of edge numbers into the notion of basic configuration means to look for asymptotic spectral 0-laws. The former results with the basic class of all graphs appear as integrated statements and thus as corollaries of spectral laws.

The figure shows the number of graphs for n = 17 according to the edge number M and indicates that connectivity and hamiltonicity happen to begin in a region given by $M(n) \approx \frac{1}{2} n \log n \approx 24$, where there are relatively few graphs with such a low number of edges

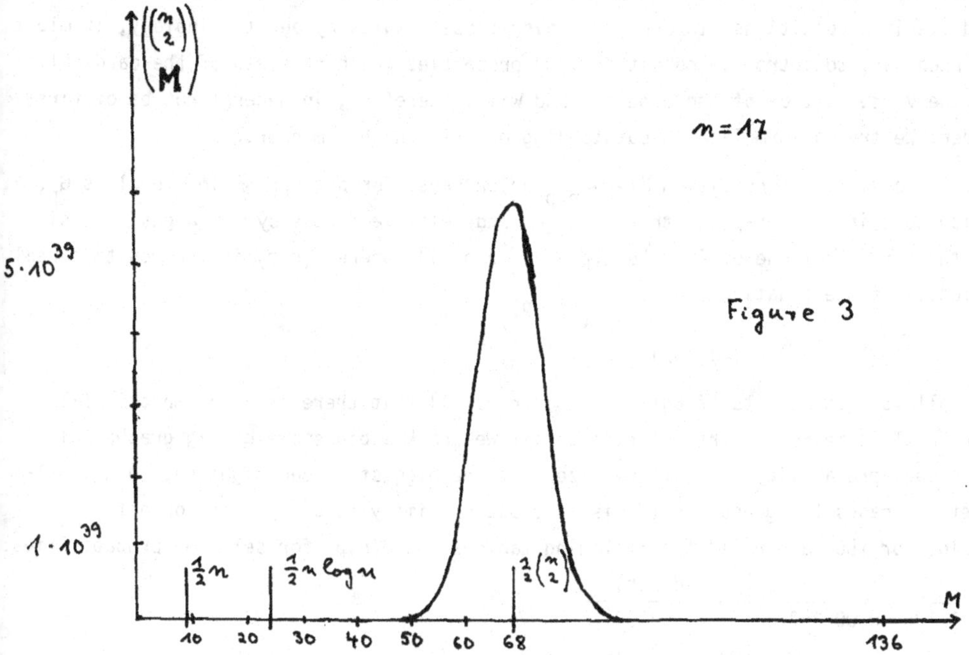

Figure 3

ooking at the present state of randam graph theory it seems that our terminology for symptotic 0-1 laws is not fine enough to express directly all the information which as been obtained by ad hoc methods. But beyond the desire for a uniform terminology e should like to obtain general explanations, for instance,for the fact of threshold hich seems to occur for all graph conditions $\mathcal{L}$, which are monotone increasing (in he sense that by adding edges to a given graph a monotone condition can never get lost).

e give an illustration of such a threshold situation by the following model, which as been described by many authors in randam graph theory analogously. Consider a big umber N of big graphs with n vertices, where n is also big. We start with all the aphs consisting only of isolated vertices. Then we add to each of the N graphsone ore edge at random. Thus all the graphs are in a process of evolution. Now we test r a fixed monotone increasing property $\mathcal{L}$ (like connectedness or hamiltonicity) and t a bell ring if this condition is fulfilled for one of the graphs. Then we should expect a theory which explains generally why, within a narrow bandwidth (depending on $\mathcal{L}$), almost all the bells begin to ring.

Finally we mention that even for conditions which are not monotone increasing, asymptotic 0-1 laws occur. The theory of random graphs again presents those results within an impressive collection of asymptotic 0-1 laws for various basic properties and specific conditions. In order to cover those results by our terminology, it might be necessary to introduce conditions and properties which make use of the cardinality of the vertex set or of the edge set and will, therefore, in general not be of first order. We try to formulate an outstanding example in this manner:

Let us consider the condition $\mathcal{L} = \mathcal{L}_{n,p}$ which says: For a graph $\mathcal{G}$ in the class $G_p(n)$, which contains all graphs with n vertices and relative frequency p of edges, the size of the maximal clique of $\mathcal{G}$ is $\lfloor d(n,p) \rfloor$ or $\lceil d(n,p) \rceil$, where d(n,p) is the positive real solution of the equation

$$\binom{n}{d(n,p)} \cdot p^{\binom{d(n,p)}{2}} = 1.$$

It follows from results of Bollobás and Erdös [4] that there is an asymptotic 1-law for $\mathcal{L}$. This means in particular, that, if we check a big store of big graphs, all with edge-probability p, what the size S ot the biggest clique might be, then in almost all cases the guess that s is in close proximity to d(n,p) is correct.
Again, for the case n = 17 the following table shows d(n,p) for selected probabilities:

p	d(n,p)
1/136 = 0.007353	2.00
0.1	2.91
0.25	3.85
0.333	4.39
0.5	5.685
0.8	9.93
0.9	12.81
0.95	14.81
1	17

Table

The most probable clique number for graphs with edge probability p and n = 17.

The methods of random graph theory are not easy to classify. But the hint that there are often techniques for approximating the binomial by the poisson distribution and for using the central limit theorem does not lead into the wrong direction.

6. Conclusion.

We do not believe that there will ever be a uniform theory for getting asymptotic 0-1 laws in combinatorics which will cover all results, which are proved sometimes with deep and laborious ad hoc methods. However, it should be an inspiring field of research for the future to develop uniform patterns of argumentation in this field

References

1. E.A.Bender, Asymptotic methods in enumeration. SIAM review 16 (1974), 485-515.

2. A. Blass and F. Harary, Properties of almost all graphs and complexes. Jour.graph theory 3 (1979), 225 - 240.

3. B. Bollobás, Graph theory. Springer, New York, Heidelberg, Berlin (1979).

4. B. Bollobás and P. Erdös, Cliques in random graphs. Math.proc.Cambridge phil.soc. 80 (1976), 419-427.

5. K.J.Compton, Application of logic to finite combinatorics. Dissertation Wesleyan Univ., Middletown (1981) (unpublished manuscript).

6. P. Erdös and A. Rényi, On the evolution of random graphs. Publ.math. inst. Hungar. acad. sciences 5 (1960), 17-61.

7. P. Erdös and J. Spencer, Probabilistic methods in combinatorics. Academic Press New York, London (1974).

8. G. Exoo, On a adjacencyproperty of graphs. Jour. graph theory 5 (1981), 371-378.

9. R. Fagin, Probabilities on finite models. Jour. Symb. Logic 41 (1976), 50-58.

10. F. Harary and E.M.Palmer, Graphical enumeration. Academic Press New York, London(1973).

11. D.J. Kleitman and B.L.Rothschild, Asymptotic enumeration of partial orders on a finite set. Trans. AMS 205 (1975), 205-220.

12. A.D. Korshunov, Solution to a problem of Erdös and Rényi on Hamilton cycles in nonoriented graphs. Soviet Math. Doklady 17 (1976), 760-764.

13. J.F. Lynch, Almost sure theories. Ann.math.logic 18 (1980), 91-135.

14. J.W.Moon, Almost all graphs have a spanning cycle. Canad.math.bull. 15 (1972), 39-41.

15. L. Moser and M. Wyman, An asymptotic formula for the Bell numbers. Trans. roy. Canad. soc. 49, ser. III (1955), 49-54.

16. W. Oberschelp Kombinatorische Anzahlbestimmungen in Relationen. Math. Annalen 174 (1967), 53-78.

17. W. Oberschelp, Monotonicity for structure numbers in theories without identity.
In: Lecture notes in math. 579 (1977), 297-308.

18. W. Oberschelp, Asymptotische 0-1 Anzahlformeln für prädikatenlogisch definierte Modellklassen.
Manuscript unpublished.
Abstract in: Deutsche Math.Vereinigung, Vorträge Univ. Dortmund 1980, p.162.

19. E.M.Wright, Graphs on unlabelled nodes with a given number of edges
Acta Math. 168 (1971), 1-9.

Generalized Block Designs as Approximations for Optimal Coverings

Jörg Remlinger
Lehrstuhl für Angewandte Mathematik
RWTH Aachen, Templergraben 64
5100 Aachen, Federal Republic of Germany

Abstract

Let V be a set of cardinality v, $v \in \mathbb{N}$. We are looking for the minimal number of k-sets (i.e. subsets of V having cardinality k), such that every t-set of V, $t \leq k$, is covered by at least λ of these k-sets. This special covering problem is called the generalized block design problem with parameters v,k,t,λ. It is equivalent to the problem of Turân [16] and also to the generalized covering problem [4]. Therefore, the known bounds for these two equivalent problems are also bounds for the generalized block design problem and vice versa.
Using some type of greedy algorithm, we will compute an approximative solution for an optimal generalized design with arbitrary parameters. The number of blocks in such an approximation will be at most $(1+\log\binom{k}{t})$-times the optimal number of blocks. This result depends essentially on a theorem of Lovâsz [11].

Introduction

Let A and B be finite sets and R a binary relation, $R \subseteq A \times B$. The triple (A,B,R) is called a covering structure or incidence structure. A subset A' of A with the property that there exists for every $b \in B$ an $a \in A$ such that $(a,b) \in R$ is called a "cover of B". The aim is to find a so-called optimal cover of B, this is a cover of minimal cardinality.
For example, the set-cover-problem [3] and the lottery-problem [12] are such covering problems. There are several similar problems in the theory of Information Retrival [13] and Operations Research [19].
A special case of these general covering problems are incidence structures of the form $([V]^k, [V]^t, R)$. Here $V = \{1,\ldots,v\}$, $t \leq k \leq v$, $[V]^k$ and $[V]^t$ are the sets of all k-sets and t-sets respectively of V, and a t-set b is incident with a k-set a if and only if b is contained in a. A k-graph A' is a subset of $[V]^k$ and an element of A' is called a block. A k-graph A' is called a block design with parameters (v,k,t,λ) if and only if $v,k,t,\lambda \in \mathbb{N}$, $v > k > t$, and every t-set of v is contained in precisely λ blocks. For further details see Hall [5].

A recent survey is contained Lindner and Rosa [10]. For an existence theory for the case $t=2$ see R.M.Wilson [17].

Generalized Block Designs

Definition 1 Let $V=\{1,\dots,v\}$, $t,v,k \in \mathbb{N}$ with $t \le k \le v$. A k-graph $A' \le [V]^k$ is called a generalized block design with parameters (v,k,t,λ) if and only if every t-set of V is contained in at least λ blocks of A'. Thus every block design is a generalized block design.

Definition 2 Let $t \le k \le v$, and let $c(v,k,t,\lambda)$ denote the class of all k-graphs G over V. The problem: "Compute $c(v,k,t,\lambda) = \min\{|G| : G \in T(v,k,t,\lambda)\}$" is called the generalized covering problem (GCP).

Definition 3 Let $t \le k \le v$. $T(v,k,t,\lambda)$ is the class of all t-graphs G over V, so that there exist at least λ edges $T_1,\dots,T_\lambda$ of G with $T_i \subseteq K$, $1 \le i \le \lambda$, for any k-set $K \in [V]^k$. The problem: "Compute $\tau(v,k,t,\lambda) = \min\{|G| : G \in T(v,k,t,\lambda)\}$" is called the generalized Turân problem (GTP).

We have:

(1) The generalized block design problem (GBDP): "Find an optimal GBD" is equivalent to the GCP.

(2) $c(v,k,t,\lambda) = \tau(v,v-t,v-k,\lambda)$.

This means that all inequalities and bounds for Turân- or covering-numbers are also approximations for the numbers of blocks in an optimal GBD. Therefore, we can deduce some bounds for the number of blocks in an optimal GBD from the following inequalities:

(3) (Schönheim [15])

$$\tau(v,k,t,\lambda) \ge \left\lceil \frac{v}{v-t} \left\lceil \frac{v-1}{v-t-1} \left\lceil \cdots \left\lceil \frac{k+2}{k-t+2} \left\lceil \frac{\lambda(k+1)}{k-t+1} \right\rceil\right\rceil \cdots \right\rceil\right\rceil\right\rceil,$$

where $\lceil\ \rceil$ denotes the upper Gaussian brackets.

(4) (Katona, Nemetz, Simonovitz, see [4])

$$\binom{v}{t}/\binom{k}{t} \le \tau(v,k,t,1) \le \binom{v}{t}\cdot\left(\frac{k-1}{t-1}\right)^{1-t}.$$

(5) (Spencer, see [4]).

$$\tau(v,k,t,\lambda) \ge \left(\frac{v}{t}\right)^t \cdot \left(\frac{k-1}{t-1}\right)^{1-t}.$$

Further intensifications can be found, for instance, in Gutschke [4].

In most cases it is impossible to find an optimal GBD (or a generalized covering etc.). Therefore, we are looking for a good approximation. There are many possibilities to find an approximative algorithm for the problems mentioned. However, not all of them run "fast enough".

We try to find an algorithm which runs in polynomially, in v, bounded time and is best possible.

Some alternatives are perhaps:

(i) "brutal algorithm" : Check all possible GBD's and take the optimum (not polynomially bounded);

(ii) Choose the "nearest" exact block design having parameters (v',k,t,λ), $v' \geq v$ or $v' \leq v$, and introduce dummies (if necessary);

(iii) A kind of greedy algorithm with a special optimization function.

We will follow the third possibility, since it looks quite simple and there is a good approximation for the number of blocks of the approximative design.

Remarks:

(1) Algorithms of type (ii) and certain types of (iii) can be found in [13]. In all the computations I did, the type-(iii)-algorithm which I will present in the next chapter produced the best results.

(2) Perhaps one can associate with the GBD-problem a special NP-complete problem and it may be possible to reduce it to such a problem. This may be a justification for choosing an approximative algorithm for the GBD-problem. I looked for such a reduction, but I did not find an approximate NP-complete problem. In [14] one can find a lot of arguments and motivation for solving the GBD-problem with an approximative algorithm. There is also mentioned another criterion of intractability which is applicable to the GBD-problem (as well as to Ramsey-numbers etc.).

A Greedy Algorithm for the Generalized Block Design Problem

The general greedy heuristic can be described as follows:

The greedy algorithm computes stepwise a "nearly" optimal solution. In each step it chooses the best possible "subsolution" by a given optimization criterion. The solution is thus the union of all subsolutions.

Remark 3: If the class of the considered problems has a matroid-(or, more generally, a greedoid-) structure [7], then the greedy algorithm produces an optimal solution. Unfortunately the class of the GBD-problems has not such a matroid-structure.

The greedy cover algorithm (for hypergraphs):
Let H be a hypergraph. We denote the set of vertices by V(H), the set of edges by E(H).
Problem: Compute a minimal subset $C \subseteq V(H)$ with the following property :
for all $e \in E(H)$ there exists a $c \in C$ such that $c \in e$.
"Solution": In each step choose a point v in V(H) having maximal degree relative to the rest of the hypergraph (maximal rest-degree = optimization criterion). Eliminate this point and all edges incident with it. The algorithm stops if the rest-graph is empty.

Now consider a special kind of hypergraph H_b:
Let the set of vertices $V(H_b)$ be $[V]^k$. The set of edges is defined as follows:
Two k-sets of $V(H_b)$ are in the same edge if and only if their intersection contains a t-set $T \in [V]^t$.
It is clear that there is a 1-1-correspondence between the edges of H_b and the set of all t-sets. We have, therefore, the following

Theorem 1. The greedy cover algorithm, applied to the hypergraph H_b, produces a GBD with parameters (v,k,t,1). For a hypergraph H_b with multiple edges (each multiplicity = λ) it produces a GBD with parameters (v,k,t,λ).

For analysing the greedy cover algorithm we will use a theorem due to Lovász [11]. For this reason we introduce the notion of p-matching.

Definition 4: Let H = (V(H), E(H)) be a hypergraph and let p be any integer.

(i) A p-matching M of H is a (multi-)set of edges, such that each vertex x belongs to at most p edges of M.
We write $\nu_p(H) = \max\{\#\text{edges in M: M is p-matching of H}\}$.

(ii) A p-Matching M of H is called simple if and only if every edge occurs at most once in M.
We write $\tilde{\nu}_p(H) = \max\{\#\text{edges in M : M is a simple p-matching of H}\}$.

Now we can consider the following result of Lovász.

Theorem 2. Given a hypergraph H. The greedy algorithm may produce b covering points Then

$$b \leq \sum_{i=1}^{d(H)} \frac{\tilde{\nu}_i}{i(i+1)} + \frac{\tilde{\nu}_{d(H)}}{d(H)} ,$$

where d(H) is the maximum degree of the vertices of H.

Proof: Let b_i be the number of steps in which the algorithm chooses a point of maximum rest-degree i.
(Note that the rest-degree of the point which is selected in the j-th step is

greater than or equal to the rest-degree of the one which is chosen in the (j+1)-st step).

Let $c_i = \sum_{j=i+1}^{d} b_j$, $d = d(H)$,

and let $E_i(H)$ be the set of all those edges of H which are not covered by any of the prevailing chosen points. Finally, let $V(H_i)$ be the set of points incident with the edges of $E_i(H)$ and define

$$H_i = (V_i(H), E_i(H)).$$

By the construction it is clear that the maximum degree $d(H_i)$ of H_i is at most i, since the point which is chosen in the (c_i+1)-st step of the greedy cover algorithm has maximum rest degree i (compare the definition of c_i).
Therefore, we have

(6) $|E(H_i)| \le \tilde{\nu}_i$.

$|E(H_i)| > \tilde{\nu}_i$ would imply that $E(H_i)$ contains more edges than a maximal i-matching of H. This is a contradiction, because each edge of $E(H_i)$ contains at most i points. In each of the next b_i steps the algorithm selects a point which covers i new (i.e. not yet covered) edges of $E(H_i)$. In the following b_{i-1} steps each of the selected points covers exactly i-1 new edges and so on. These arguments imply

(7) $|E(H_i)| = ib_i + (i-1)b_{i-1} + \dots + 2b_2 + b_1$

and, by (6), we have

$ib_i + \dots + 2b_2 + b_1 \le \tilde{\nu}_i$, $1 \le i \le d$.

Multiplying the i-th of these inequalities, $1 \le i \le d-1$, by $1/i(i+1)$ gives

(8) $\frac{i}{i(i+1)}b_i + \dots + \frac{2}{i(i+1)}b_2 + \frac{1}{i(i+1)}b_1 \le \frac{\tilde{\nu}_i}{i(i+1)}$, $1 \le i \le d-1$;

and by multiplying the d-th inequality by $1/d$ we obtain

(9) $\frac{1}{d}db_d + \dots + \frac{2}{d}b_2 + \frac{1}{d}b_1 \le \frac{\tilde{\nu}_d}{d}$.

The summation over all the left sides of (8) and the left side of (9) yields

$$\sum_{i=1}^{d-1} \left(\sum_{j=1}^{i} \frac{j}{i(i+1)} b_j \right) + \sum_{j=1}^{d} \frac{j}{d} b_j \le \sum_{j=1}^{d-1} \frac{\tilde{\nu}_j}{j(j+1)} + \frac{\tilde{\nu}_d}{d} .$$

This is equivalent to

$$b_1 \left(\frac{1}{1\cdot 2} + \frac{1}{2\cdot 3} + \ldots + \frac{1}{d(d-1)} + \frac{1}{d} \right) + b_2 \left(\frac{2}{2\cdot 3} + \frac{2}{3\cdot 4} + \ldots + \frac{2}{d(d-1)} + \frac{2}{d} \right) +$$

$$\ldots + b_i \left(\frac{i}{i(i+1)} + \frac{i}{(i+1)(i+2)} + \ldots + \frac{1}{d(d-1)} + \frac{i}{d} \right) + \ldots + b_d$$

$$\leq \frac{\tilde{\nu}_1}{1\cdot 2} + \frac{\tilde{\nu}_2}{2\cdot 3} + \ldots + \frac{\tilde{\nu}_{d-1}}{d(d-1)} + \frac{\tilde{\nu}_d}{d} .$$

Since the coefficients of the b_i's are all equal to 1 (by using associativity), we finally obtain

$$b = b_1 + \ldots + b_d \leq \frac{\tilde{\nu}_1}{1\cdot 2} + \ldots + \frac{\tilde{\nu}_{d-1}}{d(d-1)} + \frac{\tilde{\nu}_d}{d} .$$

This is exactly the assertion of Theorem 2. □

In the case of the GBD-problem (see above) we have for the special hypergraph H_b:

(10) $d(H_b) = d = \binom{k}{t}$, and

(11) for a simple p-matching the equality

$$|M| \leq p\cdot\binom{v}{k} / \binom{v-t}{k-t} = p\cdot\binom{v}{t} / \binom{k}{t} \text{ holds.}$$

Consequently it follows that

$$\tilde{\nu}_p \leq p\cdot\binom{v}{k} / \binom{k}{t} .$$

Now Theorem 2 implies

$$b \leq \binom{v}{t} / \binom{k}{t} \cdot \sum_{i=1}^{d} \frac{1}{i} \leq \binom{v}{k} / \binom{k}{t} \cdot \left(1 + \log\binom{k}{t}\right) ,$$

i.e., $\frac{b}{b_{opt}} \leq 1 + \log\binom{k}{t}$.

Using the equalities

(12) $\tilde{\nu}_p = \binom{v}{k} - \tau\left(v,k,t,\binom{v-t}{k-t}-p\right)$ and

(13) $\tilde{\nu}_p = \binom{v}{k} - c\left(v,v-t,v-k,\binom{v-t}{v-k}-p\right)$, respectively,

we can derive better evaluations for $\tilde{\nu}_p$, and also for b, if we replace the estimations for the $\tilde{\nu}_p$'s by the inequalities (3), (4) or (5) (or perhaps by better bounds for Turán- or covering-numbers).

Remark 4

a) The algorithm and the results above are still valid for $\lambda > 1$. This follows by a simple transformation.

b) The result $\frac{b}{b_{opt}} \le 1 + \log \binom{k}{t}$ gives an upper bound for the number of covering blocks in the greedy cover algorithm. The average case behaviour of the algorithm may be much better, but it is not easy to compute. Several computations showed that the solutions of the greedy cover algorithm diverge at most by 50% from the optimal solution for a GBD-problem.

c) The complexity of the greedy cover algorithm is in $O(v^{t+k})$, thus polynomially bounded in v when t and k are fixed.

Generalizations and Concluding Remarks

Definition 5 A generalized partial block design (GPBD) with parameters $(v,f,t,\underset{\sim}{\lambda})$, $\underset{\sim}{\lambda} = (\lambda_1,\ldots,\lambda_{\binom{v}{t}})$, $\lambda_i \in \mathbb{N}$, $1 \le i \le \binom{v}{t}$, is a set of k-sets of $V = \{1,\ldots,v\}$, called blocks, such that for every $j \in \{1,\ldots,\binom{v}{t}\}$ the t-set of V labelled j is contained in at least λ_j blocks.

Analogous considerations as in the case of $\underset{\sim}{\lambda} = (\lambda,\ldots,\lambda)$ yield the following result

$$(14) \qquad b \ge \sum_{j=1}^{\binom{v}{t}} \frac{\lambda_j}{\binom{k}{t}} .$$

In this case upper bounds for the value of b are very hard to compute because they depend essentially on the parameters λ_i, $1 \le i \le \binom{v}{t}$. Some results are known in the special case, where $\lambda_i \varepsilon \{\lambda,\lambda+1\}$ for some $\lambda \in \mathbb{N}$, for all $i = 1,\ldots,\binom{v}{t}$. This is the case of the regular generalized partial block designs (see, for instance, [2]).

Let us finally illuminate this generalized problem from another point of view. Sometimes it may be useful to describe the GPBD-problem as an LP-problem (Linear Programming-problem).
Label the $\binom{v}{t}$ t-sets by $T_1,\ldots,T_{\binom{v}{t}}$ and the $\binom{v}{k}$ k-sets by $K_1,\ldots,K_{\binom{v}{k}}$.

Define the incidence-matrix A between the t-sets and the k-sets by

$$A = (a_{ij})_{i=1,j=1}^{n\quad m}, \quad a_{ij} = \begin{cases} 1 \; ; \; T_i \subset K_j \\ 0 \; ; \; \text{otherwise} \end{cases}$$

With this notation the GPBD-problem can be described as follows.

Find an optimal cover $\underset{\sim}{x} = (x_1,\ldots,x_{\binom{v}{t}})$, $x_i \in \mathbb{N}_0$, $1 \le i \le \binom{v}{t}$, such that $A\underset{\sim}{x} \ge \underset{\sim}{\lambda}$

$$\text{i.e.}\quad \left.\begin{array}{ll} \min & \sum_{i=1}^{\binom{v}{t}} x_i \,, \\ \text{s.t.} & \begin{array}{l} A\underset{\sim}{x} \ge \underset{\sim}{\lambda} \\ \underset{\sim}{x} \ge \underset{\sim}{o}\,,\ \underset{\sim}{x}\ \text{integer}\,. \end{array} \end{array}\right\}\quad (P)$$

If there is an optimal solution for (P), we can find one by using methods of Operations Research (cf. [19]). Unfortunately the required algorithms are in general, not polynomially bounded in v.

Another point of view may also be of interest.

Consider the dual problem of (P):

$$\left.\begin{array}{ll} \max & \sum_{i=1}^{\binom{v}{k}} \lambda_i y_i \\ \text{s.t.} & \begin{array}{l} A^t\underset{\sim}{y} \le \underset{\sim}{\lambda} \\ \underset{\sim}{y} \ge \underset{\sim}{o}\,,\ \underset{\sim}{y}\ \text{integer}\,. \end{array} \end{array}\right\}\quad (DP)$$

(DP) is a weighted generalization of a problem of Brown, Erdös and Sôs [1] and equivalent to a generalized packing problem (see, for instance, [4]). If we replace $\underset{\sim}{\lambda}$ by $(\lambda,\ldots,\lambda)$, $\lambda \in \mathbb{N}_0$, we have (P) as the GBD-problem and (DP) as the original problem of Brown, Erdös and Sôs. With a duality theorem in LP we obtain the following

Theorem 3. If the problem (P) (this is the GPBD-problem) has an optimal solution, then (DP), the weighted form of the B-E-S-problem, has the same optimal solution, and vice versa.

This means that the problems are essentially the same.

References

1. G.W.Brown, P.Erdös, V.Sôs — Some extremal problems on r-graphs, in: Harary,F, New directions in the theory of graphs. Acad.Press, New York, 1973, 53-63.

2. N.Gaffke — Optimale Versuchsplanung für lineare Zwei-Faktor-Modelle. Dissertation, RWTH Aachen, 1978.

3. M.R.Garey, D.S.Johnson — Computers and intractability. W.H.Freeman, San Francisco, 1979.

4. K.-U.Gutschke — Untersuchungen zu einer Klasse kombinatorischer Extremalprobleme. Dissertation, RWTH-Aachen, 1974.

5. M.Hall — Combinatorial Theory. Blaisdell, Waltram-Toronto-London, 1967.

6. H.Hanani — The existence and construction of BIBD's. Ann.Math.Statist., 32, 1961, 361-386.

7. B.Korte, L.Lovâsz — Mathematical structures underlying greedy algorithms, in: Fundamentals of Computation Theory, Lect.Notes in Comp. Sci. 117, 1981, 205-209.

8. O.Krafft — Lineare statistische Modelle.Vandenhoek & Ruprecht, Göttingen, 1978

9. E.Lawler — Combinatorial Optimization: Networks and Motroids. Holt-Reinehart-Winston, New York, 1976.

10. C.Lindner, A.Rosa (eds.) — Topics on Steiner Systems. Ann. of Discrete Math., 7, 1980.

11. L.Lovâsz — On the ratio of optimal integral and fractional covers. Discrete Math. 13, 1975, 383-390.

12. W.Oberschelp — Lotto Garantiesysteme und Blockpläne. Math.-Phys.-Semesterberichte 19, 1972, 55-67.

13. W.Oberschelp, D.Wille — Mathematischer Einführungskurs für Informatiker. Diskrete Strukturen, Teubner, Stuttgart, 1976.

14. J.Remlinger — Verallgemeinerte Blockpläne als Approximation optimaler Überdeckungslösungen, Diplomarbeit, RWTH-Aachen, 1980.

15. J.Schönheim — On coverings, Pacific J.Math. 14, 1405-1411, 1964.

16. P.Turân — On the theory of graphs, Colloq.Math. 3, 19-30, 1954.

17. R.M.Wilson — An existence theory for pairwise balanced designs III: Proof of the existence conjectures, Journ.Comb.Theory (A) 18, 1975, 71-79.

18. E.Witt — Über Steinersche Systeme, Abh.Math.Sem.Univ.Hamburg 12,1938 265-275.

19. H.-J.Zimmermann — Einführung in die Grundlagen des Operations Research, München, 1971.

A Graphic Theory of Associativity and Wordchain Patterns

Dov Tamari
175 W 76 Street New York, N.Y. 10023, U.S.A.

Abstract

The problem of deciding whether a partial binary operation, a "bin"[1],can be embedded in a semigroup is the associativity problem (for general bins). It is known that it is equivalent to the word problem for (semi)groups and thus unsolvable, even for the class of finite bins. This paper establishes a close association between bins and their wordchains and 3-connected 3-regular planar graphs, or, equivalently convex 3-regular polyhedral nets (skeletons). This permits a constructive approach revealing the combinatorial depth of the associativity problem in detail and leads to a naturally enumerable hierarchy of standard wordchain patterns, of universal bins, and of associative laws. Each bin is a superposition of homomorphic images, i.e. "colourings" of edges, of universal bins. One side result is a purely algebraic equivalent of the 4-colour-theorem. The obtained results open further ways for an efficient search by computer for simplest non-associativity contradictions. It is hoped that they lead to solutions of the associativity problem for further subclasses of bins, further insight into the structure of partial binary operations and of polyhedra and will yield precise measures of presentations for associative systems and their classifications.

0. Introduction

The complexity of the general concept of associativity of partial binary operations could hardly be better hidden than by its collapsing into the simple elementary formula $(xy)z = x(yz)$ for the all-important, yet still very special case of closed (i.e. complete) operations. Furthermore, the veil of deceptive simplicity is not lifted by the first encounters while reconnoitring the wilderness of partial operation However, some dents have been lately made in this "terra incognita", and, hopefully, some headway may be started here. As further motivation for this pursuit may serve

---

1) The term bin has been proposed by K. Osondu in his thesis (Buffalo N.Y., 1974) and can be used, when wanted, with "partial" or "full", similarly to "partial" or "linear" order.

the fact that partial operations have come in their own right in recent electronics circuitry when parts of the function (mathematical and technical) are prescribed while others are left open (oral remark by Professor Dexter).

1. Roots of this Research

The following is a brief, admittedly subjective, account going back to the problem of extending a cancellation semigroup S to a group (indeed rather rings without zero-divisors to fields or division rings).

It was discovered

a) that this problem is best understood as the ordered superposition of two distinct ones, the first of which is the (usually easier) symmetrisation problem leading in general, but not always, to a partial bin. The symmetrisation sym(S) is followed by the usually more difficult problem of completion of sym(S) to a group, and

b) that sym(S) need not be associative and that its associativity is the sufficient (and of course necessary) condition that it can be completed to a group.

Since the semigroup generated by sym(S) (or even by sym(P), where P is any presentation of S=S(P)) is by itself already a group, it seemed advantageous to look at the problem as a special case of the more general one of embedding bins in semigroups. It is indeed much more general in view of the fact that any presentation of semigroups or groups can be standardized to a bin. However, in spite of so much more generality the new problem turns out to be not more difficult. Indeed, it is simpler in the sense that the new "associative laws" A^n_m are simpler, more immediate in form and in concept, than the famous n. and s. conditions of Malcev for the embedment of semigroups into groups. A posteriori, these are but a special-purpose adaption for a special case; and so too are, more than a decade later, the simultaneous but independent results of Lambek and Tamari, precursors of this present work. Malcev of course, was the precursor of all. (A more objective report would have to start with Hamilton, Cayley and continue with Dyck, Thue, Dehn, Magnus, Etherington, Bruck, Coxeter, Moser, Lyndon, Schupp and many others).

For more details about basic concepts underlying this theory and for historical background the reader is referred to [6] (in particular §§ 1, 2) and to [2] (§ 1).

2. Basic Concepts

A bin B (or partial binary operation, partial groupoid, monoid, multiplication table with "holes") is essentially a conjunction of ternary relation statements $(l_i, r_i; p_i)$, usually written $l_i\, r_i = p_i$ $(l_i, r_i, p_i \in B)$. B is called <u>associative</u> if it can be embedded in a semigroup S; or, more specifically, if the canonical map $k: B \to S(B)$ from B into the semigroup S(B) generated by B as a set of generators and defining relations, in brief as a presentation, is injective, or in other words, if the distinct generators (i.e. the elements of B) will still represent distinct elements of the semigroup S ("generator problem").

Algebraists describe the "generation" of S(B) as a "quotient construction" $S(B)=F(B)/E_B$, where F(B) is the free semigroup over the alphabet B, i.e. the set of words with concatenation as multiplication, E_B the equivalence relation induced in F(B) by the relations of B (E_B is in fact a congruence in F(B)); k assigns to each element $b \in B$ the equivalence class of words containing the one-letter word "b".

This is unfortunately, but unavoidably, in general, an ineffective definition. No suitable collections of semigroups nor details of their generation (even not that of single elements as infinite equivalence classes of words) are generally accessible for constructive inspection to see if there exists an embedding semigroup, or that no distinct elements of B become equivalent (mod E_B) under k - except in special cases. Indeed one knows that the decision problem of associativity for the class of finite bins is equivalent to the word problem for finitely presented (semi)groups and, therefore, unsolvable. Furthermore, every finite presentation of a (semi)group can be standardized to a finite bin. Thus bins are universal standard presentations of binary operations (even non-associative ones). (For details of standardization see, for instance, [6](§2)).

Still, these equivalence classes of words have some general and yet definite, genuinely constructive features, namely so-called wordchains. These are finite, linearly ordered sets of words, each one obtained from its neighbour by one of the substitutions of the given presentation (an equation representing two substitutions), in our case a bin. Each equivalence class is a set union of its chains.

Wordchains "progress" or transform inside equivalence classes from one word to the next by the standard bin substitutions

(Ø): $l_i\ r_i \rightarrow p_i$, the binary multiplication, contraction or fusion $\emptyset_i$,

(S): $p_i \rightarrow l_i\ r_i$, the binary factorisation, expansion or split S_i,

one only at each step. Hence at each step the length of a word changes by +1 or -1. These wordchains are referred to as standard wordchains. If such a wordchain begins and ends with a single letter word, say "a" and "z", it is called a special standard wordchain, denoted by C_z^a. The inverse chain of C_z^a is denoted by C_a^z. The total number of their constraction steps, say n, must equal that of their expansion steps, and thus the number of all steps is 2n, that of all words, including "a" and "z", 2n + 1. At each step three letters "act" making a total of 6n individual actions; each letter, except a ana z, acts twice: appears and disappears in the chain at distinct steps. Thi makes a total of 3n+1 letters in the chain, 3n-1 "full-life" letters called "edges" and the two "half-edges" a and z.

3. Wordchain Patterns and their Associated Prototypes P^n

Every wordchain can be visualized as "physically" written on a sheet of paper and thus as a planar figure. More specifically, one identifies in successive words letters repeated without any other action on them as successive parts of one and the same segment or edge, or, if one wants so, as the repeated name of such an edge. One further identifies the changes by substitution (i.e. applying the multiplication table) even when preserving one or the other letters, as vertices into which edges enter from former words and end (= disappear), and from which new edges originate and proceed to later words. As there would be no point in repeating the same whole word one has exactly one vertex between each pair of successive words. Thus the standard wordchain has become a planar 3-valent or 3-regular graph except for its ends. However, special standard wordchains can be closed if one can identify the two end letters, a=z, to become one edge - otherwise one has a "contradiction" to associativity. The whole figure becomes a planar 3-regular graph or, equivalently, a 3-regular division of the sphere or polyhedral net. From Euler's formula (Descartes's rule) one obtains v=2n, e=3n, f=n+2 for their number v of vertices, e of edges and f of faces, where n = (1) , 2, 3, .. is a parameter called degree. There is a minor point of beauty well fitting the system in keeping n=1, graphically as well as algebraically, by starting with general ternary relations; see A^1, the first in the list of examples. Some statements, however, will obviously only hold for n >1. (There is perhaps even a point for starting at n=0.)

Each vertex has 3 edges as well as 3 faces, each edge 2 vertices and 2 faces, and each face f^g (g-gon, g ≥ 2) g vertices and g edges (= sides). Vertices or faces with a common edge, or edges with a common vertex, are called neighbours; vertices or edges with a common face are "vertices and sides respectively of that face". For n >1 each vertex has 3 vertices as neighbour, each edge 4 edges as neighbour and each g-gon g faces as neighbour (g > 2).

In the construction of special standard wordchain patterns and their associated associative laws, or in the search for "contradictions" to associativity, one progresses in natural order with the parameter n from shorter to longer chains. To avoid trivialities, juxtapositions repeating already encountered cases of wordchain patterns, one imposes also 3-connectedness on the graphs. i.e. their separation into two disjoint graphs requires the (omission) of at least 3 edges. We shall refer to planar (or spherical) 3-regular and 3-connected graphs (or nets) as prototypes P^n of degree n. (The term "prototype" comes from the author's thesis (Paris 1951) where it is used in a less general context, while Lambek uses "polyhedral condition" with a somewhat different meaning). By a classical theorem of Steinitz they are indeed equivalent to the nets of convex 3-regular polyhedra. One has thus associated with each essentially new special standard wordchain pattern a convex 3-regular polyhedron P^n with all edges directed and one distinguished as the closure edge.

4. The Converse Construction

Conversely, every prototype P^n with vertices $v_1,\ldots,v_{2n}$ and edges $e_0,\ldots,e_{3n}=e_0$ can be associated with a special standard wordchain and with couples of so-called universal bins (A^n,B^n) by a judicious directing and labelling of the edges which become the elements of a bin A^n as well as, with a slight modification , those of a bin B^n. This is done by linearly ordering the 2n vertices such that

1) v_1 and v_{2n} are vertices of one edge $e_0=e_{3n}=a=z=(v_1,v_{2n})$, and

2) each vertex v_i, $1<i<2n$, has at least one of its 3 edges coming from an earlier vertex and one going to a later one. This means that all edges become naturally directed by the indices of their endpoints-except perhaps (v_1,v_{2n}) - and that the neighbour relation of vertices in the plane (or on the sphere) is preserved to some degree by this projection on the index line "i": a vertex v_i($1< i< 2n, 1<n$) remains surrounded by most of its neighbours by falling between some couples of neighbours. The ordering and labelling of the vertices, of the edges with their induced directions and of the faces can be done in a finite number of distinct ways as follows:
First one obtains an open net N^n from a prototype P^n by choosing any edge, say $e_0=e_{3n}=(v_1,v_{2n})$, cutting it into 2 halfedges or sticks "a" and "z", and pulling them apart, say a to the top, z to the bottom. It has n bounded and 2 unbounded faces. It is convenient to visualize N^n as spread out in the plane from left to right and from "a" at the top to "z" at the bottom. N^n remains 3-connected in its interior. However its "ends", i.e. a with v_1 and z with v_{2n}, are only 2-connected, and so are parts of N^n containing an end. No P^n contains a P^m, $m<n$, since P^m would be disconnected (i.e. 0-connected "with") from the remainder of P^n. Nor does P^n contain any N^m, $m<n$, since N^m would be only 2-connected to the remainder of P^n through its half-edges "a" and "z". Hence no N^n contains any N^m, $m<n$. Thus no P^n nor N^n, $n>1$ contains a "2-side" (i.e. digon), nor, for $n>2$, two adjacent triangles because they constitute the bounded faces of an N^1, an N^2, etc. respectively.

For convenience of reference the following construction proceeds in a Cartesian number plane. N^n becomes an ordered (open) net or wordchain model or pattern O^n as follows: Stretch N^n between v_1 and v_{2n} giving the vertices v_1 distinct natural number ordinates $y_i=i$, $i=1,2,\ldots,2n$, such that each vertex v_i, $1<i<2n$, gets (at least) one neighbour vertex with lower index and one with higher index. This means that the neighbourhood relation between vertices remains reflected in the indices by projection on the vertical line as an "interior" or "between" relation. Thus all edges descend strictly, i.e. pass any ordinate (i.e. horizontal line) at most once and no edges ever meet except in vertices. Thus each vertex becomes a tripod with one edge to one side, either up or down, and 2 edges to the other. This permits 1) to distinguish the two arrows simultaneously entering or simultaneously leaving the same vertex as a left and a right factor, 2) to order all edges after $a=e_0$ as $e_1, e_2, \dots, e_j, \dots, e_{3n}=z$ in this linear order by their starting vertices and when needed from left to right. A different way would be 3) to order all edges meeting one and the same ordinate from left to right as letters of one word, and 4) to order, similarly as in 2), the faces $f_0, f_1, \dots, f_k, \dots, f_n, f_{n+1}$, f_0 the unbounded "polygon" to the left, f_{n+1} to the right, both derived from the two faces of the "cut" edge, while each other polygon f_k has a distinct vertex v_{i_k} at its "top", i_k a strictly increasing function of k and, dually, another at its "bottom", as well as a left and a right "side" (=sequence of edges).

Each vertex v_i, $i=1,2, \dots, 2n$, is either a <u>split</u> (factorisation, expansion, or top of face) into which a so-called product arrow $e_j=p_i$ enters and from which 2 factor arrows, a left one $e_{j*}=l_i$ and a right one $e_{j*+1}=r_i$, $j<j*$, exit; or a <u>fuse</u> (multiplication, contraction, or bottom of face) into which 2 factor arrows l_i and r_i enter and from which one product arrow p_i exits. Denote by $s_i(t_i)$ the number of splits (fuses) among $v_1, v_2, \dots, v_i$ and observe that $s_i > t_i$ for $i < 2n$, but $s_{2n}=t_{2n}=n$, $s_i+t_i=i$, $s_1=1$, and for $n \geq 2$ $s_2=2$, etc.

The edges, including the sticks, presented by segments or arcs meeting any ordinate at most once, are also called <u>letters, elements, generators, variables or indeterminates</u>. The sequence of letters encountered by any ordinate between v_i and v_{i+1} is well determined and is the word

$$W_i = e_{j_{io}} e_{j_{i1}} \dots e_{j_{i\lambda_i}} \quad \text{of length } \lambda_i+1, \text{ where } \lambda_i=s_i-t_i.$$

The sequence of 2n+1 words $W_o=a, W_1, \dots, W_{2n}=z$ associated with the ordered model O^n is its <u>wordchain</u>, the sequence λ_i its <u>profile</u>:

$$\lambda_o=0=\lambda_{2n}, \ \Delta\lambda_i = \lambda_{i+1}-\lambda_i = \pm 1, \quad \lambda_i \geq 1 \text{ for } 1 \leq i \leq 2n-1.$$

5. The Universal Bins

To each vertex v_i belongs a triple $(l_i, r_i; p_i)$ of letters "active" in the <u>transformation</u> or <u>transition</u> $T_i: W_{i-1} \to W_i$ with

$$T_i: p_i \to l_i r_i \text{ or } T_i: r_i l_i \to p_i, \text{ depending on } v_i \text{ being a split or a fuse.}$$

The collection of these triples written as a binary oparation relation $l_i r_i = p_i$, in particular, their tabulation into a partial multiplication table, defines the <u>free</u> or <u>universal bin B^n</u> associated with O^n.

Once an ordering has been fixed each of the 3 edges, or rather half-edges at each vertex receives a unique natural interpretation as one of the three components - left factor, right factor or product - of 2n ternary relations or table entries of a bin A^n or B^n. Thus each of the 3n edges of A^n and the 3n-1 edges of B^n respectively plays two roles which may or may not differ - one at its beginning vertex, one at its end vertex; but the half-edges $a \neq z \in B^n$ play only one product role. Thus the bin B^n is the same as the bin A^n, except for the chosen "start-end" or closure edge of P^n $e_0 = e_{3n}$, which will be "cut" with the result that $p_1 = a \neq z = p_{2n}$ in B^n. Thus A^n is obtained from B^n by identifying a and z; in other words one has the "near-identity" epimorphism $\varepsilon: B^n \to A^n$. B^n is, evidently , non-associative, while A^n is easily proven to be associative. Thus A^n is the greatest associative homomorphic image of B^n. An A^n may belong to several B^n; a P^n to several A^n; the number of P^n rises steeply with larger n.

6. A Homological Definition of Associativity

One can now state the following

Proposition. A bin B is associative if and only if every morphism $\beta: B^n \to B$ splits into the epimorphism $\varepsilon: B^n \to A^n$ and a morphism $\alpha: A^n \to B$.

This statement can also serve as a "homological" definition of associativity. It can be turned constructively to supply an enumerable hierarchy of independent associative laws. The totality of these can be expressed in a "metaformula" of implications

$$(\mathbf{A}^n) \ : \qquad B^n \Rightarrow a=z \ ,$$

where the hypothesis B^n is considered the conjunction of the 2n bin (ternary relation) statements of B^n, for all universal bins.

Anticipating later results (section 15) subclassifying the A^n and B^n for $n > 1$ one writes with more detail

$$(\mathbf{A}^n_m) \ : \ B^n_m \Rightarrow a = z, \ \text{where } 1 \leq m < n .$$

Here m is the number of letters of A^n_m which possess two factorizations, i.e. they appear twice as products inside the multiplication table including the one special letter $a = z \in A^n_m$, while the "primes" are those letters which have no decomposition into factors, i.e. they do not appear inside the multiplication table.

A^n_m admits a more compact and more explicit "normal" form valid for $n > 1$

$$(\mathbf{A}^n_m) : M_{q_k} = M'_{q_k} \ , \quad 1 \leq k < m < n \, , \ \Rightarrow M_a = M_z \ ,$$

where the hypothesis of k-1 equations is empty if m=1, and where the M are monomials, i.e. full binary bracketings in the n+m primes of the universal bins. These primes serve as general variables of indeterminates like, e.g. x, y, z in (xy)z=x(yz) to express the ordinary associative law, which is just the first instance of an A^n_m, namely $A^2 = A^2_1$. The $q_1, \ldots, q_{m-1}$, are the m-1 twice directly factorizable letters common to an associated couple of universal bins A^n_m, B^n_m, while $q_m = a = z$ in A^n_m only.

The monomials M_{q_k}, M'_{q_k} indexed by the q_k are the couple of their prime factor decompositions derived from the couple of their entries in the multiplication table, M_a and M_z the unique prime factor decompositions of a and z.

Corresponding to the m monomial equations above the polyhedral net belonging to A^n_m decomposes into m regions, each one a binary double tree - a pair of binary trees with a common root, rather like a natural tree - generated from the root by successive binary factoring (splitting) as long as possible, i.e. till one is stopped when all last components present are primes. Each prime belongs to the extremes of two distinct such double-trees, except the case that it may belong to M_a and M_z. The totality of primes constitutes the common boundary regions of these double-trees whose common roots are just these elements with double factorization. The corresponding open nets and wordchain patterns belonging to B^n_m decompose in the same way, but rather into m+1 regions: m-1 double trees $M_k = M'_k$, $1 \le k < m$, and the 2 trees M_a and M_z with roots a and z. The two appearances of each of the primes are in the same order, but no couple of monomials has any pair of brackets in common. This is the equivalent of 3-connectedness for monomial systems.

7. Types and Characters of Letters in Wordchains and Universal Bins

Each letter, except "a", originates (starts, appears, begins) in a vertex v_i, and each one, except "z", ends (disappears) in a vertex $v_{i'}$, $i < i'$, in one and only one of three possible ways: either as a left, or as a right factor, or as a product. This yields the nine types indicated by the following self-explanatory symbols: (u,v), $u,v \in \{l,r,p\}$, u being the letter character at the start v_i, v that at the end $v_{i'}$.

Each letter has a natural number $L = i'-i$ as its life-span in the chain during which it participates in the wordchain.

The universal bins A^n and B^n are rather "lean" and very special. Both have only 2n entries in their multiplication table of size $(3n)^2$ and $(3n+1)^2$ respectively, the diagonals are empty. All elements are used exactly twice as left (l) or right (r) factors (f), or as products (p), except the two special distinct ones ($a \neq z$) in B^n. This produces 9 possible types (u,v) of elements with $u,v = l,r$, or p; one may say that "a" is of type (-,p) and "z" (p,-). The nine, or even eleven types are conveniently grouped into three principal types p^i, $i = 0,1,2$, where i is the frequency of the character p in their type (u,v). Thus p^0 comprises four types without p, the already mentioned primes, of which there are altogether n+m, the same in A^n_m and B^n_m; p^1 also comprises four types (u,v), those with exactly one single p of which there are altogether 2(n-m) elements in A^n_m and 2(n-m)+2 in B^n_m adding the two special elements a and z; finally p^2 comprises one single type, the already mentioned twice factorizable elements of which there are, by definition, m in A^n_m, but only m-1 in B^n_m.

(I) $m \geq 1$: In A^n there is at least one p^2-element $a = z$.

(II) $|\{p^o\text{-elements of } A^n_m\}| = |\{p^o\text{-elements of } B^n_m\}| = n + m$:
In A^n_m 2m of the 2n entries in the multiplication table are of the m p^2-elements; there remain $2(n-m)$ unique entries of p^1-elements. Therefore, the number of primes is $3n - 2(n-m) - m = n + m$ in A^n_m.
In B^n_m one has only one p^2-element less, but two p^1-elements $a \neq z$ more, with no change of primes.

(III) All three principal types must be present in a universal bin except that p^2 may lack in B^n (namely in B^n_1): It suffices to prove that $m < n$, i.e. that elements p^1 must always be present. Indeed, the first two substitutions must be splits "s", i.e. $b=e_1$, or $c=e_2$ of type ss (see below), therefore p^1. There must be at least two distinct p^1-elements and at least one ending in z which is of type øø (see below).

Denoting by ø the fuse type of a vertex and s the split type one gets 2x2=4 other edge types. Only pp becomes øs, all primes become sø and the p^1 become either ss or øø.

8. Normalization

The classification of vertices in a chain as splits s and fuses ø suggests classifications of edges into 4 classes ss, sø, øs, and øø. As a chain is equivalent to its dual by inversion of directions which interchanges the s and ø characters of the vertices, the edge characters øs and sø are each one invariant, indeed identical with p^2=øs and p^o=sø, and thus ss ∪ øø = p^1. So far nothing is new. However, the s-ø characterization is useful for the normalization of wordchain patterns, to weed out some irrelevant but "annoying" vertex order changes by delaying all fusions until afte execution of all already destined future splits of all letters present at any time of this process, i.e. including also iterated splits. In other words, one gives absolute priority to expansion as long as possible. One could not do this with fusing because the cofactor for fusing a present letter need not yet exist in the chain and will only be produced by a later split. However, there is total symmetry between splits anc fuses as they must finally balance out, fuses becoming prominent in the second half of the chain.

Among simultaneously possible splits one could fix a priority, say from left to right in a word, but need not; one could also shorten chains by decreeing multiple simultaneous splits as far as possible, but does not. The already introduced partial normalization has certain consequences: in each "narrow" or "bridge", i.e. a word of locally minimal length in a chain, there must be a p^2-letter with lifespan 1, which means that it appears in this minimal word only. One does not need to go into further detail because the normalization adopted here is automatically taken care of by the monomial equations presentation (see sections 6. and 15.) which takes also care of the essentials of the vertex ordering and which will be studied further.

9. General Discussion of Remaining Problem

The preceding description of the systems of monomial equations, the universal bins, and the standard wordchains are not yet sufficient for their complete direct algebraical-combinatorial construction and enumeration independent from the construction of their polyhedral graphs. This can certainly not be an easy problem in the general case because its solution would, conversely, resolve the long outstanding problem of an effective closed construction and enumeration of convex polyhedra to start with the 3-regular ones, or at least a recursive construction not requiring individual inspection of each newly constructed polyhedron for identification. However, some dents have been made in this problem as we show in the following section.

10. The Case $m = 1$

This is the so-called contraction-associativity. It has been completely resolved in earlier work, including the enumeration of the associative laws A_1^n by a complicated formula of recurrence; for $n = 2, 3, 4, 5, 6$ their numbers are 1, 5, 34, 273, 2436 respectively. For further details the reader is referred to [4] (Resumé p. 70, and § 8, p. 80), where these numbers are denoted by $D(P_{n-1})$ (D for "diagonals"). Although the enumeration is by recurrence, the construction of the A_1^n themselves is quite explicit and simple. This is also evident from the monomial form of (A_m^n): a single unconditional identity in 2 disjoint bracketings over the same sequence of n+1 letters However, cases with $m > 1$ seem never to have been considered before; but even for m=1 one has not yet evaluated what could be learned about convex polyhedra from what one knows about the A_1^n.

11. The General Case, the Successor Operation

For general 3-connected planar graphs, including especially 3-regular ones, only relatively little seems to be known. This is quite surprising considering the fact that simple 3-connected polyhedra, in particular 3-regular ones, are the most elementary "furniture" of ordinary 3-space in which we live, "the stuff from which things small and large are made", from which Descartes had to start to build his material world. Thus one can know little from this source about ordered nets, standard wordchain patterns, universal bins, and associative laws. However, there is a simple recursive construction leading from prototypes P^n to all protottypes P^{n+1} and, therefore, to the whole infinity of prototypes P, starting with the single P^1, the trihedron, or if one wants with the single P^2, the tetrahedron, or even with the single P^3, the pentahedron, better known as the triangular prism. The construction of the P^{n+1} from the P^n is by insertion of just a new edge dividing a face and any two distinct edges of it. This creates two new vertices, three new edges and one new face, resulting often, but not always, in several distinct P^{n+1} depending on the face and its edges chosen to be divided. The big differencefrom the successor construction of the naturals **N** which can also start indifferently , say with 0, or 1, or 2, etc., is of course that after n=3 the **P**-construction bifurcates and then "polyfurcates" more and more, and what is worse, in both directions; i.e. the binary (predecessor - successor) relation becomes highly many-to-many with growing n. As the **P**-construction becomes so quickly impractical and is certainly well-known it will not be treated here further. Its algebraical interpretation is probably new, but this by itself will not change the situation and will be treated elsewhere with a closer look at the details of the successor operation.

12. The Closed Ordered Model

The closure condition $B^n \Rightarrow a=z$, also called an associative law $\mathbf{A}^n$ of degree n, reproducesfrom a given ordered model O^n its parent prototype P^n enriched with a cyclic ordering of its vertices, which may or may not be a Hamiltonian circuit, and, more importantly, with a consistent labelling and "colouring" of the edge with their character types giving them and the vertices a meaningful algebraical interpretation. Or, conversely, one has provided a combinatorial-geometrical meaning of general associativity. The prototypes thus enriched are called closed ordered models. The principal letter types p^i, i=0, 1, 2, are preserved under the dualities of top-bottom and left-right direction inversion which were arbitrarily determined by the choices of v_1, and thus of v_{2n}, of the cut-edge, and of e_1 and e_2 as left and right. Each edge will still get one of the above 9 types. The closure edge $e_0 = a = z = e_{3n}$ will be of type p^2, or, equivalently, $\emptyset$s, whatever its direction.

13. The Uniqueness of the Cut-Edge

This remains preserved under closure even if the closed O^n has several letters of type p^2, i.e. m > 1. To see this consider the directions of the sticks a and z and the characters of the vertices v_1 and v_{2n}. It was understood that v_1 is a split with the product "a" directed into it, while v_{2n} is a fuse and the product "z" directed away from it. This gives the closure edge the direction $v_{2n} \to v_1$. The formerly unbounded faces f_0 and f_{n+1} have now become bounded by this common edge and cyclically oriented, say f_0 clockwise and f_n+1 anti-clockwise. All other faces $f_1, \dots, f_n$ keep their top and bottom vertices and retain their left and right sides ("sides" in the sense of"sequences of adjacent edges") parting at the top and meeting at the bottom. Thus $(\overrightarrow{v_{2n},v_1})$ is singled out as the only edge between the only two cyclically oriented faces. One can imagine a global map of "ocean currents" parting and meeting around "islands" in $f_1, f_2, \ldots, f_n$, but circling the two "continents" in f_o and f_{n+1} in opposite directions which thus are joined in the "channel" along the closing edge.

One could, however, decide otherwise. Indeed, one must invert the direction of the current in the "continental channel" making v_1 the only <u>source</u> (all 3 edges going out) and v_{2n} the only <u>sink</u> (all 3 edges coming in) in order to preserve generalisability of (equational) associativity to (quasi-ordered) semi-associativity, (see for instance, the references [3] to [6] in [4], or [3], [32] in [6]) in which transitivity without symmetry imposes direction $v_1 \rightarrow v_{2n}$ for the closure edge. This again singles out this couple of vertices (and their edge) among all others which remain splits or fuses.

14. The Vertex Ordering

The hypothesis B^n of $\mathbf{A}^n$ is a conjunction of bin statements. Therefore, it has to be independent of their order. From this point of view the vertex ordering is only a convenient auxiliary construction for the derivation of the associated wordchain. On the other hand, the vertex ordering induces the arrowing of the edges instrumental in obtaining B^n. What really matters is the characterization of the three halfedges at each vertex as l, or r, or p in a manner consistent with the general flow picture described above. This could "a priori" be achieved on the polyhedron in various ways. Once this has been done and the universal bin B^n has been constructed it is not difficult to recover the associated vertex ordering except for irrelevant indeterminacies. This will be done in the next section by the method of monomials. One will thus also obtain conciser and more familiar expressions for the associative laws $\mathbf{A}^n$.

15. Monomials and Binary Trees, the Prime Factor Decomposition

One can considerably reduce the number of letters and statements required for an A^n by using the classical bracket notation or any equivalent device. This will also eliminate parasitic bugs in the vertex ordering. One pays by complicating the statements. One starts with the "extremes" a and z, the only elements used only once in the multiplication table. Substitute according to the multiplicate table of B^n a by its factorization e_1e_2 and, similarly, z by e_je_{3n-1} or $e_{3n-1}e_j$, where e_j is the cofactor of e_{3n-1}. For each p-type letter among the e_1, e_2, e_j, e_{3n-1} substitute its factorization included in a pair of brackets according to the applicable formula $p_i \rightarrow (l_i r_i)$. Repeat this procedure as long as possible, i.e. as long as there are stil p-letters in the compounded expression. At the end all letters present must be of p^o-type, i.e. primes. Denote by M_a and M_z the thus obtained final monomials (the complete binary bracketing expressions). $M_a = M_z$ will serve as the new consequence replacing $a = z$, while one has erased in the multiplication table all used statements. Note that in the process no pp-letters have been used as each substituting letter had to have an l-type or r-type.

If this exhausts B^n one has finished, and A^n is reduced to the unconditional identity $M_a = M_z$. Both are monomials without common brackets in the same n+1 letters, namely all the primes of B^n, appearing on both sides in the same order, while the eliminated 2n letters were all the entries inside the multiplication table. That M_a and M_z are two disjoint bracketings over the same word follows from the 3-connectednes and the fact that each p^o-letter "born" in $a \to bc \to \dots \to M_a$ must join its equal in M_z in order to disappear in the contracting chain $M_z \to \dots \to z$. In fact, one has done nothing other than reconstituted the wordchain leading from a to z, perhaps with some "improvement", namely normalization which in this case reveals the simple "1-mountair profile which may not have shown itself in the original form.

The graphic equivalent of this procedure is the "growing" of 2 binary trees from the roots a and z in opposite directions by following up all uninterrupted sequences of splits starting from a and similarly for fuses starting from z (indeed coming from z the original fuses appear as splits). If this exhausts B^n then the crowns of these two trees completely overlap and form a zone composed of primes only. One may call these trees prime-leaved trees and denote them by their monomials M_a and M_z which determine the trees completely. But what if there remain entries in B^n?

Indeed, each letter (edge) of B^n, say e, is root of its well determined binary prime-leaved tree M_e. If e is a prime $M_e \equiv e$ is its own such tree and monomial. If e is a p^1-letter it also uniquely determines its prime-leaved tree by binary factorization. The first one is uniquely determined and so are the later ones since all appearing letters come from factorizations, thus they cannot be p^2, and are, therefore, either primes or p^1-letters.

Moreover, as e is p^1 it must have at the "other" end a cofactor, say d. Then $de = g$ or $ed = g$, where g is a p-letter, too, d either a prime or a p^1-letter, i.e. also having a uniquely determined tree M_d. One has, therefore, a tree $M_g = M_d M_e$ or $M_e M_d$ and M_g is larger than M_e. If g is p^2 it cannot have a cofactor and M_g is a maximal tree; if not it is p^1 and itself contained in a larger tree. Continuing with comultiplication as long as one can one must finally be stopped at a p^2-type letter determining a maximal tree, say M_q. Of course if e was already p^2 one would already have such a q. Thus among the remaining entries there must be at least one such q and, therefore, even at least two entries. Hence one can conclude:

If B^n is not yet exhausted there must be entries and relations of the form $e\,e^* = q = q_1 = e'e''$ for $q = q_1$ and, possibly, other p^2-letters $q_2, q_3, \dots$, each one to be treated twice as a and z have been treated before. This means that e, e*, e', e" have to be treated as e_1, e_2, e_j, e_{3n-1} (or e_{3n-1}, e_j) before, by replacing each still present p-letter by its two direct factors enclosed in brackets, till one is stopped when one has arrived at two prime-leaved trees M_q and M'_q, which may be of different length. Similarly, one obtains M_{q_2}, M'_{q_2}, etc. till B^n is completely exhausted. One has arrived at a collection of monomial equations

$$C^n: M_{q_1} = M'_{q_1}, M_{q_2} = M'_{q_2}, \dots ;$$

their conjunction constitutes the new hypothesis and the implication $A^n: C^n \Rightarrow M_a = M_z$ the monomial equations form of an associative law. One can provide a further subscript m indicating the total number of equations. This system is, in general, non-homogeneous: its monomials may have different degrees; they must have all letters different in each equation except $M_a = M_z$; they form a completely disjoint system of brackets, i.e. all submonomials are distinct, although each prime, and primes only, just appear twice in the whole system . The associated wordchain profile is more pitted with "valleys", each one having under its bottom a short-lived p^2-letter. M_a and M_z are now monomials over distinct words, namely the first and the last "peaks" (= λ_i having a local maximum) of the wordchain, although they may have some common subwords. Each letter of M_a, M_z, M_{q_k}, M'_{q_k} must finally join its equal in some other monomial in a consistent order inducing also a well defined direction in each double-tree. The whole wordchain will be recomposed from these pieces like some picture puzzles or a planar wiring system from such subsystems by joining end-wires without overcrossing to their single correspondents in other subsytem terminals. It will be easy because they correspond in whole segments in which words overlap.

16. Homomorphic Images of Bins or "Colourings"

By closing the special standard wordchain into a circle, or rather on a cylinder mantle, i.e. by imposing A^n, the 2 extremal pieces M_a and M_z melt into one double-tree M_z—...—z=a —...— M_a and one may consider $a=z=q_m$. By imposing A^n_m on its B^n_m one obtains a new universal bin A^n_m, a homomorphic image of B^n_m, indeed its greatest associative homomorphic image with $|A^n_m| = 3n$: every associative homomorphic image of B^n_m is also a homomorphic image of this A^n_m.

One is also interested in the non-associative homomorphic images of B^n, i.e. those which still "separate" a and z.

The concept of homomorphism for general bins, even that of universal bins, needs special attention. This is an important topic, indeed a crucial one for further applications of this theory, which one must leave open here for a later occasion, except for some simple but enlightening examples. By calling the image elements under a bin homomorphism "colours" one puts in evidence that the concept of bin homomorphism generalises that of edge colouring in a certain sense. In such "colourings" adjacent edges may, or may not have the same colour, but colours will compose according to a bin, i.e. constitute a "colour-bin".

The following interesting remark gives a new, purely algebraical equivalent to the famous 4-colour conjecture. (Or should one rather say "theorem" after the recent "proof by computer"?)

It is well-known that the general 4-colouring of faces (F_4) reduces to that for **P** and is equivalent to 3-colouring of edges (E_3), (see, for instance, [1] pp.267; the author is obliged to Professor G. Dirac for calling his attention to this theorem). However, E_3 is obviously equivalent to the existence of surjective homomorphisms

$\epsilon^n : A^n \rightarrow C$, C the "colour bin" of colours, say X,Y,Z, with the relations XY=YX=Z, XZ=ZX=Y, YZ=ZY=X. (One may remark that C is a "truncated" 4-group, i.e. with its identity element excised.) One has thus

$$F_4 \Longleftrightarrow E_3 \Longleftrightarrow \epsilon \quad .$$

17. The Prototypes P^n

As already mentioned, there is only one prototype for each of the lower degrees n = 1, 2, and 3. They and some of higher degree are wellknown.

n = 1: the trihedron P^1 has three digons (i.e. biangles or twosides).
Think of three meridians trisecting the globe. Its two vertices are the poles. It is 3-valent and 3-connected.

Remark: No digon ($=f^2$) can appear for n > 1 because of 3-connectedness.

n = 2: the tetrahedron P^2 has four 3-angles: $4f^3$.

n = 3: the pentahedron P^3, better known as triangular prism, has two 3-angles and three 4-angles ($2f^3 + 3f^4$).

n = 4: There are two hexahedra P^4:
(a) the 4-angular prism, for brief the "cube" $6f^4$, and
(b) the pentagonal "half-prism" $2f^3 + 2f^4 + 2f^5$ of two pentagons with one common edge and two triangles and two quadrangles between them.

Remark: The tetrahedron is a triangular half-prism, the triangular prism a quadrangular half-prism. For $n \geq 4$ the n-gonal prism and the (n+1)-gonal half-prism are distinct.

n = 5: There are five heptahedra P^5, two successors of the cube (which admits only two distinct stroke operations):
(a) the pentagonal prism $5f^4 + 2f^5$,
(b) truncated cube (truncated by cutting off a tetrahedral corner) $1f^3 + 3f^4 + 3f^5$

We note that (a) and (b) are also successors of the pentagonal half-prism, while the 3 following ones, each with hexagons, are successors of the pentagonal halfprism only :

(c) the hexagonal halfprism $2f^3+3f^4+2f^6$, the two f^6 with a common edge

and 2 less wellknown polyhedra

(d) $2f^3+2f^4+2f^5+1f^6$, and

(e) $3f^3+3f^5+1f^6$.

With increasing n the number of prototypes increases steeply: card P^n for $n=6,7,8$ are 14,50,233. (The next two numbers (for $n=9,10$) are not sure and beyond they are unknown [3]).The "polynomial" notation $\Sigma c_i f^i = (c_3, c_4, c_5 \ldots)$ becomes insufficient because of the existence of allomorphic polyhedra, i.e. combinatorially distinct polyhedra, with equal numbers c_i for all f^i, the first instance being also the only one for $n=6$, namely two distinct octahedra (2,2,2,2). One has two adjacent 4-gons separating two 5-gons, the other one has the two 5-gons adjacent and the two 4-gons separated and different and lesser symmetry. For larger n allomorphy becomes the usual thing. Similarly, completely asymmetric polyhedra become more and more frequent.

18. The First Associative Laws and some Higher Degree Examples

For the associative laws the situation is similarly deceptively simple for degrees 1 and 2 - no surprises because essentially well-known:

$n=1$: There is only one 0^1, the digon with its two sticks. It yields only one A^1,

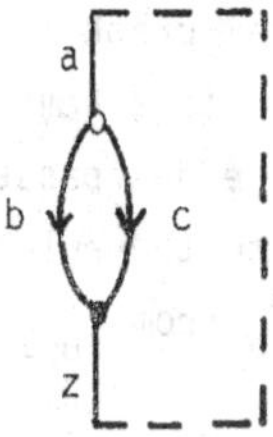

namely $(b,c;a)$, $(b,c;z) \Rightarrow a=z$, i.e. the well-known law of uniformity of binary operations; in other words A^1 singles out bins among ternary relations. One can now take A^1 for granted and can write, as usual, the relations as equalities. Should a digon appear in a wordchain it will be replaced by a simple segment and the number of vertices reduced by two, the number of edges by three and the number of faces by one.

$n=2$: All edges of the tetrahedron are homologuous and there is only one open net,

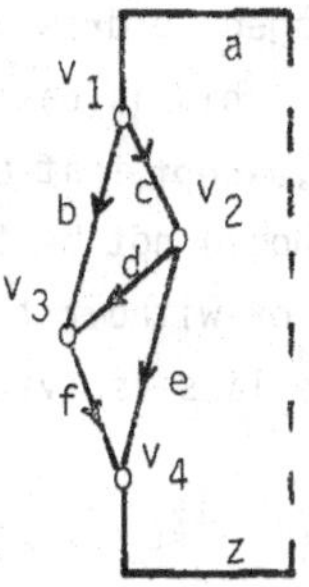

namely two adjacent triangles with two sticks, or, even simpler, divide the 0^1 above by a "stroke" of the successor operation and obtain 0^2. Either v_2 is at the left and v_3 at the right or vice versa, depending on the inclination of the stroke: or even better, as one wishes to direct the stroke. Because of duality this does not matter and one obtains only one associative law:

A^2: $a=bc$, $c=de$, $bd=f$, $fe=z \Rightarrow a=z$, or, by substitution $b(de)=(bd)e$, the ordinary associative law. From now on we take A^2 for granted, and replace two adjacent triangles in a wordchain by a segment reducing the number of vertices, edges and faces by four, six and two respectively.

$n=3$: P^3 has two kinds of edges , six 3-gon sides and three edges separating 4-gons yielding two open nets N^3: either (a) a 4-gon between two 3-gons, or (b) a

3-gon between two 4-gons; or, even quicker, by the "stroke" operation applied to N^2, say to the upper triangle (this suffices by symmetry) to obtain N^3; or similarly, a conveniently arrowed stroke applied to an O^2 yielding an O^3. The reader may draw the two N^3 (a) and (b) and their five orderings O_1^3 yielding the well-known five $\mathbf{A}_1^3$ written out in former work. We remark that the unique self-dual $\mathbf{A}_1^3$: $((fg)e)c = f(g(ec))$ derived from $N^3(a)$, the four others from $N^3(b)$.

New are the first two instances of $m > 1$. They are derived from $N^3(a)$ which yields two dual O_2^3 and the corresponding $\mathbf{A}_2^3$:

$$(\mathbf{A}_2^3)\qquad \begin{aligned} bd = gh &\Rightarrow b(de) = g(he) \\ ec = gh &\Rightarrow (de)c = (dg)h. \end{aligned}$$

The difference between $O_1^3(a)$ and the two $O_2^3(a)$ is in the arrowing of the quadrangle between the two triangles: In $O_1^3(a)$ its left and its right side (= sequences of edges) have both edges. In O_2^3 one side has 3 edges, the other one edge; the exchange of left and right yields the two dual cases. Again, all 7 A^3 now granted, no N^3, i.e. adjacent sequences # Δ # or Δ # Δ, will be admitted for $n > 3$.

= 4: There is no need to write out the 34 $\mathbf{A}_1^4$ listings the 34 couples of disjoint bracketings out of the 14 binary bracketings over 5 letters. One observes that both P^4 admit cycles entering and leaving each country f_k once and only once through distinct edges, starting, say, in f_0 and returning to it by passing from f_{n+1}. The "tour" corresponds to a Hamiltonian circuit in the dual graph (a simplicial or triangulated polyhedron) and divides the "globe" into 2 hemispheres, say a northern and a southern. It induces an orientation in all edges passed, say from north to south, and also in the remaining edges according to the rules. It distinguishes also the northern end of $e_0 = e_{3n}$ separating f_0 from f_{n+1} as the "North Pole", the other as the "South Pole".

For the remainder the reader is referred to the listed examples of associative laws with indication of their prototypes. The reader is encouraged to draw lettered figures of the ordered models and to tabulate their bins. This is easily done from their monomial expressions following the conventions, except that one has replaced for convenience $e_1, e_2, e_3, \ldots$ by $b, c, d\ldots$. It should not be too hard to complete some of these lists of associative laws, with or without the help of prototypes. Indeed, one can first construct associative laws as systems of monomial equations and then derive their prototypes.

Two $\mathbf{A}_1^4$: $((ij)(fg))c = i((jf)(gc))$ (0,6) (="cube")

$(((jk)h)e)c = j(k(h(ec)))$ (2,2,2) (="pentagonal halfprism")

Two $\mathbf{A}_2^4$: ef = ij ⇒ (de)(fg) = (di)(jg) (0,6)

ef = ij ⇒ d(e(fg)) = ((di)j)g
Two $\mathbf{A}_3^4$: ec = gh, dg = jk ⇒ (de)c = j(kh)
bd = gh, he = jk ⇒ b(de) = (gj)k
} (2,2,2)

One $\mathbf{A}_2^5$: gc = (lm)k ⇒ (d(fg))c = ((df)l)(mk) (2,2,2,1)

One $\mathbf{A}_3^5$: gc = ij, fi = lm ⇒ (d(fg))c = d(l(mj))
One $\mathbf{A}_4^5$: bd = gh, he = jk, gj = mn ⇒ b(de) = m(nk)
} (2,3,0,2)(hexagonal halfprism)

One $\mathbf{A}_3^6$: b(df) = jk, kg=(op)n ⇒ b(d(fg))=((jo)p)n $(2,2,2,2)_1$
Six $\mathbf{A}_4^6$: bd=g(ij), je=lm, g(il)=pq ⇒ b(de)=P(qm) $(2,2,2,2)_2$
] allomorphic polyhedra, both with f^6/f^6 "cut".

bd=(jk)h, he=lm, kl=op ⇒ b(de)=j(o(pm)) (2,2,3,0,1)

ec=gh, dg=(mn)k, kh=op ⇒ (de)c=m((no)p)
ec=gh, dg=jk, kh=m(op) ⇒ (de)c=(j(mo))p
} (2,3,1,1,1)

ec=gh, dg=j(lm), l(mh)=pq ⇒ (de)c=(jp)q
ec=g(lm), dg=jk, (kl)m=pq ⇒ (de)c=(jp)q
} (3,1,2,1,1)

The last two P^6 are the smallest (completely) asymmetrical planar simple (i.e. 3-valent) and 3-connected graphs (octahedral) in which no two edges are homologuous. This was first remarked by R. Frucht (Compositio Math. 6 (1938), 239-250); communicated by A. Hill, London). Thus each of their 18 edges leads to a distinct N^6.

Acknowledgements: Thanks are due to the hearers of my talks on this subject this last half year at various universities, in particular to Prof. M. Perles and his Seminar (Jerusalem) and Prof. A. Ginzburg (Tel Aviv), for their interest and valuable remarks; also to Mr. A. Hill, an artist in London, for showing me some (completely) asymetric graphs and polyhedra. Particular thanks are due to the organizers of the Combinatorics Colloquium at Rauischholzhausen, the Mathematisches Institut der Universität Giessen, for hospitality and for permitting to present this paper in spite of its too late entry; also to the Matematisk Institut Aarhus for its facilities and hospitality, in particular due to the efforts of Prof. G.Dirac; this paper profited from many conversations with him. The final exposition was considerably improved following the remarks of the referee; the author and the readers alike owe him thanks. Last but not least the author feels grateful to his friends, Professors Ken and Saris Magill, for their hospitality and unfailing friendship. They all made this endeavor possible at this late stage of the author's life, in spite of obstructing difficult circumstances.

References

1. C. Berge, Graphs and Hypergraphs. North Holland (2nd edition) 1976.

2. Paul W. Bunting, Jan van Leeuwen and Dov Tamari, Deciding Associativity for Partial Multiplication Tables of Order 3. Mathematics of Computation, 32 (1978), 593-605.

3. Branko Grünbaum, Convex Polytopes. Interscience Publ.1967, Ch.13 and Table 1, p.424.

4. Danièle Huguet et Dov Tamari, La Structure Polyédrale des Complexes de Parenthésages. J. of Combinatorics, Information and System Sciences, 3 (1978), 69-81.

5. Kevin E. Osondu, Symmetrisations of Semigroups. Semigroup Forum, 24 (1982), 67-75.

6. Dov Tamari, The Associativity Problem for Monoids and the Word Problem for Semigroups. In "Word Problems", North Holland Publ.Co., Amsterdam 1973, 591-607.

Elusive Properties

E. Triesch,
Lehrstuhl für Angewandte Mathematik
RWTH Aachen, Templergraben 64
5100 Aachen, Federal Republic of Germany

1. Introduction:

We consider properties P of subsets X of a t-element set $T (t \in \mathbb{N})$. Imagine that such a property P is given and that there are two players A("Algy") and C("Constructor") playing the following game: A asks C questions of the form: "Is $x \in X$?" about a hypothetical set $X \subseteq T$. The prupose of those questions is to determine whether or not the evolving set X has property P. A wants to minimize the number of questions he asks and C wants to force A to ask as many questions as possible by providing very inconvenient answers. The number of questions which are asked in the game if both players play optimally is called the complexity of P and we denote it by c(P). P is called elusive if $c(P) = t$. If you think of P as a Boolean function you recognize immediately that "c(P) is a lower bound on the time any algorithm recognizing P must take in the worst case, on any model of machine where no two operations can take place at the same time"[RV]. A very important special case occurs if T is the set of two-element-subsets of an n-element set V, that is, the set of edges of a complete graph K_n. We are then asking for the complexity of graph properties. Most graph properties which are investigated in practice contain with a graph G each isomorphic copy of G. I want to call such graph properties invariant. (In what follows we always identify a property with the set of sets $X \subseteq T$ having this property.) We call a property P monotonic if P or its complement contain with a set X all its subsets. P is non-trivial if $P \neq \emptyset$ and $P = \mathcal{P}(T)$. For the rest of this paper we assume every property to be non-trivial. At any moment the situation of the game can be characterized by a pair (E,N) with $E \subseteq T$, $N \subseteq T$ and $E \cap N = \emptyset$. We think of E as the set of elements of T which are known to be elements of X, and of N as the set of elements of T which are known not to be elements of X. An algorithm φ is a function which chooses for each such pair (E,N) with $E \cup N \neq T$ an element of $(E \cup N)^c$, the complement of $E \cup N$, which we interpret as the next probed element. A strategy ψ is a function, which assigns to each pair ((E,N),x) with (E,N) as above and $x \in (E \cup N)^c$ one of the pairs $(E \cup \{x\},N)$ and $(E,N \cup \{x\})$.

For an algorithm φ and a strategy ψ we denote by $c(P;\varphi,\psi)$ the number of questions which are asked in the game if A uses algorithm φ and C strategy ψ. Let $c(P;\psi) = \min\{c(P;\varphi,\psi) \mid \varphi \text{ algorithm}\}$. Obviously, we have $c(P) = \max\{c(P;\psi) \mid \psi \text{ strategy}\}$

2. Duality:

For any property P we define the dual property P^* by $P^* = \{X \subseteq T \mid X^c \in P\}$. (Notice that this definition differs from the notion of duality in [BBL]). We obviously have $P^{**} = P$. Furthermore, the following lemma holds.

Lemma: $c(P^*) = c(P)$.

Proof: For any strategy ψ define a strategy ψ^* by

$$\psi^*((E,N),x) = (E \cup \{x\},N) \Leftrightarrow \psi((E,N),x) = E,N \cup \{x\}).$$

Let φ be any algorithm and assume that the game for testing P ends with the pair (E,N) if A uses φ and C uses ψ. We have $c(P;\varphi,\psi) = |E \cup N|$ and

$$\forall X\ (E \subseteq X \subseteq N^c \Leftrightarrow X \in P) \qquad (1)$$

or

$$\forall X\ (E \subseteq X \subseteq N^c \Rightarrow X \notin P) \qquad (2)$$

By using φ and ψ^* in the game for testing P^*, we get the following situation after $|E \cup N|$ questions: Our characterizing pair is (E',N') with E' = N and N' = E. Now suppose that $E' \subseteq Y \subseteq N'^c$. Taking complements we get $N' \subseteq Y^c \subseteq E'^c$ or $E \subseteq Y^c \subseteq N^c$. If (1) is true we get

$$\forall Y\ (E' \subseteq Y \subseteq N'^c \Rightarrow Y \in P^*)$$

and if (2) is true we have

$$\forall Y\ (E' \subseteq Y \subseteq N'^c \Rightarrow Y \notin P^*).$$

In any case the game is finished. So

$$c(P;\varphi,\psi) \geq c(P^*;\varphi,\psi^*)$$

for all C and ψ and, therefore,

$$c(P) \geq c(P^*) \geq c(P^{**}) = c(P).$$

3. Some important Results:

In 1973 Rosenberg [R] conjectured that there is a $\gamma > 0$ such that for all (non-trivial) invariant graph properties P

$$c(P) \geq \gamma \cdot n^2.$$

In [BBL] you find some counterexamples to this conjecture the first of which was constructed by Aanderaa (for directed graphs). Aanderaa and Rosenberg then formulated together the following conjecture:

There is a $\gamma > 0$ such that for all monotone, invariant graph properties P we have

$$c(P) \geq \gamma \cdot n^2.$$

Their conjecture was proved in 1975 by Rivest and Vuillemin [RV] with $\gamma = 1/16$. Kleitman and Kwiatkowski [KK] improved the value of γ to 1/9 if n is large. It is conjectured that all monotone, invariant graph properties are elusive ([BBL],p.4). There is even a more general conjecture of Rivest and Vuillemin [RV]: If $P \subseteq p(T)$ is invariant under a transitive permutation group on T and $\{\emptyset,T\} \nsubseteq P$ and $\{\emptyset,T\} \nsubseteq P^c$ then P is elusive.

Rivest and Vuillemin proved their conjecture for $|T|$ a prime power. In fact, this result is the most important step in their proof of the Aanderaa-Rosenberg-conjecture. There are also some results proving the elusiveness of some special properties: A theorem of Hopcroft and Tarjan shows that planarity is elusive ([BBL],p.7). A result of Bollobás [B1] shows the elusiveness of the properties "$cl(G) \geq r$" and "$\chi(G) \geq r$", where $cl(G)$ and $\chi(G)$ are the clique number and the chromatic number of G, respectively.

4. The "Simple Strategy":

Let us define a strategy ψ_o by $\psi_o((E,N),x) = (E \cup \{x\},N) \Leftrightarrow \exists X \in P(E \cup \{x\} \subseteq X \subseteq N^c)$. In [MW 1,2] and [B2], pp.4o6-4o7, this strategy is discussed and it is claimed that for $T \notin P$ ψ_o is winning strategy for C (i.e. $c(P;\psi_o) = t$) if and only if P satisfies the following condition:

$$\forall X \in P \quad \forall x \in X \quad \exists y \in X^c \quad \exists Y \in P \quad ((X \setminus \{x\}) \cup \{y\} \subseteq Y) \qquad (3)$$

Now this condition is indeed sufficient for the elusiveness of P but not necessary. We have to change it a little in order to obtain necessity, too. The right condition is:

$$\forall X \in P \quad \forall x \in X \quad ((X \setminus \{x\}) \in P \Rightarrow \exists y \in X^c \quad \exists Y \in P((X \setminus \{x\}) \cup \{y\} \subseteq Y)) \qquad (4)$$

The restriction that $T \notin P$ is unnecessary.

Theorem 1: ψ_o is a winning strategy for C iff P satisfies condition (4).

Proof. The proof is almost the same as the one given in [B2].

Sufficiency: Let us suppose that the game ends after $s < t$ steps with the characterizing pair (E,N). Then one of the conditions (1) and (2) must hold. Since P is not empty it follows immediately from the definition of ψ_o that (1) must hold. So, if we define $X = N^c$ and choose $x \in X \setminus E$ we have that $X \in P$ and $X \setminus \{x\} \in P$. Now condition (4) implies that there are $y \in N$ and $Y \in P$ with $(X \setminus \{x\}) \cup \{y\} \subseteq Y$. But then it is impossible that ψ_o chose y to be a non-element, a contradiction.

Necessity: Let us suppose that (4) does not hold. We get $X \in P$ and $x \in X$ with $X \setminus \{x\} \in P$ such that for all $y \in X^c$ and all $Y \in P$ $(X \setminus \{x\}) \cup \{y\} \nsubseteq Y$. Now Algy first probes the elements of $X \setminus \{x\}$ and then those of X^c. This results in the pair $(E,N) = (X \setminus \{x\}, X^c)$ after t-1 steps. Obviously the game (and therefore the proof) is finished.

We also want to write down the dual form of Theorem 1 which is not completely obvious. We have:

$$\psi_o^*((E,N),x) = (E,N \cup \{x\}) \Leftrightarrow \exists X \in P(E \subseteq X \subseteq (N \cup \{x\})^c).$$

Theorem 1^*: ψ_0^* is a winning strategy for C if and only if P satisfies the following condition:

$$\forall X \in P \; \forall y \in X^c (X \cup \{y\} \in P \Rightarrow \exists x \in X \; \exists Y \in P((X \cup \{y\}) \setminus \{x\} \supseteq Y)). \quad (5)$$

We can use Theorem 1^* to prove the following generalization of a theorem of Bollobás and Eldridge [BE]:

Theorem 2: Let P be invariant under a transitive permutation group on T. If P satisfies the following condition (6) then P is elusive:

$$\forall X_1, X_2, X_3 \in P(X_1 \subseteq X_2 \subseteq X_3 \Rightarrow |X_2 \setminus X_1| \neq 1 \text{ or } |X_3 \setminus X_2| \neq 1) \quad (6)$$

Proof: Let us assume that our strategy ψ_0^* fails for otherwise the assertion holds. Negating condition (5) implies that there exist $X \in P$ and $y \in X^c$ with $X \cup \{y\} \in P$, but $(X \cup \{y\}) \setminus \{x\}$ does not contain a $Y \in P$ for all $x \in X$. Since P is invariant under a transitive permutation group we can assume that the first probed element is y. We choose y to be not an element and from now on choose each probe z to be an element if and only if $z \in X$. It is now easy to see that (by condition (6)) the constructor wins.

5. 2-connectedness is elusive:

In [B2], Theorem 1,2 (vi), it is claimed that P = "2-connectedness" is an elusive graph property and that this could be proved by applying ψ_0 to P or its complement. It is easy to see that this not the case. It is the purpose of this final paragraph to present a strategy showing that P is elusive. Here it is:

$$\psi_1((E,N),x) = (E, N \cup \{x\}) \Leftrightarrow x \text{ closes a cycle in the graph with edge-set } E \cup \{x\}$$

in which not all diagonals have been probed yet.

Theorem 3: ψ_1 is a winning strategy for the constructor C.

Proof: Suppose on the contrary that C looses in time $c = c(P; \psi_1)$, (E,N) being the characterizing pair of the game at that time.

(i) If condition (1) holds, the graph G with edge set E is 2-connected. Let $e = \{u,v\}$ be an unprobed edge. It follows from the 2-connectedness of G that there is a cycle in G containing u and v. Hence e is an unprobed diagonal in that cycle which is impossible by the definition of ψ_1.

(ii) So let us assume that condition (2) holds. It is easy to see that G is connected. Note also that no block of G contains unprobed edges by the same kind of reasoning as in (i). Consider two blocks B and B' of G having the same cutvertex w. Denote by V(B) and V(B') the vertex sets of B and B', respectively. We show that there is an unprobed edge joining nodes in $V(B) \setminus \{w\}$ and $V(B') \setminus \{w\}$. This shows that we can add edges from $(E \cup N)^c$ to G to obtain a 2-connected graph, a fact which obviously contradicts condition (2). Suppose that no such edge exists. Then all edges between $V(B) \setminus \{w\}$ and $V(B') \setminus \{w\}$ are in N. Let $e = \{u,v\}$ be the last probed edge in $V(B) \cup V(B')$. (All edges in this set are probed at time c).

By symmetry there are only the following two essentially differend cases:

Case 1: $u \in V(B) \setminus \{w\}$, $v \in V(B') \setminus \{w\}$. ψ_1 would choose e to be an edge, a contradiction to $e \in N$.

Case 2: $u, v \in V(B)$. Since e is the last probed edge in V(B), ψ_1 chooses e to be an edge. Now imagine that Algy asks the edges in the same order as before but that he omits e. A moment's thought shows that with this modification the characterizing pair (E',N') at time c-1 satisfies

$$N' \supseteq N \text{ (and therefore } E' \subseteq E \setminus \{e\}).$$

But then the game is finished which contradicts the minimality of $c = c(P; \psi_1)$ The proof is complete.

References:

[BBL] Best, M.R., van Emde Boas, P and Lenstra, Jr. H.W.:
"A sharpened version of the Aanderaa-Rosenberg conjecture",
Math. Centrum, Amsterdam, 1974.

[B1] Bollobás, B.:
"Complete Subgraphs are elusive",
J.Comb.Th. Ser.B. 21 (1976), 1-7.

[B2] Bollobás, B.:
"Extremal graph theory".
Acad.Press, London, 1978.

[BE] Bollobás, B. and Eldridge, S.E.:
"Problem", in: "Proc.Fifth British Combinatorial Conference",
(Nash-Williams, C.St.J.A. and Sheehan, J., eds.),
Utilitas Math., Winnipeg, 1976, 689-691.

[KK] Kleitman, D.J. and Kwiatkowski, D.J.:
"Further results on the Aanderaa-Rosenberg conjecture"
J. Comb.Th. Ser.B. 28 (1980), 85-95.

[MW1] Milner, E.C. and Welsh, D.J.A.:
"On the computational complexity of graph theoretical properties",
Univ. of Calgary, Res.Paper No.232, June, 1974.

[MW2] Milner, E.C. and Welsh, D.J.A.:
"On the computational complexity of graph theoretical properties",
in: "Proc.Fifth Brithish Combinatorial Conference".
(Nash-Williams, C.St.J.A. and Sheehan, J., eds.),
Utilitas Math., Winnipeg, 1976, 471-487.

[RV] Rivest, R.L. and Vuillemin, J.:
"On recognizing graph properties from adjacency matrices",
Theor.Comput.Sci. 3 (1976/77), 371-384.

. 817: L. Gerritzen, M. van der Put, Schottky Groups and Mumford ırves. VIII, 317 pages. 1980.

I. 818: S. Montgomery, Fixed Rings of Finite Automorphism Groups Associative Rings. VII, 126 pages. 1980.

I. 819: Global Theory of Dynamical Systems. Proceedings, 1979. ited by Z. Nitecki and C. Robinson. IX, 499 pages. 1980.

I. 820: W. Abikoff, The Real Analytic Theory of Teichmüller Space. , 144 pages. 1980.

I. 821: Statistique non Paramétrique Asymptotique. Proceedings, 79. Edited by J.-P. Raoult. VII, 175 pages. 1980.

. 822: Séminaire Pierre Lelong–Henri Skoda, (Analyse) Années 78/79. Proceedings. Edited by P. Lelong et H. Skoda. VIII, 6 pages, 1980.

. 823: J. Král, Integral Operators in Potential Theory. III, 171 ges. 1980.

I. 824: D. Frank Hsu, Cyclic Neofields and Combinatorial Designs. 230 pages. 1980.

I. 825: Ring Theory, Antwerp 1980. Proceedings. Edited by F. van staeyen. VII, 209 pages. 1980.

I. 826: Ph. G. Ciarlet et P. Rabier, Les Equations de von Kármán. 181 pages. 1980.

I. 827: Ordinary and Partial Differential Equations. Proceedings, 78. Edited by W. N. Everitt. XVI, 271 pages. 1980.

I. 828: Probability Theory on Vector Spaces II. Proceedings, 1979. ited by A. Weron. XIII, 324 pages. 1980.

I. 829: Combinatorial Mathematics VII. Proceedings, 1979. Edited R. W. Robinson et al.. X, 256 pages. 1980.

I. 830: J. A. Green, Polynomial Representations of GL_n. VI, 118 ges. 1980.

I. 831: Representation Theory I. Proceedings, 1979. Edited by Dlab and P. Gabriel. XIV, 373 pages. 1980.

I. 832: Representation Theory II. Proceedings, 1979. Edited by Dlab and P. Gabriel. XIV, 673 pages. 1980.

I. 833: Th. Jeulin, Semi-Martingales et Grossissement d'une tration. IX, 142 Seiten. 1980.

I. 834: Model Theory of Algebra and Arithmetic. Proceedings, 79. Edited by L. Pacholski, J. Wierzejewski, and A. J. Wilkie. VI, 0 pages. 1980.

I. 835: H. Zieschang, E. Vogt and H.-D. Coldewey, Surfaces and anar Discontinuous Groups. X, 334 pages. 1980.

ol. 836: Differential Geometrical Methods in Mathematical Physics. oceedings, 1979. Edited by P. L. García, A. Pérez-Rendón, and M. Souriau. XII, 538 pages. 1980.

ol. 837: J. Meixner, F. W. Schäfke and G. Wolf, Mathieu Functions nd Spheroidal Functions and their Mathematical Foundations rther Studies. VII, 126 pages. 1980.

ol. 838: Global Differential Geometry and Global Analysis. Pro- edings 1979. Edited by D. Ferus et al. XI, 299 pages. 1981.

ol. 839: Cabal Seminar 77 – 79. Proceedings. Edited by A. S. echris, D. A. Martin and Y. N. Moschovakis. V, 274 pages. 1981.

ol. 840: D. Henry, Geometric Theory of Semilinear Parabolic Equa- ons. IV, 348 pages. 1981.

ol. 841: A. Haraux, Nonlinear Evolution Equations - Global Behaviour Solutions. XII, 313 pages. 1981.

ol. 842: Séminaire Bourbaki vol. 1979/80. Exposés 543–560. , 317 pages. 1981.

ol. 843: Functional Analysis, Holomorphy, and Approximation heory. Proceedings. Edited by S. Machado. VI, 636 pages. 1981.

ol. 844: Groupe de Brauer. Proceedings. Edited by M. Kervaire and . Ojanguren. VII, 274 pages. 1981.

Vol. 845: A. Tannenbaum, Invariance and System Theory: Algebraic and Geometric Aspects. X, 161 pages. 1981.

Vol. 846: Ordinary and Partial Differential Equations, Proceedings. Edited by W. N. Everitt and B. D. Sleeman. XIV, 384 pages. 1981.

Vol. 847: U. Koschorke, Vector Fields and Other Vector Bundle Morphisms – A Singularity Approach. IV, 304 pages. 1981.

Vol. 848: Algebra, Carbondale 1980. Proceedings. Ed. by R. K. Amayo. VI, 298 pages. 1981.

Vol. 849: P. Major, Multiple Wiener-Itô Integrals. VII, 127 pages. 1981.

Vol. 850: Séminaire de Probabilités XV. 1979/80. Avec table générale des exposés de 1966/67 à 1978/79. Edited by J. Azéma and M. Yor. IV, 704 pages. 1981.

Vol. 851: Stochastic Integrals. Proceedings, 1980. Edited by D. Williams. IX, 540 pages. 1981.

Vol. 852: L. Schwartz, Geometry and Probability in Banach Spaces. X, 101 pages. 1981.

Vol. 853: N. Boboc, G. Bucur, A. Cornea, Order and Convexity in Potential Theory: H-Cones. IV, 286 pages. 1981.

Vol. 854: Algebraic K-Theory. Evanston 1980. Proceedings. Edited by E. M. Friedlander and M. R. Stein. V, 517 pages. 1981.

Vol. 855: Semigroups. Proceedings 1978. Edited by H. Jürgensen, M. Petrich and H. J. Weinert. V, 221 pages. 1981.

Vol. 856: R. Lascar, Propagation des Singularités des Solutions d'Equations Pseudo-Différentielles à Caractéristiques de Multipli- cités Variables. VIII, 237 pages. 1981.

Vol. 857: M. Miyanishi. Non-complete Algebraic Surfaces. XVIII, 244 pages. 1981.

Vol. 858: E. A. Coddington, H. S. V. de Snoo: Regular Boundary Value Problems Associated with Pairs of Ordinary Differential Expressions. V, 225 pages. 1981.

Vol. 859: Logic Year 1979–80. Proceedings. Edited by M. Lerman, J. Schmerl and R. Soare. VIII, 326 pages. 1981.

Vol. 860: Probability in Banach Spaces III. Proceedings, 1980. Edited by A. Beck. VI, 329 pages. 1981.

Vol. 861: Analytical Methods in Probability Theory. Proceedings 1980. Edited by D. Dugué, E. Lukacs, V. K. Rohatgi. X, 183 pages. 1981.

Vol. 862: Algebraic Geometry. Proceedings 1980. Edited by A. Lib- gober and P. Wagreich. V, 281 pages. 1981.

Vol. 863: Processus Aléatoires à Deux Indices. Proceedings, 1980. Edited by H. Korezlioglu, G. Mazziotto and J. Szpirglas. V, 274 pages. 1981.

Vol. 864: Complex Analysis and Spectral Theory. Proceedings, 1979/80. Edited by V. P. Havin and N. K. Nikol'skii, VI, 480 pages. 1981.

Vol. 865: R. W. Bruggeman, Fourier Coefficients of Automorphic Forms. III, 201 pages. 1981.

Vol. 866: J.-M. Bismut, Mécanique Aléatoire. XVI, 563 pages. 1981.

Vol. 867: Séminaire d'Algèbre Paul Dubreil et Marie-Paule Malliavin. Proceedings, 1980. Edited by M.-P. Malliavin. V, 476 pages. 1981.

Vol. 868: Surfaces Algébriques. Proceedings 1976–78. Edited by J. Giraud, L. Illusie et M. Raynaud. V, 314 pages. 1981.

Vol. 869: A. V. Zelevinsky, Representations of Finite Classical Groups. IV, 184 pages. 1981.

Vol. 870: Shape Theory and Geometric Topology. Proceedings, 1981. Edited by S. Mardešić and J. Segal. V, 265 pages. 1981.

Vol. 871: Continuous Lattices. Proceedings, 1979. Edited by B. Bana- schewski and R.-E. Hoffmann. X, 413 pages. 1981.

Vol. 872: Set Theory and Model Theory. Proceedings, 1979. Edited by R. B. Jensen and A. Prestel. V, 174 pages. 1981.

Vol. 873: Constructive Mathematics, Proceedings, 1980. Edited by F. Richman. VII, 347 pages. 1981.

Vol. 874: Abelian Group Theory. Proceedings, 1981. Edited by R. Göbel and E. Walker. XXI, 447 pages. 1981.

Vol. 875: H. Zieschang, Finite Groups of Mapping Classes of Surfaces. VIII, 340 pages. 1981.

Vol. 876: J. P. Bickel, N. El Karoui and M. Yor. Ecole d'Eté de Probabilités de Saint-Flour IX – 1979. Edited by P. L. Hennequin. XI, 280 pages. 1981.

Vol. 877: J. Erven, B.-J. Falkowski, Low Order Cohomology and Applications. VI, 126 pages. 1981.

Vol. 878: Numerical Solution of Nonlinear Equations. Proceedings, 1980. Edited by E. L. Allgower, K. Glashoff, and H.-O. Peitgen. XIV, 440 pages. 1981.

Vol. 879: V. V. Sazonov, Normal Approximation – Some Recent Advances. VII, 105 pages. 1981.

Vol. 880: Non Commutative Harmonic Analysis and Lie Groups. Proceedings, 1980. Edited by J. Carmona and M. Vergne. IV, 553 pages. 1981.

Vol. 881: R. Lutz, M. Goze, Nonstandard Analysis. XIV, 261 pages. 1981.

Vol. 882: Integral Representations and Applications. Proceedings, 1980. Edited by K. Roggenkamp. XII, 479 pages. 1981.

Vol. 883: Cylindric Set Algebras. By L. Henkin, J. D. Monk, A. Tarski, H. Andréka, and I. Németi. VII, 323 pages. 1981.

Vol. 884: Combinatorial Mathematics VIII. Proceedings, 1980. Edited by K. L. McAvaney. XIII, 359 pages. 1981.

Vol. 885: Combinatorics and Graph Theory. Edited by S. B. Rao. Proceedings, 1980. VII, 500 pages. 1981.

Vol. 886: Fixed Point Theory. Proceedings, 1980. Edited by E. Fadell and G. Fournier. XII, 511 pages. 1981.

Vol. 887: F. van Oystaeyen, A. Verschoren, Non-commutative Algebraic Geometry, VI, 404 pages. 1981.

Vol. 888: Padé Approximation and its Applications. Proceedings, 1980. Edited by M. G. de Bruin and H. van Rossum. VI, 383 pages. 1981.

Vol. 889: J. Bourgain, New Classes of $\mathcal{L}^p$-Spaces. V, 143 pages. 1981.

Vol. 890: Model Theory and Arithmetic. Proceedings, 1979/80. Edited by C. Berline, K. McAloon, and J.-P. Ressayre. VI, 306 pages. 1981.

Vol. 891: Logic Symposia, Hakone, 1979, 1980. Proceedings, 1979, 1980. Edited by G. H. Müller, G. Takeuti, and T. Tugué. XI, 394 pages. 1981.

Vol. 892: H. Cajar, Billingsley Dimension in Probability Spaces. III, 106 pages. 1981.

Vol. 893: Geometries and Groups. Proceedings. Edited by M. Aigner and D. Jungnickel. X, 250 pages. 1981.

Vol. 894: Geometry Symposium. Utrecht 1980, Proceedings. Edited by E. Looijenga, D. Siersma, and F. Takens. V, 153 pages. 1981.

Vol. 895: J.A. Hillman, Alexander Ideals of Links. V, 178 pages. 1981.

Vol. 896: B. Angéniol, Familles de Cycles Algébriques – Schéma de Chow. VI, 140 pages. 1981.

Vol. 897: W. Buchholz, S. Feferman, W. Pohlers, W. Sieg, Iterated Inductive Definitions and Subsystems of Analysis: Recent Proof-Theoretical Studies. V, 383 pages. 1981.

Vol. 898: Dynamical Systems and Turbulence, Warwick, 1980. Proceedings. Edited by D. Rand and L.-S. Young. VI, 390 pages. 1981.

Vol. 899: Analytic Number Theory. Proceedings, 1980. Edited by M.I. Knopp. X, 478 pages. 1981.

Vol. 900: P. Deligne, J. S. Milne, A. Ogus, and K.-Y. Shih, Hodge Cycles, Motives, and Shimura Varieties. V, 414 pages. 1982.

Vol. 901: Séminaire Bourbaki vol. 1980/81 Exposés 561–578. III, 299 pages. 1981.

Vol. 902: F. Dumortier, P.R. Rodrigues, and R. Roussarie, Germs of Diffeomorphisms in the Plane. IV, 197 pages. 1981.

Vol. 903: Representations of Algebras. Proceedings, 1980. Edited by M. Auslander and E. Lluis. XV, 371 pages. 1981.

Vol. 904: K. Donner, Extension of Positive Operators and Korovkin Theorems. XII, 182 pages. 1982.

Vol. 905: Differential Geometric Methods in Mathematical Physics. Proceedings, 1980. Edited by H.-D. Doebner, S.J. Andersson, and H.R. Petry. VI, 309 pages. 1982.

Vol. 906: Séminaire de Théorie du Potentiel, Paris, No. 6. Proceedings. Edité par F. Hirsch et G. Mokobodzki. IV, 328 pages. 1982.

Vol. 907: P. Schenzel, Dualisierende Komplexe in der lokalen Algebra und Buchsbaum-Ringe. VII, 161 Seiten. 1982.

Vol. 908: Harmonic Analysis. Proceedings, 1981. Edited by F. Ricci and G. Weiss. V, 325 pages. 1982.

Vol. 909: Numerical Analysis. Proceedings, 1981. Edited by J.P. Hennart. VII, 247 pages. 1982.

Vol. 910: S.S. Abhyankar, Weighted Expansions for Canonical Desingularization. VII, 236 pages. 1982.

Vol. 911: O.G. Jørsboe, L. Mejlbro, The Carleson-Hunt Theorem on Fourier Series. IV, 123 pages. 1982.

Vol. 912: Numerical Analysis. Proceedings, 1981. Edited by G. A. Watson. XIII, 245 pages. 1982.

Vol. 913: O. Tammi, Extremum Problems for Bounded Univalent Functions II. VI, 168 pages. 1982.

Vol. 914: M. L. Warshauer, The Witt Group of Degree k Maps and Asymmetric Inner Product Spaces. IV, 269 pages. 1982.

Vol. 915: Categorical Aspects of Topology and Analysis. Proceedings, 1981. Edited by B. Banaschewski. XI, 385 pages. 1982.

Vol. 916: K.-U. Grusa, Zweidimensionale, interpolierende Lg-Splines und ihre Anwendungen. VIII, 238 Seiten. 1982.

Vol. 917: Brauer Groups in Ring Theory and Algebraic Geometry. Proceedings, 1981. Edited by F. van Oystaeyen and A. Verschoren. VIII, 300 pages. 1982.

Vol. 918: Z. Semadeni, Schauder Bases in Banach Spaces of Continuous Functions. V, 136 pages. 1982.

Vol. 919: Séminaire Pierre Lelong – Henri Skoda (Analyse) Années 1980/81 et Colloque de Wimereux, Mai 1981. Proceedings. Edité par P. Lelong et H. Skoda. VII, 383 pages. 1982.

Vol. 920: Séminaire de Probabilités XVI, 1980/81. Proceedings. Edité par J. Azéma et M. Yor. V, 622 pages. 1982.

Vol. 921: Séminaire de Probabilités XVI, 1980/81. Supplément: Géométrie Différentielle Stochastique. Proceedings. Edité par J. Azéma et M. Yor. III, 285 pages. 1982.

Vol. 922: B. Dacorogna, Weak Continuity and Weak Lower Semicontinuity of Non-Linear Functionals. V, 120 pages. 1982.

Vol. 923: Functional Analysis in Markov Processes. Proceedings, 1981. Edited by M. Fukushima. V, 307 pages. 1982.

Vol. 924: Séminaire d'Algèbre Paul Dubreil et Marie-Paule Malliavin. Proceedings, 1981. Edité par M.-P. Malliavin. V, 461 pages. 1982.

Vol. 925: The Riemann Problem, Complete Integrability and Arithmetic Applications. Proceedings, 1979-1980. Edited by D. Chudnovsky and G. Chudnovsky. VI, 373 pages. 1982.

Vol. 926: Geometric Techniques in Gauge Theories. Proceedings, 1981. Edited by R. Martini and E.M.de Jager. IX, 219 pages. 1982.

ol. 927: Y. Z. Flicker, The Trace Formula and Base Change for L (3). XII, 204 pages. 1982.

ol. 928: Probability Measures on Groups. Proceedings 1981. Edited y H. Heyer. X, 477 pages. 1982.

ol. 929: Ecole d'Eté de Probabilités de Saint-Flour X – 1980. roceedings, 1980. Edited by P.L. Hennequin. X, 313 pages. 1982.

ol. 930: P. Berthelot, L. Breen, et W. Messing, Théorie de Dieudonné ristalline II. XI, 261 pages. 1982.

ol. 931: D.M. Arnold, Finite Rank Torsion Free Abelian Groups nd Rings. VII, 191 pages. 1982.

ol. 932: Analytic Theory of Continued Fractions. Proceedings, 1981. dited by W.B. Jones, W.J. Thron, and H. Waadeland. VI, 240 pages. 982.

ol. 933: Lie Algebras and Related Topics. Proceedings, 1981. dited by D. Winter. VI, 236 pages. 1982.

ol. 934: M. Sakai, Quadrature Domains. IV, 133 pages. 1982.

ol. 935: R. Sot, Simple Morphisms in Algebraic Geometry. IV, 6 pages. 1982.

ol. 936: S.M. Khaleelulla, Counterexamples in Topological Vector paces. XXI, 179 pages. 1982.

ol. 937: E. Combet, Intégrales Exponentielles. VIII, 114 pages. 82.

ol. 938: Number Theory. Proceedings, 1981. Edited by K. Alladi. , 177 pages. 1982.

ol. 939: Martingale Theory in Harmonic Analysis and Banach paces. Proceedings, 1981. Edited by J.-A. Chao and W.A. Woyzyński. VIII, 225 pages. 1982.

ol. 940: S. Shelah, Proper Forcing. XXIX, 496 pages. 1982.

ol. 941: A. Legrand, Homotopie des Espaces de Sections. VII, 2 pages. 1982.

ol. 942: Theory and Applications of Singular Perturbations. Proedings, 1981. Edited by W. Eckhaus and E.M. de Jager. V, 363 ages. 1982.

ol. 943: V. Ancona, G. Tomassini, Modifications Analytiques. IV, 0 pages. 1982.

ol. 944: Representations of Algebras. Workshop Proceedings, 80. Edited by M. Auslander and E. Lluis. V, 258 pages. 1982.

ol. 945: Measure Theory. Oberwolfach 1981, Proceedings. Edited D. Kölzow and D. Maharam-Stone. XV, 431 pages. 1982.

ol. 946: N. Spaltenstein, Classes Unipotentes et Sous-groupes de orel. IX, 259 pages. 1982.

ol. 947: Algebraic Threefolds. Proceedings, 1981. Edited by Conte. VII, 315 pages. 1982.

l. 948: Functional Analysis. Proceedings, 1981. Edited by D. Butvić, H. Kraljević, and S. Kurepa. X, 239 pages. 1982.

ol. 949: Harmonic Maps. Proceedings, 1980. Edited by R.J. Knill, . Kalka and H.C.J. Sealey. V, 158 pages. 1982.

ol. 950: Complex Analysis. Proceedings, 1980. Edited by J. Eells. , 428 pages. 1982.

ol. 951: Advances in Non-Commutative Ring Theory. Proceedings, 981. Edited by P.J. Fleury. V, 142 pages. 1982.

ol. 952: Combinatorial Mathematics IX. Proceedings, 1981. Edited y E. Billington, S. Oates-Williams, and A.P. Street. XI, 443 pages. 982.

ol. 953: Iterative Solution of Nonlinear Systems of Equations. Proeedings, 1982. Edited by R. Ansorge, Th. Meis, and W. Törnig. I, 202 pages. 1982.

Vol. 954: S.G. Pandit, S.G. Deo, Differential Systems Involving Impulses. VII, 102 pages. 1982.

Vol. 955: G. Gierz, Bundles of Topological Vector Spaces and Their Duality. IV, 296 pages. 1982.

Vol. 956: Group Actions and Vector Fields. Proceedings, 1981. Edited by J.B. Carrell. V, 144 pages. 1982.

Vol. 957: Differential Equations. Proceedings, 1981. Edited by D.G. de Figueiredo. VIII, 301 pages. 1982.

Vol. 958: F.R. Beyl, J. Tappe, Group Extensions, Representations, and the Schur Multiplicator. IV, 278 pages. 1982.

Vol. 959: Géométrie Algébrique Réelle et Formes Quadratiques, Proceedings, 1981. Edité par J.-L. Colliot-Thélène, M. Coste, L. Mahé, et M.-F. Roy. X, 458 pages. 1982.

Vol. 960: Multigrid Methods. Proceedings, 1981. Edited by W. Hackbusch and U. Trottenberg. VII, 652 pages. 1982.

Vol. 961: Algebraic Geometry. Proceedings, 1981. Edited by J.M. Aroca, R. Buchweitz, M. Giusti, and M. Merle. VI, 500 pages. 1982.

Vol. 962: Category Theory. Proceedings, 1981. Edited by K.H. Kamps, D. Pumplün, and W. Tholen, XV, 322 pages. 1982.

Vol. 963: R. Nottrot, Optimal Processes on Manifolds. VI, 124 pages. 1982.

Vol. 964: Ordinary and Partial Differential Equations. Proceedings, 1982. Edited by W.N. Everitt and B.D. Sleeman. XVIII, 726 pages. 1982.

Vol. 965: Topics in Numerical Analysis. Proceedings, 1981. Edited by P.R. Turner. IX, 202 pages. 1982.

Vol. 966: Algebraic K-Theory. Proceedings, 1980, Volume I. Edited by R. K. Dennis. VIII, 407 pages. 1982.

Vol. 967: Algebraic K-Theory. Proceedings, 1980. Volume II. VIII, 409 pages. 1982.

Vol. 968: Numerical Integration of Differential Equations and Large Linear Systems. Proceedings, 1980. Edited by J. Hinze. VI, 412 pages. 1982.

Vol. 969: Combinatorial Theory. Proceedings, 1982. Edited by D. Jungnickel and K. Vedder. V, 326 pages. 1982.